D0812164

FISH HANDLING

AND PROCESSING

Fish Handling & Processing

Editors

G. H. O. Burgess C. L. Cutting (*in part*)
J. A. Lovern J. J. Waterman

Illustrated by

Keith Atkinson

CHEMICAL PUBLISHING COMPANY, INC.
New York 1967

First American Edition
1967
CHEMICAL PUBLISHING CO., INC.
New York N. Y.

Printed in United States of America

Contents

Contents

Acknowledgements

Thanks are due to Messrs. J. M. Dent & Sons, publishers of *The Face is Familiar*, by Ogden Nash, for permission to use the quotation from the poem 'The Fish' at the head of Chapter 12; to Dobson Books Ltd., publishers of *The Little Pot Boiler,* by Spike Milligan, for permission to quote from the poem 'Contagion' at the head of Chapter 5; and to Messrs. Macmillan & Co., publishers of *Wrack at Tidesend*, by Sir Osbert Sitwell, for permission to use the quotation from 'Mrs. Busk' at the head of Chapter 3.

Foreword

Food Science and Technology are relatively new subjects. Most food research laboratories in the world were established after the 1914–18 war, and only recently has it become possible to study food science in British Universities. The Torry Research Station in Aberdeen was set up by the Department of Scientific and Industrial Research in 1929 and was one of the first institutions in the world concerned entirely with the problems of fish handling, processing, distribution and storage. The staff of the Station have always been available to give advice, so far as current knowledge would allow, to firms in the industry, and the annual course at Torry for people engaged in handling and processing fish, both in this country and abroad, has been widely patronized since it was first begun ten years ago.

The need for a convenient and readable handbook, written especially for people employed in the British fish industry, has been apparent for a very long time, and it is therefore with great pleasure that I see this work completed.

It has in many instances proved difficult to decide how much attention to give to a specific point, but an attempt has been made to give greater attention to those theoretical and practical points of most importance to workers in the industry. Furthermore, to quote the author of a practical manual in another field, it is believed '. . . that it is more important to be nearly right and understandable than to be academically accurate and incomprehensible'.

Many workers at the Torry Research Station and Humber Laboratory have contributed to this book. It was felt, however, that the authorship should remain anonymous, especially since the editors, in the interests of uniformity of style, have rewritten large sections of the text.

If this book helps those engaged in the many day-to-day operations of the fish industry to do their jobs better, and so makes a contribution towards the increasingly rapid technical development that is now occurring, all those who have helped in its production will be well satisfied.

TORRY RESEARCH STATION
ABERDEEN

G. A. REAY

Illustrations

I

About this book

'Begin at the beginning,' the King said, gravely, and go on till you come to the end, then stop.'

LEWIS CARROLL

ALTHOUGH a number of books on scientific aspects of fish handling and processing have been written by scientists, these have in the main been written for other scientists and technologists. Few attempts, and certainly none in Britain, have been made to write a work specially for those with wide practical knowledge of fish handling and processing but with little or no scientific training. Such an attempt has been made here. The products and processes used in other countries have not been discussed and even so far as the British industry is concerned, the coverage is not exhaustive. Attention has rather been concentrated on the major processes about which staff at Torry have gained first-hand knowledge through their own researches. Further work may modify some of the ideas discussed here, but it is believed that this constitutes a reasonably accurate and authoritative account of current knowledge.

Scientists, when they write for each other, use terms that are largely incomprehensible to the layman and sometimes, indeed, to other scientists working in different fields. So far as is possible, scientific terms have been avoided here. But they have not been avoided entirely. The scientist, when he attempts to write non-technical English, immediately becomes aware of the paralysing effect of other scientists looking over his shoulder. It is easy to poke fun at technical phrases, but in certain fields they are indispensable whilst in others they are difficult to do without. Deprived of them altogether, the scientific writer finds his sentences bursting forth into such a profusion of qualifying clauses and parenthetical asides that he begins to despair of his task altogether. Clearly, some compromise is necessary.

Here, two different approaches have been tried. A conscious effort has been made *not* to write for other scientists and technologists. The hypnotizing effect of the thought that every word on the printed page may damage a scientific reputation has, it is hoped, been avoided by the anonymity of the authorship of each chapter. Where it has seemed to suit the purpose and where slight inaccuracy has seemed of little moment, straightforward non-technical explanation has been given. But such a course could not always be adopted and in consequence chapters 12 to 16 deal with some basic scientific principles involved in fish technology. It is hoped that, by this means, the man who wishes to look up a particular point can obtain the information he requires whilst anyone reading the book as a whole will not be infuriated by constant repetition of the same basic facts.

The editors have felt that the text should, so far as is possible, be easily read without numerous references to other parts of the book or to other work. In consequence no cross-references have been given in the text; instead, a very full index has been provided, which will, it is hoped, prove sufficient for people using the book. No references to original papers have been given, primarily because this book is for people who are handling and processing fish and not for other scientists, but also partly because the relevant literature is so considerable that if it were to be covered adequately, it would greatly increase the length and cost of the book.

Specialized terms have generally been printed in italics when first introduced, and where necessary have been defined. Rigid scientific definitions of terms have been given only when this has seemed essential. Where abbreviations have been used, these conform as far as practicable to *British Standard* 1991: Part 1: 1954.

All temperatures have been given in degrees Fahrenheit. The Fahrenheit scale is used universally throughout the British fish industry;

standards for quick freezing are based on it and so is all measuring equipment used in kilns and cold stores. Although it may well be that the Centigrade scale will eventually replace the Fahrenheit scale in the fish industry, there is at present no sign whatsoever of this occurring and the editors therefore decided that there was little point in giving the Centigrade equivalent of every temperature. The method of converting one scale to the other is given in Chapter 14 for those who wish to know the equivalent.

The editors have attempted to achieve some uniformity of style, but slight differences will still, no doubt, be noted by the discerning reader; these are partly due to the differences of treatment unavoidable when discussing two such divergent topics as, for example, Chemistry and Retailing.

2

The handling of fish at sea

The ice was here, the ice was there,
The ice was all around:

S. T. COLERIDGE
The Ancient Mariner

GOOD handling of fish at sea should ensure that the catch retains its initial freshness, so far as is possible, until landing. The important requirements are to chill the fish rapidly as soon as it is caught, to prevent it warming again, and to maintain a high standard of cleanliness, both on deck and in the fish room.

A good design of vessel makes the proper handling of the catch easier, but few fishermen will be lucky enough to sail in an ideal ship; they all, however, should handle the catch properly. Good handling and stowage practice can help greatly to maintain the freshness of the catch, even in vessels that are in some respects badly designed, whereas bad handling, even on a well-designed vessel, can only result in poor quality fish.

The importance of good practice at sea cannot be over-emphasized. Fish begins to spoil immediately it is caught. Even if well gutted, washed and stowed at once in plenty of ice, cod and similar species become inedible within sixteen days. Herring is not normally gutted at sea; even if it is well iced, it may be inedible within three days. Neglect, even of apparently trifling details, may render fish stale after only a day or two. The time between catching and landing fish is often many days longer than the time between landing it and selling it in the shop. The fisherman consequently bears much of the responsibility for the state of freshness in which his catch reaches the consumer.

Port wholesalers and processors are paying increasing attention to the signs that show whether or not the catch has been properly handled and stowed at sea. For example, quality control schemes are being introduced in some ports; the intention of these is to withdraw fish from the auction that, although not bad enough to be condemned by Port Health Inspectors, yet does not reach a certain standard of freshness. In some fisheries, the fear that fish may be withdrawn from the sale, rather than the size of the hold, limits the length of the voyage. The industry in general is increasingly prepared to pay more for a good quality product and this in itself should be an incentive to the whole catching section to adopt the highest standards of handling and stowing at sea.

There are many marketable species of fish and many ways of catching them; British fishing ports vary considerably not only in the weight, species and average size of fish landed at each of them, but also in the lengths of trip of particular types of craft and the methods of catching, handling and stowing employed on them. In this chapter the principles of good handling and stowage are explained and illustrated, mainly by reference to deep sea trawling and partly also to drifting. The principles that are set out are, however, applicable to other types of vessel and methods of fishing.

METHODS OF HANDLING AND STOWING

WHITE FISH

White fish is generally gutted at sea. On seine net vessels, however, when fishing is heavy, it is practically impossible to gut the entire catch, especially if it is composed mainly of small fish such as small haddock, mackerel or whiting. On trawlers all fish are gutted except redfish or soldiers (*Sebastes*), which are rarely gutted at sea on account of the spiny nature of the fins and gill covers. Fish caught by the smallest inshore vessels are not always gutted; indeed, many vessels of this class do not even carry any ice. Quantities of ice carried vary from 100 tons or more on the big distant water trawlers to a few hundredweights on some inshore vessels.

After gutting and washing, the catch is usually stowed with ice. Methods of stowage vary but consist mainly of boxing, bulking and shelfing.

(1) *Bulking*. This is the standard method of stowage employed on distant water trawlers. Fish are stowed, intimately mixed with ice,

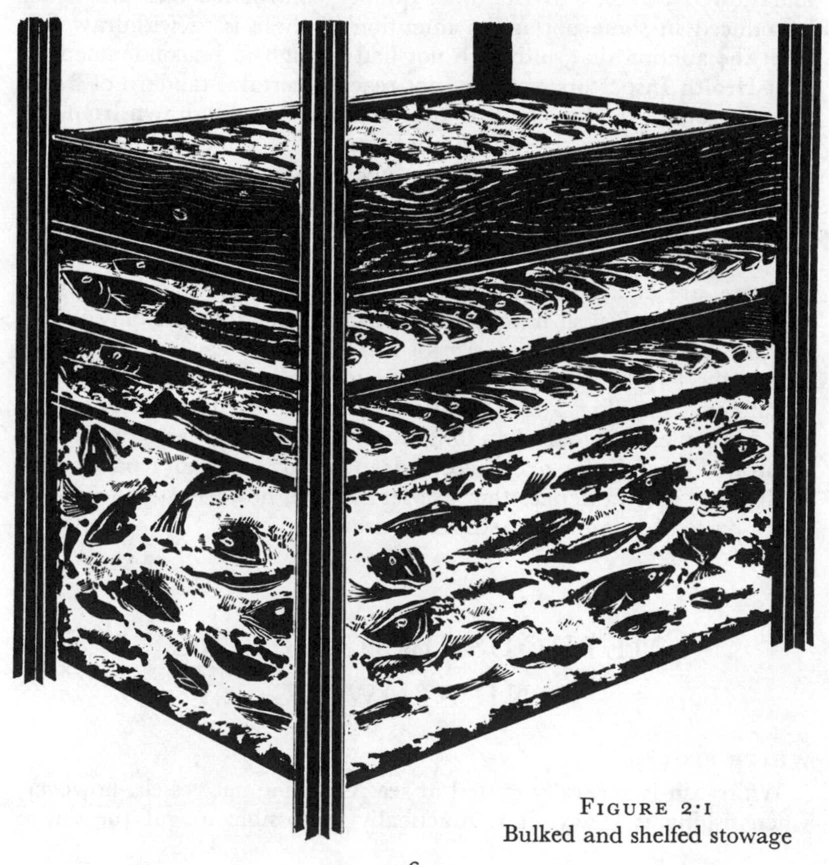

FIGURE 2:1
Bulked and shelfed stowage

in *pounds*. Pounds are built up with removable pound boards, fixed stanchions, and divisions, some of which are removable and some fixed athwartships; battens are provided, either loose or fixed to certain of these pound boards, and provide support for a horizontal shelf made up of boards which thus form the supports for the layers of fish and ice (see Figure 2·1). In shallow bulking, layers of fish and ice do not exceed about 18 inches in depth. In deep bulking, common on distant water trawlers just after the war but now fortunately rare, the depth of fish and ice was much greater, up to 4 feet or even more.

(2) *Shelfing*. Single-shelfed fish are laid belly downwards side by side and head to tail upon a bed of crushed ice supported by a shelf, with little or no ice over the fish, and an air space between them and the shelf above (see Figure 2·1). *Semi-shelfing* or, as it is sometimes called, *bulk shelfing*, is similar to single shelfing except that one or more layers of fish are alternately laid belly uppermost and belly downwards upon the first layer and a layer of crushed ice finally shovelled over the top.

(3) *Boxing*. Boxing, if carried out properly, is one of the best ways of stowing iced fish; so far it has not been generally adopted in the deep sea trawling fleets. On British fishing vessels a wide variety of boxes is used. These differ both in design and size, and many have seen service in other capacities before they reach the fish industry. Fish is often stowed with ice in boxes, although even those boxes made specially for use on the vessel do not always allow sufficient space for the requisite volume of fish and of adequate ice to cool it and keep it cool. In view of the wide differences in boxing practice, it is not possible here to be more specific about current methods.

HERRING

Herring, whether it is caught by drifter or ring netter, is never gutted at sea. In the past, herring was usually stowed in bulk, but since the war, and partly as a result of official encouragement, boxing has become increasingly common. Drifters now carry ice and whenever possible box and ice their fish at sea.

(1) *Bulk stowage*. This is the way in which all herring was formerly stowed. Herring, as it is shaken or brailed out of the nets, according to the method of capture, falls into the hold and accumulates. Apart from putting in boards to reduce pressure on the fish and to prevent it moving about with the motion of the vessel, no attempt is made to preserve the quality of the catch.

(2) *Boxed stowage*. Herring is so perishable that even an hour or two at summer temperatures may do irreparable harm to its quality; it is

so easily damaged that herring at the bottom of a pile 18 inches deep, a depth that is commonly used in white fish stowage, may well be useless for anything but fish meal. Boxing is an attempt to reduce pressure on the fish and, when used in conjunction with enough ice in the right places, is almost an ideal solution to the stowage problem.

Boxes commonly in use contain 5 or 6 stones of fish, that is, six to the cran and five to the cran respectively, and are increasingly made of aluminium. In practice, about two shovelfuls of ice are put into the bottom of a box, which is then filled with fish, and a further two or three shovelfuls of ice put on top.

THE PRINCIPLES OF GOOD STOWAGE

All fish, if it is to be sold as wet fish on landing, should be rapidly chilled. Quick and efficient chilling of the catch to as low a temperature as is practicable without actually freezing the flesh is essential if spoilage is to be kept to a minimum. This means stowage in ice.

Although pure ice melts at 32°F, the temperature of fish stowed in crushed ice may be only 31°F or even slightly lower. This effect is due to salts present in the flesh and sea water on the surface of the fish, which cause some of the ice to melt and at the same time absorb heat from the surroundings. The ice not only cools the fish, however, but also the boards and stanchions, or boxes, and linings and, if enough is used in the right places, absorbs heat entering the hold and therefore prevents the fish warming again until it is landed. The actual process of spoilage also produces some heat which should be absorbed by the ice. It is therefore essential to distribute sufficient ice in the right places.

It is also important that fish should be in contact only with ice and not even with other fish. Fish touching one another do not cool as rapidly as they do when each is buried completely in ice. But, apart from this, if a fish such as cod is stowed so that it is in contact over an appreciable area with a smooth surface such as a pound board or another fish, then air may be excluded. Some of the spoilage bacteria inevitably present, when they are deprived of air, rapidly produce foul-smelling substances which spread throughout the flesh. Herring are also occasionally affected by this form of spoilage. Crushed ice contains innumerable pockets of air and therefore fish properly stowed in ice does not become spoiled in this way. The Canadians call such affected fish *bilgy*; it is perhaps unfortunate that the term *stinker*, used in Britain, is applied not only to white fish affected by this type of spoilage, but also to cod that has been feeding on certain species of floating mollusc related to slugs and snails, and which has in consequence a strong smell and taste rather like seaweed or iodine.

It is the melt water, flowing over the surface of the whole fish, which does most of the cooling. In practice, cooling is more effective the higher the temperature of the surroundings. The explanation appears to be that the transfer of heat from fish to crushed ice by direct contact is a minor effect, in such fish as cod, owing to the small areas of fish and ice actually in contact. The major effect is the flow of melt water produced by contact with warm fish, warm shelves and especially with warm air. The temperature of the melt water rises slightly as it flows over the fish but, if there is enough ice amongst the fish, the water is cooled again and yet more is produced. With flake ice it may be that the shape of the flakes gives more rapid cooling because of better contact between fish and ice. A free flow of melt water is still desirable, however, to preserve the fresh appearance.

Besides contributing to the cooling of fish the flow of melt water also washes away bacterial slime, spoilage products, and remaining blood and thus helps to preserve the fresh appearance and the fresh smell. It follows, therefore, that the fish, especially small fish, should never be packed so tightly that the flow of melt water is prevented. It is important at all times to ensure that there is free drainage of melt water and to avoid immersion of fish in dirty water.

Although it is theoretically desirable to keep the fishroom temperature well above 32°F so that a rapid flow of melt water may be maintained, in practice the weight of ice required to ensure that the fish will remain covered, and also that fish temperatures will not rise, will be large if fishroom air temperatures are high. Thus fishroom temperature has to be kept reasonably low in order to avoid excessive wastage of ice.

The temperatures of single-shelfed fish usually lie somewhere between that of melting ice and that of the air in the fishroom, and it is therefore necessary to maintain an average air temperature of 34°F to 35°F if spoilage is not to be excessively rapid. On the other hand, if the air temperature is lowered very much, to 30°F for example, uncontrolled slow freezing of the fish may occur with considerable loss of quality. It is possible, however, to reduce the temperature of the contents of the fishroom to a point slightly below that obtained with melting ice and thereby extend the storage life of the fish. This carefully controlled reduction of temperature, when carried out quickly and uniformly throughout the mass of fish and ice, is known as *superchilling*. The most satisfactory temperature range has been found by experiment to be between 28°F and 30°F, and storage life of white fish in ice can be extended by about seven days by this method. There are technical problems, however, in achieving close control of temperature and rapid, uniform cooling that necessitate rather elaborate refrigerating arrangements in the fishroom. The chief disadvantage of the method is that the fish are partially frozen on discharge and need some time to thaw out before they can be filleted.

Fish should not be crushed or bruised when it is stowed. Apart from the inevitable damage and softening of flesh, there may also be a considerable loss in weight; an average of 7% of the initial gutted weight of cod stowed in bulk in layers 3 feet deep may be lost on a distant water voyage. This loss may reach 15% in the case of early-caught haddock at the bottom of a 3-foot shelf.

Although single-shelfed fish are not subjected to pressure from above, and may therefore look more attractive than fish that have been bulked or boxed, they are cooled only from below. Higher stowage temperature and slower rate of cooling result in more rapid spoilage and make it inadvisable to shelf fish near the beginning of a trip. The length of time before the smell, flavour and texture of shallow-bulked or boxed fish becomes definitely superior to that of shelfed fish is not yet known nor have all the factors involved been accurately assessed. It is certain, however, that after 12 days from catching, shelfed fish is definitely inferior to properly stowed, shallow-bulked or boxed, fish. In some instances marked differences have been observed after only three days.

Finally, much of the care exercised by the fisherman in handling and stowing the catch may be wasted if the boards or boxes and supporting structures are not cleaned before the fish is stowed. Unused ice remaining at the end of a trip should always be discarded since it has been shown that such ice, although apparently clean, may be heavily contaminated with fish-spoilage bacteria. Fish stowed in dirty ice spoil more rapidly than fish stowed in clean ice.

GOOD HANDLING AND STOWAGE

ON THE TRAWLER

(1) *On deck.* Before fishing begins, the deck of the gutting pounds, the fixed stanchions and boards and washing machine or tank should be thoroughly washed, scrubbed and steam-hosed. The gutting pounds should be washed clean again every time the gear is about to be hauled.

If fishing is on a muddy bottom, the fish may with advantage be given a preliminary wash when being roughly sorted before gutting. If fish from the previous haul remains on deck, it should be removed to a separate gutting pound so as to avoid putting fish on top of fish, and the pounds that are to receive the fresh drag prepared by washing as already recommended. Gutting should start as soon as possible after the cod-end is emptied. Guts and inedible or undersized fish and rubbish, should not be thrown back on top of edible fish, but over the side or into a separate space. A suitable space can be arranged between gutting pounds and bulwarks and, if required, also between the gutting pounds and hatch coamings.

Fish should never be trodden on, roughly handled or dropped. If the fish are to be headed, a sharp knife should be used to make angled cuts behind the head before the spine is broken; heading should not be done merely by tearing the fish across a warp. On British trawlers it is neither practicable nor worthwhile to bleed the fish as a separate operation before gutting, but this can be done with advantage on inshore line vessels to ensure attractive white flesh. The sooner gutting is carried out, the better the fish will bleed.

In gutting, the aim should be thoroughly to clean out and wash the belly cavity. For this purpose all round fish of sufficient size such as cod and haddock should be *double naped* and the belly opened up over its entire length. Special care should be taken to remove the whole of the liver.

The sooner fish is put in ice after catching, once it has been gutted, the better. Fish that lie on deck without ice begin to soften quite rapidly and the way is made easy for spoilage bacteria. A heap of fish on deck becomes spontaneously warmer, in a way rather similar to a compost heap, and this makes the softening and spoilage processes occur even faster. Furthermore, a certain amount of dirt from the intestines, which contain many bacteria, is bound to remain on some fish and should be washed off as soon as possible. All these troubles are avoided to a great extent by using a continuously operated washing machine and chute to carry the fish below.

Where there is no washing machine, one or more of the crew should begin to wash the gutted fish by hand and put them below soon after gutting has begun. In this way, most of the fish is washed and put in ice sooner than if kept in a heap on the deck and washed only after the completion of gutting; washing is also likely to be more thorough. Where there is no machine, washing should be done in a galvanized steel or light alloy tank. A continuous flow of clean water should be maintained from the pump and the fish washed in *open*-mesh baskets swirled around in the tanks. If the fish must be put below in baskets rather than by chute, the fish should be allowed to drain thoroughly before being brought to the hatch, to avoid introducing dirty water into the fishroom and so contaminating the ice and boards or boxes. Washing need be sufficient only to remove all visible blood, debris and filth from the skin and gills and the belly cavity. More intensive washing than this will do little, if any, good. After all fish are gutted and put below, the decks and pounds should be cleared and washed for the next haul. Boots and frocks should also be washed.

So long as the fish is on deck all hands on watch should continue working. Fish should never be left lying on deck for any reason other than that the ship is in immediate danger. In very heavy fishing and in warm weather, it is possible that some fish will be in *rigor mortis* or death stiffening before being stowed. Fish in stiff rigor should be handled very carefully indeed and should not be forcibly bent or straightened.

(2) *In the fishroom.* The care of the catch once it gets into the hold is largely a matter of chilling it rapidly and making sure that its temperature is kept at about 32°F until landing. It is also important to handle and stow fish so that contamination by unclean surfaces is prevented, and damage due to crushing or bruising is kept to a minimum.

Although sea and air temperatures may occasionally fall below 32°F on the Arctic fishing grounds in winter, they are usually above this temperature. Even when they are below 32°F on the fishing ground itself, higher sea and air temperatures are generally encountered on the voyage home. In consequence, heat from the water surrounding the fishroom, or from the air above, is continually leaking into the hold. This heat melts the ice and may warm the fish if it is inadequately protected with ice. In the design and construction of the ship, and especially of the fishroom, everything possible should be done to reduce this heat leak, by insulation or otherwise, to a level at which it will not melt sufficient of the outer protective layers of ice during the course of a normal trip to allow the fish to warm and spoil more rapidly.

Methods of stowage are generally governed by the custom of the port, landing facilities and similar considerations. Nevertheless, certain practices are not technologically desirable and it may be of interest to set down a few of the important considerations.

Boxing is the method recommended for herring drifters; it is also used in the smaller white-fish vessels as well as one deep sea trawling fleet. It seems from experiments that boxing and icing can also be the best method for retaining the quality of white fish. This is partly because handling during discharge becomes unnecessary and also there is less crushing in stowage and a lower temperature is maintained after unloading. It has already been mentioned that bulk-stowed white fish may lose weight during stowage; cod boxed with plenty of ice and without crushing may actually gain 1% to 2% in weight. Boxing also helps greatly in laying out the catch on the market in the order of catching. If the full advantages in quality are to be obtained, the fish should not be sorted or emptied from the boxes during discharge. Since a layer of ice should remain above the fish, and so will obscure them, some reliable code-marking of boxes is required in order to make identification easier on the quayside. The weight of fish in the box must, to some extent at least, be taken on trust by the buyer. Boxing if carried out in this way will, therefore, be suitable where there are a few large-scale buyers who are not particularly selective regarding the exact size of the fish and where fish are required to fill contract orders, such as are now being introduced in some ports.

Boxed fish should be easier to handle, and centralized cleaning of standard boxes easier to organize than that of boards. Weight for weight of fish, boxing requires more room than bulking but less than single shelfing. The stowage rates are rather less than $4\frac{1}{2}$ cubic feet to a 10-stone kit for bulked, about 6 cubic feet for boxed and 10 cubic feet for single-shelfed sprag cod. An additional difficulty in some fisheries might be that of stowing boxes in order to leave sufficient space in the hold for working and for ice. Schemes involving part-boxing and part-pounding of the catch seem wasteful of space.

The many hundreds of pound boards discharged on to the deck of a large trawler during unloading often cause acute congestion; this is one of the arguments in favour of a standardized board which can be thoroughly cleaned and maintained in a central plant.

Before fishing begins, the Mate should make sure that the fishroom is clean; he should pay particular attention to the underside of fixed battens. The slush well should be inspected and cleaned with a suitable chemical disinfectant and the bilge lines tested, if necessary by running the pump. A layer of crushed ice 6 to 12 inches deep, depending upon the weather, insulation and length of voyage, should be spread over the bottom of the fishroom. A grid of boards may be put down to provide extra insulation and to assist drainage towards the central gutterway and slush wells. For boxed stowage, battens should be laid fore and aft on the bottom of the hold to carry the boxes clear of the layer of ice. The refrigerated pipe coils, if any, should be turned off for bulked or boxed stowage.

Species such as coalfish that are likely to discolour other fish, and those such as skate and dogfish that give off ammonia during spoilage, should be stowed separately. Other species may be sorted according to demand and the custom of the port. So-called *rough* fish should not be treated roughly; much rough fish has a poorer reputation on the market than it need have simply because it is not handled and stowed with sufficient care.

The following procedure should be used for icing fish in bulk or for boxed stowage. Boards and rest angles, or boxes, should be examined for cleanliness before use. The bottom of the box or shelf should be covered with about 2 inches of crushed ice on to which a layer of fish should be placed or allowed to slide. Ice should then be shovelled over the fish to fill up the spaces and lightly cover it before the next layer is stowed. Finally, another 2-inch layer of ice should be placed at the top of the box or shelf. Fish should not be bulked in layers more than 18 inches deep between shelves. Each layer should have ice mixed right through it and a layer of ice both above and below it; ice should also be placed where it will prevent contact between fish and pound divisions or linings. The amount of ice used in direct contact with fish should be at least one part to three parts of fish by weight; up to one part to two is occasionally used in commercial practice. These figures do not include ice specially disposed to absorb heat leaking in from outside.

Pounds and boxes should never be overfilled. Where fish are placed individually, they should be stowed belly cavity downwards on a layer of ice about 2 inches deep to ensure drainage of melt water. If fish are shelfed towards the end of the trip, metal boards should be used for preference to help in keeping the air above the fish cool. Refrigerated grids, if any, should be operated as necessary to keep air temperatures above shelfed fish in the region of 33°F to 36°F.

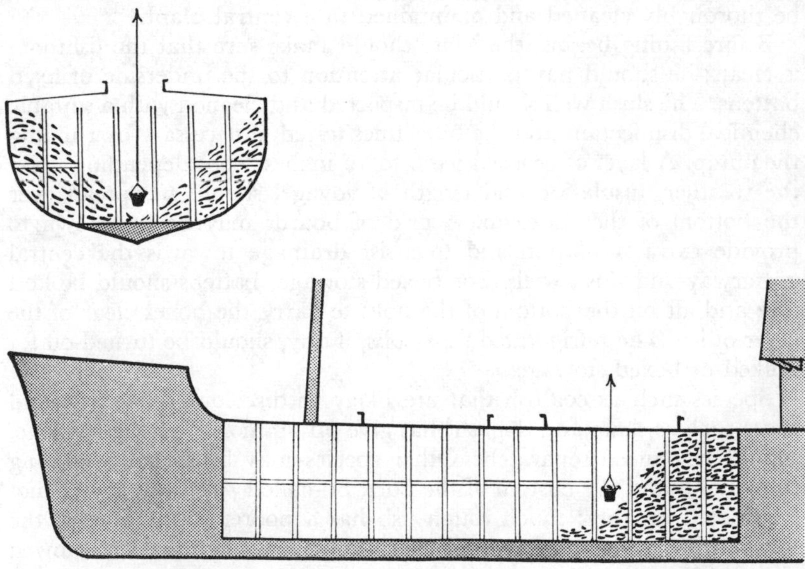

FIGURE 2·2 Mixing of catch during discharge; two ages being pulled down into common discharge point; (*top*) thwartships section of fishroom; (*bottom*) fore-and-aft section of fishroom

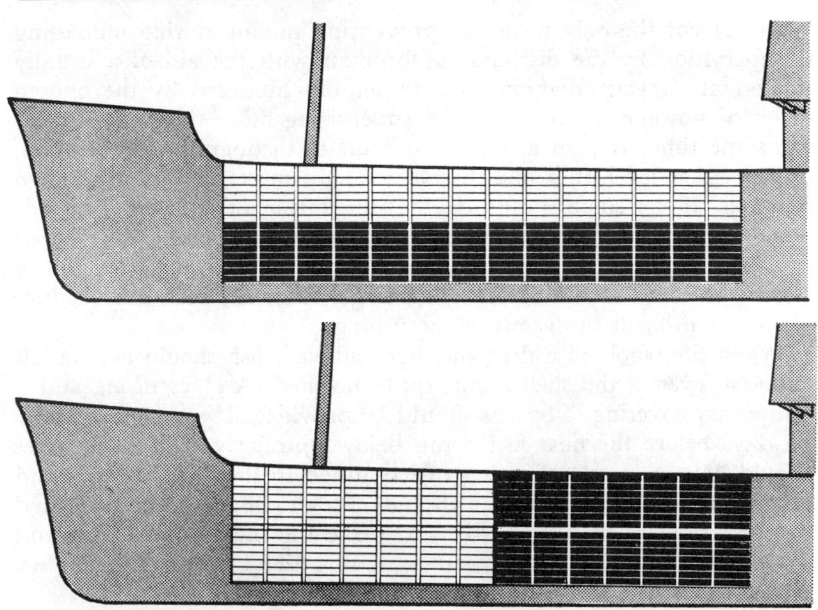

FIGURE 2·3 Fishroom: (*top*) with bulked stowage below staging, shelfed stowage above; (*bottom*) with bulked stowage aft, shelfed stowage forward

Mention should also be made of the problem of the mixing of fish of different ages during unloading. When fish are being discharged from pounded stowage, the lumpers tend to work so that the fish stowed outboard and high up slide towards the centre line and are discharged simultaneously with those inboard and low in the stowage (see Figure 2·2, *top*). In the case of an aftermost pound under a winch, they will tend to work so that the fish slide forward into a pound directly below a hatch (see Figure 2·2, *bottom*). These facts should be borne in mind when deciding the plan of stowage. While considerations of trim and stability and of keeping fishroom work to a minimum are of prime importance, an attempt should be made to stow the fish so that the possibility of qualities becoming mixed during discharge is reduced. It is perfectly feasible in most large trawlers to stow all bulk fish below the staging, with shelfed fish above, rather than bulk aft and shelfed forward (see Figure 2·3). A stowage record should be kept, showing the whereabouts and boundaries of each day's catch, and to make this easier, a slightly more elaborate diagram than is usual should be provided. It should at least show the correct number of stanchions and pound divisions and show the staging at the correct relative height.

At present the only means of preventing mixing during unloading is supervision by the discharging foreman with the aid of a usually inadequate stowage diagram and sometimes hindered by the chosen order of stowage; he may also be supervising more than one ship at the same time. It is in any case difficult to distinguish staging level, or any other level, in a half-discharged ship merely by looking down a hatch. If, by some simple device, the lifting of a layer of boards could be prevented at will until the whole layer was completely cleared of fish, this would be one answer. Another would be the adoption of boxing or some similar means of stowage, although such a procedure would be difficult in distant water fishing.

When the whole of a drag has been stowed, fish should not be left exposed, even if the shelf is not quite finished. A layer of ice and a temporary covering of boards should be provided. It may occasionally be days before the next fish come below. Similarly, half-filled boxes should have a couple of inches of ice added to them. All boards and boxes should fit snugly and, although fish should never be subjected to pressure, care should be taken to make the individual shelves and divisions or boxes reasonably air-tight in order to exclude draughts. Nevertheless, free drainage must not be prevented.

It cannot be over-emphasized that in most British vessels it is essential to place an extra layer of ice against the linings at the sides and bulkheads. Where fish is stowed in boxes, the space between the ends of the tiers and the linings should be filled with ice and, where in doubt, a box should be omitted rather than squeezed in with no protecting wall of ice. Ice should also be placed between the aftermost row of boxes and the aft bulkhead; vertical battens on the bulkhead will help in doing this. When shelfing, a heap of ice should be piled up against the lining where each shelf meets it, to cover the lining as far as possible with ice and keep down the air temperature.

Similarly, a layer of ice 6 to 9 inches deep should be placed between bulked or boxed fish and the deck. This absorbs heat penetrating the deck and from warm air and lights. Even when refrigerated grids are still operating, the layer of ice is advisable for two reasons. First, it will reduce the extent of evaporation of water from the fish due to the drying effect of the grids. Secondly, the risk of partial freezing, resulting from radiation of heat from the fish to the grids, which are usually below 20°F, is avoided. A shelf carrying nothing but a layer of ice is required above shelfed stowage for the same reasons.

To prevent entry of warm air, hatches should be left open for as short a time as possible and never more than one at a time (see Figure 2·4). If at all possible, only a manhole should be opened; this is feasible where a washing machine and chute are used. Lights should be switched off when not needed. Fishroom air temperatures, sea water temperatures and external air temperatures, also jacket air temperature if a jacketed hold is fitted, should be recorded twice daily. In abnormally hot weather, the deck awning should be erected and the hose arranged to keep a film of water running over the deck. The slush well should be pumped regularly, twice a watch, and the strainers maintained in good condition to avoid the necessity for blowing back. The layer of ice over the slush well should be maintained.

Ice can be contaminated by bacteria from dirty boots and tools and wooden and cement surfaces, as well as from dripping fish baskets. Fish stowed in dirty ice will spoil more rapidly than that in clean ice. Care should therefore be taken to keep ice clean and, so far as is possible, to use ice which has not been directly in contact with wooden or cement surfaces.

During discharge of a new ship, opportunity should be taken to note the places where sufficient ice seems to remain and where the deckhead grid has caused it to congeal into unmelted slabs. These observations should be compared with the stowage plan and temperature records until a thorough knowledge and experience of that particular ship has been built up.

A pound weight of ice varies in refrigerating value according to how much is really ice and how much is water, but differences are not likely to be great once the fishing grounds are reached. If ice then appears slushy, the cause should be sought first in excessive heat leak into the hold. Apart from varying wetness, the refrigerating value of ice depends on its weight; the apparent quantity as judged by volume is a bad guide to the refrigerating value of different ices. Thus a cubic foot of flake ice and a cubic foot of crushed ice differ markedly in weight. Ice made from hard water tends to congeal rather more easily but in general ice is ice, provided it is ice and not a mush of ice

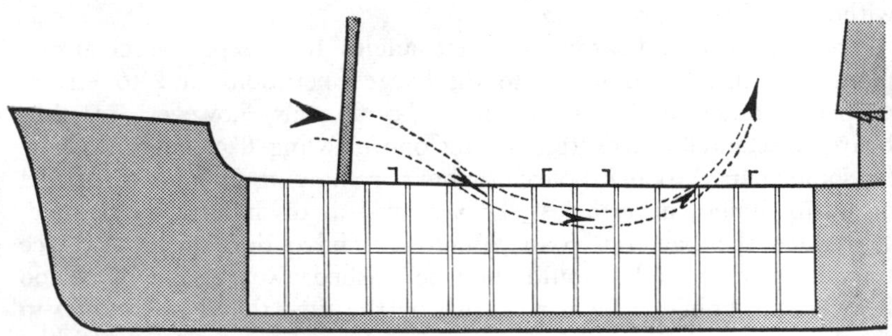

FIGURE 2·4 Air flowing through the fishroom with two hatches open

and water. The fact that some types of ice are delivered at temperatures below 32°F means that there is no liquid water present and the customer receives only ice, which may remain without any melting for some time and be easy to handle. A pound of ice at 22°F, however, has only about 3% more cooling power than a pound of ice at 32°F. Ice which 'runs away too fast', it should be noted, is probably doing its job well and cooling the fish and fittings more rapidly than the ice with which it is compared. In general, the quality of the ice supplied should be the last possible excuse for a poor quality trip. The explanation is much more likely to be the quantity of ice used in relation to the weather, length of trip and insulation of the ship, or some other defect in handling and stowage practice.

Much of the care exercised by the crew may be rendered useless by carelessness during discharge. Lumpers should, for example, stand on shelves and not, as frequently happens, lift the boards and trample on the next layer of fish. Similarly, bulked fish in the centre staging should not be used as a shock absorber for empty baskets thrown down the hold. Bruised fish goes bad much more quickly than undamaged fish. Damage to fish is also caused by hooks and by carelessly-used shovels. Rubbing of fish, often attributed to the trawl, is in fact frequently caused by ice shovels. Some alternative to hooks is desirable that will damage neither the fish nor the finish on boards; perhaps boxing at sea or something similar is the best solution.

As far as possible the tipping of fish from one container to another during unloading should be avoided. Kits or market boxes should have drain holes and these should be so arranged that melt water from one box does not drain into the box below it in a stack but tends to flow away over the sides or ends of the box below. The rate of warming of fish is slowed if it is in a compact stack of boxes. Top icing may also help, although only the top layers of fish will be greatly affected by this procedure. At all events, discharge should be rapid, and everything possible done to reduce the time that the fish remains without ice before processing.

The removal of boards and rest angles to a separate cleansing plant is desirable, to facilitate discharge operations and to ensure proper cleaning and maintenance. To do this, however, requires boards of standard sizes and in fact one trawling fleet has had these for some years. Wooden boards or boxes remaining unused at the end of a trip should, nevertheless, be removed at discharge and cleaned, to reduce the bacterial load which will have developed. Any ice remaining in the hold should then be washed away as it will be too heavily contaminated for use on fish by the time the vessel returns to the fishing grounds. In the case of vessels making shorter trips, the same procedure could be carried out once a week.

The fishroom should then be thoroughly washed out in harbour using clean mains water according to the following procedure: The fishroom should first be washed down with clean cold water, preferably containing a suitable detergent. A hose delivering water at a pressure of at least 20 lb/in^2 is the most effective method. The detergent may be applied to all surfaces which are then hosed down, or it may be introduced into the water supply at a convenient point by means of a suitable dispenser. A stiff brush should be used if necessary to remove any obvious dirt and accumulated fish slime and debris. The whole fishroom should then be thoroughly hosed with clean water to remove all the remaining detergent. Great care must be exercised in the choice of detergents to avoid subsequent tainting of the fish. In selecting detergents for this job, the manufacturers should be consulted and their recommendations followed.

Every opportunity should be taken to dry out and repaint or revarnish wooden fishrooms; the same applies to boards and boxes. Insulation should be inspected at annual surveys, particularly in the region of the bilges. Thermometers should be checked at least once a year at ordinary air and ice temperatures. Refrigerating plant should be inspected and overhauled by the makers at the same time.

Maintenance of fishrooms is much easier if this is made the main responsibility of a member of the shore staff. A suitable man is a retired Mate with a reputation for turning out good fish. He may also be responsible for instructing prospective Mates and fishroom hands in the principles and practice of good handling and stowage. At least one large trawling company employs such a person. The existence of such a member of staff does not, of course, absolve the management from their responsibility for seeing that all other members of their staff and crews are properly aware of their own special part in producing fish of the best possible quality, and of ensuring that they have the necessary knowledge and skill and are provided with the best possible equipment and supplies for the care of the catch.

ON THE DRIFTER

In general, the comments made in the previous section, on the care of the trawler's catch, apply to the care of herring on drifters or, indeed, of any relatively small fish caught in large quantities by any method. But there are certain additional considerations. When fishing is light it is possible to handle the catch correctly; but when it is heavy it is practically impossible to do so. It may be possible only to handle a small fraction of the catch correctly, while the rest must be bulked perhaps even without ice.

Gutting of herring on the vessel is impossible, as it is also for other species, such as whiting and mackerel, when caught in large quantities.

Ungutted fish spoils more rapidly than gutted and this fact, together with the intrinsically more rapid spoilage rate of fatty fish such as herring, explains why, even under the best conditions of handling and stowage, these species remain edible for a much shorter time than gutted white fish such as cod. Herring itself, for example, remains edible only for a week or less even if buried in ice.

The fundamental problem is the same here as on the trawler, to cool the fish rapidly and keep it cool at about 32°F. To do this properly requires the use of about one part of ice to two parts of fish by weight, and careful distribution of ice throughout the fish. When fishing is light it might be possible to ice on this scale, although it would be difficult when fishing is heavy. Tests have proved, however, that a temperature approaching that of the ice can readily be reached by covering the whole of the bottom of the box with a good layer of ice before the box is filled with herring, and putting another layer on top of the fish.

The conventional six-to-the-cran box is sufficiently large to hold one-sixth cran of herring plus a $1\frac{1}{2}$-inch layer of ice at the bottom and 1 inch of ice at the top. Obviously, the depths of these layers cannot be measured precisely while the boxes are being handled and filled but, with suitably crushed ice, it should be practicable to get very close to the quantities indicated, giving a fish to ice ratio of slightly more than three to one. This should be sufficient for herring that are to be landed within eight hours of taking from the nets.

A clear way should be arranged for the water from the melting ice to drain from around the base of each stack of filled boxes. The stacks of filled boxes should be surrounded on all sides by ice or by a refrigerated air space and the bottom tiers of boxes should therefore rest on battens. Vertical battens should similarly be provided at the sides and bulkhead.

It is preferable not to disturb the herring between the time when they are boxed and iced at sea and the time when they reach the processor's premises. If further ice is seen to be necessary when the boxes are put ashore, it should be placed on top of the herring. Should the herring have to be turned into buyers' boxes at the quayside, a layer of ice should first be put into these. If insufficient ice is found on the top of the fish after it has been tipped into the buyer's box, more should be added.

DECK AND FISHROOM DESIGN AND EQUIPMENT

DECK DESIGN AND EQUIPMENT

This section deals only with minor improvement of the decks of existing trawlers; space does not permit of a discussion of the advantages or otherwise of, for example, stern trawling or the shelter-deck which, in any event, would be outside the scope of this book.

Immediately the cod-end comes over the side, the care of the catch becomes the responsibility of the crew. Their job is made easier by attention on the part of both owner and builder to certain relatively minor details. Fish should come into contact only with surfaces that can be frequently and thoroughly cleaned. To this end the deck in the region of the gutting pounds should be sheathed either in light alloy sheet embossed with an anti-slip tread or in some equivalent material. The bulwark with its frames and stiffeners is not suitable as a wall for the gutting pounds and a portable wall of boards should be erected 12 or 18 inches inboard. The spaces between wall and bulwarks can then be used as a dumping place for guts and rubbish preliminary to putting these overboard. For easy cleaning, both deck stanchions and boards should be portable and light. Seven-foot squares can be constructed from 3 ft 6 in fishroom boards if suitable intermediate stanchions are provided; the boards are then easily carried and cleaned.

A fish-washing machine, or at least a washing tank, should be provided. A convenient tank can be made from an extra hatch cover of galvanized steel or light alloy, of sufficient size to cover the main hatch coaming completely, and provided with cleats for lashing it in place to the deck. For use as a washing tank, this cover is turned upside down. Any baskets used for fish should be easily cleaned and sterilized and water should be able to drain away easily. Wire or plastic baskets are preferable to wicker ones.

The use of a washing machine has several advantages. It can be arranged so that rehandling of the fish from one side of the deck to the other is avoided, and it can help to ensure that fish is put below as soon as possible and at a steady rate instead of in intermittent larger quantities, thus permitting easy and proper handling by the fishroom crew. To do this, the washing machine is usually connected with portable metal chutes which guide the fish leaving the machine through the hatch and thence to the point of stowage. The first part of the chute usually consists of a roller conveyor which assists drainage of the fish. The standard British washing machine consists of a large open tank supported some 4 feet above deck and on the line of the hatches. A continuous supply of water overflows at a weir at one end, carrying the fish with it. The washing and overflow action are assisted by the motion of the ship (see Figure 2·5). A more elaborate type of washer

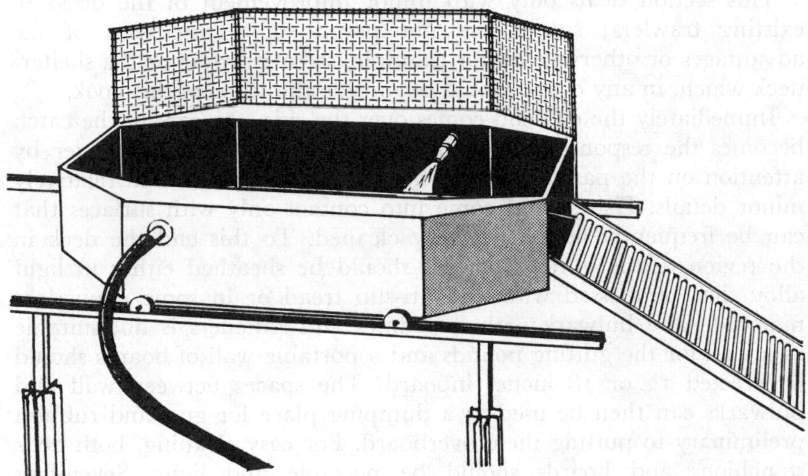

FIGURE 2·5 Standard British washing machine for trawlers

in use in German-built ships is of the rotating-drum type. The drum carries longitudinal wooden slats, is placed on a slight slope, and is electrically driven. Fish are fed in at the upper end and in the first half of the drum a series of water sprinklers washes them, the surplus water being drained off in the second half of the drum.

Washing machines thus eliminate the need to handle fish in baskets. Baskets involve extra labour and are difficult to keep clean. It is true that more sorting needs to be done in near water vessels and middle water vessels than on distant water trawlers, but there are several ways in which this problem might be overcome. With the usual type of British washer, sorting might be done below, or the washer might be arranged to deliver fish to a sorting station on deck beside the hatch, or the rotary type of washer might be used to enable sorting to take place before washing. The procedure just mentioned is not practicable with the ordinary overflow type of washer since the fish do not necessarily leave it in the order in which they are fed into it. One important advantage of a washing machine with portable chutes is that the manhole type of opening is large enough and the main hatch need not be opened at all at sea (see Figure 2·6). This reduces air exchange between fishroom and the outside.

Other deck equipment should include a steam hose to help in cleaning the deck battens and other fittings, an awning for use in strong, continuous sunshine, and a discharge nozzle for the deck-washing hose to spread a film of running water over the deck, again for use in exceptionally warm weather. Finally, adequate lighting is essential to ensure good handling of the catch on deck at all times.

FISHROOM DESIGN AND EQUIPMENT

(1) *Insulation.* The heat entering the fishroom from various sources, such as the surrounding sea or the engine room bulkhead, should all be absorbed by ice or refrigeration or it will warm the fish. The quantity of ice, or size of refrigeration plant required, depends upon the amount of heat entering the hold.

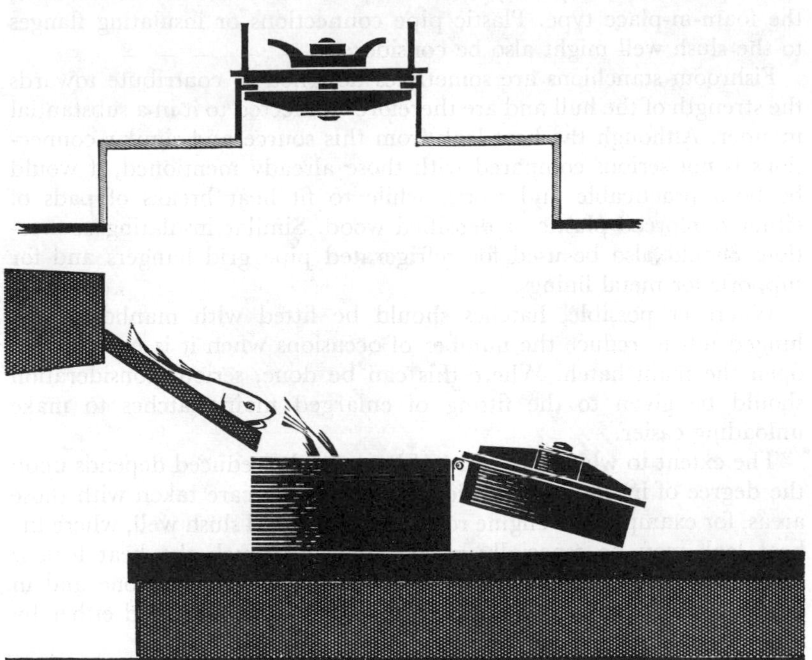

FIGURE 2·6 Manhole in fishroom hatch

Certain fairly obvious steps may be taken during building to reduce heat leak. Fishroom bulkheads should be sheathed with wood or lined and insulated on the fishroom side. The tops of double-bottom tanks must be wood-sheathed, covered with portable, well-fitting heavy wooden covers, or permanently insulated and lined, for example with high-density expanded ebonite or PVC covered either with concrete reinforced by steel mesh or with granolithic cement. These materials are best able to withstand the impact of dropped axes and boards and, although both wood and cement can harbour spoilage bacteria, good stowage practice should ensure that fish are kept well off the bottom of the hold. Similar steps should be taken to provide a moderate degree of insulation in vessels without double-bottom tanks but with cement ballast.

Slush wells may also be a source of trouble. Not only can heat enter through them but, unless care is taken in their design and construction, dirty melt water can collect in pockets in and around the well and produce offensive smells in both the fishroom and engine room. Consideration should be given to fitting a conventional type of slush well with an inner tank, possibly of reinforced plastic separated from the ship's structure by a layer of waterproof insulation of, for example, the foam-in-place type. Plastic pipe connections or insulating flanges to the slush well might also be considered.

Fishroom stanchions are sometimes designed to contribute towards the strength of the hull and are therefore connected to it in a substantial manner. Although the heat leak from this source and similar connections is not serious compared with those already mentioned, it would be both practicable and worth while to fit heat breaks of pads of either reinforced plastic or densified wood. Similar insulating connections should also be used for refrigerated pipe grid hangers and for supports for metal linings.

Wherever possible, hatches should be fitted with manholes with hinged lids to reduce the number of occasions when it is necessary to open the main hatch. Where this can be done, serious consideration should be given to the fitting of enlarged main hatches to make unloading easier.

The extent to which the entry of heat can be reduced depends upon the degree of insulation provided and upon the care taken with those areas, for example, the engine room bulkhead and slush well, where the heat leak may be especially great. However much the heat leak is reduced, it cannot be entirely eliminated by insulation alone and in order to avoid harm to the fish this heat must be absorbed either by ice or by mechanical refrigeration.

There is obviously a range of choice between, on the one hand, relatively light insulation with considerable reliance upon ice or mechanical refrigeration to absorb heat entering the hold before it affects the catch, and, on the other hand, heavy insulation with correspondingly less reliance on ice or mechanical refrigeration. The choice adopted will depend upon the fishery and the construction of the ship, and in many cases also upon the skill and attitude of the crews.

Steel Arctic trawlers may, for example, be operated successfully from Britain merely with 2-inch wooden fishroom linings and no other insulation. The heat leak into the hold of such vessels if operated from the warmer waters of Spain or Portugal would, however, be too great to be absorbed by practicable means and would therefore need to be

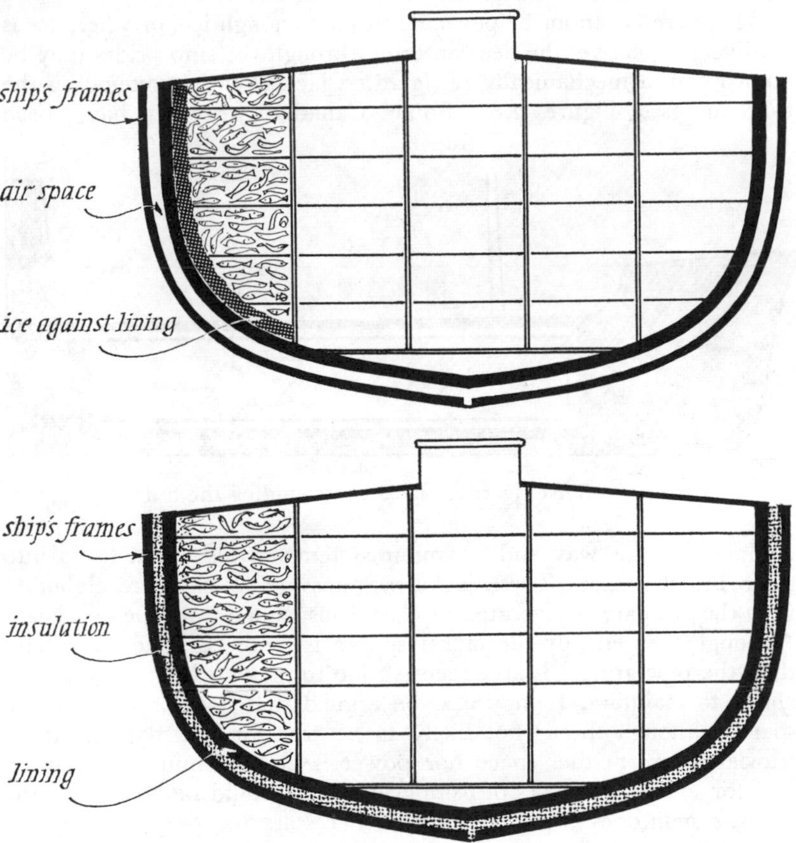

FIGURE 2·7 The value of insulation: (*top*) Uninsulated fishroom with wall of ice against lining; (*bottom*) Insulated fishroom

reduced by careful insulation. Within wide limits, however, the choice between insulation or cooling is difficult to make simply on objective economic grounds because of several technical and operational factors that must also be taken into account. The cheapest and most effective way of absorbing heat entering the uninsulated Arctic trawler with wooden linings, operated from Britain, is to put a sufficiently deep layer of crushed ice all round the catch. This means, however, relying completely on the skill and sense of responsibility of those stowing the fish; the Mate may be held responsible for the condition of the catch, but he cannot continually superintend stowage throughout fishing. Some insulation, even where insulation is apparently not strictly necessary, may therefore be regarded as a worth while insurance against the results of ignorance or carelessness.

Where crews cannot be persuaded to use enough ice, or where ice is relatively expensive, the heat entering through the ship's sides may be absorbed in a mechanically refrigerated jacket-space surrounding the fishroom (see Figure 2·8). Some Canadian trawlers have been

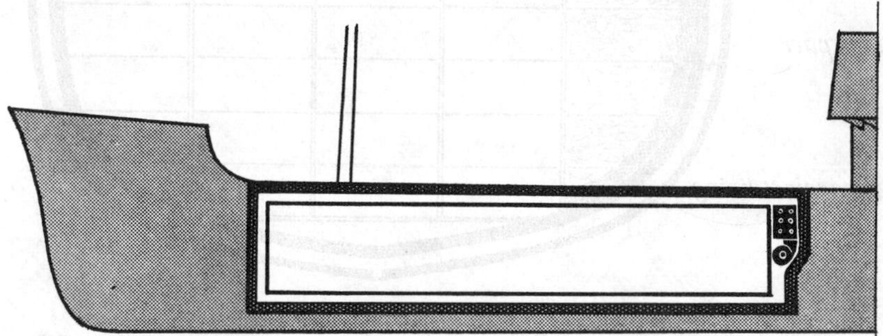

FIGURE 2·8 Refrigerated jacket surrounding the fishroom

equipped in this way and a similar system has also been fitted into some British vessels. These jacket systems are much more elaborate than the ordinary refrigerated pipe grid suspended from the deckhead. Although theoretically the jacket system is more effective and useful than the pipe grid, it is an expensive job to make it both reliable and cheap to maintain. It may also be argued that it takes up valuable space; although this is not likely to be true for the modern Arctic trawler where neither space nor power is at a premium, it may be true for some other types of fishing vessel. It should be noted that the jacket system does not make any less necessary the proper stowing of fish with ice in order to cool it down rapidly. Where ice is cheap, the simplest choice for the owner is to rely on ice, combined with a moderate amount of insulation.

Insulation is essential for fishrooms lined with metal or plastic, except, perhaps, where the metal is merely a sheathing fixed over a substantial wooden lining. The best types of insulating material for the holds of fishing vessels are those whose properties are not affected by rot or by water, even when fully or partly immersed. Leakage of some water into the insulation is almost inevitable; there may be leaky rivets or seams, fishroom linings are rarely if ever watertight and also some condensation cannot be avoided. Naturally-occurring materials, such as cork, are not ideal because they absorb water and also because their properties vary to some extent from sample to sample. Although materials made of glass fibre or mineral fibre and of aluminium foil and plastic materials of open-cell construction do not rot, their insulating values can be reduced when water and slime fill their air spaces. The best materials, therefore, seem to be those of closed-cell construction; foamed plastic and expanded ebonite are examples. Foamed-in-place, closed-cell plastic insulation is now available; in order not to exceed an economic thickness and also to allow external hull repairs, this may have to be applied between two linings. One disadvantage of plastic cellular insulation materials compared with glass wool or mineral wool is their greater susceptibility to heat and fire.

Air movement, mainly by convection, increases considerably the transmission of heat through vertical insulated walls, such as ship's sides, to above the theoretical value indicated by the thermal conductivity of the material; this is true even of walls insulated with slab materials. A great deal of care and supervision is necessary during erection to eliminate as far as possible gaps in joints between slabs where air can circulate.

Since complete air-tightness can never be achieved, and it is not likely that the layer of insulation and the fishroom lining will be equally leaky, there is more than a possibility that moisture from the atmosphere will condense between lining and insulation or between insulation and ship's side. Although at most times the hold will be colder than the sea, it is possible that when in dock, and during outward voyages in winter, the water may be colder than the hold. This condensation can occur on either surface of the lining or the insulation, as well as on the ship's plating. It is therefore essential to protect all wood grounds by suitable pressure impregnation and, if a wood lining is used over a closed-cell plastic insulation, it is preferable to leave an air space, ventilated into the hold, between the two. In wooden ships, an air space between the outer surface of the insulation and the wood of the ship's structure is also essential. This air space should be vented to the atmosphere and all structural wood well coated with preservative.

Choice of lining material lies mainly between wood and metal, although cement is being used by some owners, and plastics of various kinds are being tried from time to time. Wood is tough, resilient, and withstands ice-axes and dropped boards, but as normally fitted it harbours bacteria and is difficult, although by no means impossible, to keep clean. Metal is easier to keep clean and the additional expense of light alloy linings and boards can also be justified where the last fraction of a knot is important. Metal linings are usually copper-free aluminium alloy in the half-hard condition. In metal ships, they involve the expense of insulation, whereas this is not absolutely necessary with wooden linings in Arctic trawlers. Another slight disadvantage in large trawlers is the massiveness of light alloy stanchions of sufficient strength.

Considerable care has to be taken to prevent corrosion due to galvanic action caused by contact of dissimilar metals in the presence of water, especially sea water. These form what is virtually an electric battery; a small electric current is produced and corrosion is enormously accelerated. Where joints between dissimilar metals are inevitable they should be treated with an approved paste or be made electrically non-conducting. Galvanic action can also be responsible for decay of wood where this is in contact with metal.

A practical advantage of light alloy linings is that lighting and visibility in the hold are better than in wooden fishrooms. This advantage is also offered by cement linings which, however, seem to have little to recommend them in other respects. They are slightly porous and harbour bacteria, they are heavy and are liable to crack and to be damaged by axes.

Wood might possibly be the ideal material for fishroom linings and boards, if it could be impregnated to a sufficient depth with a suitable plastic at an economic price. Plastic reinforced with fibrous materials such as glass fibre has been tried experimentally, but it seems difficult to obtain sufficient rigidity and resistance to abrasion at a reasonable cost. It must be borne in mind that plastics reinforced with glass fibre have no great heat-insulating value. The real insulating medium in glass wool, as in all other heat-insulating materials, is the entrapped air. It is good practice to insulate steel deckheads even if these are covered with wood decking. Where light alloy linings are used deckhead linings are almost essential in order to avoid galvanic action.

The amount of heat entering fishrooms is not accurately known, because the structure and conditions of operation of fishing vessels are so variable and complex that calculation is difficult and so is precise measurement. Experiment has shown, however, that in an uninsulated, wood-lined, steel Arctic trawler making a voyage to Greenland at the time of year when the water is warmest, a layer of crushed ice 9 inches thick against the side linings was more than adequate to absorb the heat entering the hold through the sides (see Figure 2·7). A 6-inch layer would probably be sufficient for similar vessels in such conditions. Where insulation is used, a practical thickness on both sides and deckhead is probably 2 inches of expanded ebonite or PVC of low-density grade, or its equivalent. For vessels operating regularly in the southern North Sea or even further south, a somewhat greater thickness of up to 4 inches might be worth while, depending upon the price of ice and other considerations.

The sea is usually a much more important source of heat than the air. The hulls of most fishing vessels, especially in the region of the fishroom, are well immersed even in calm weather. Also, water has a very much greater heat capacity than air. Furthermore, the relatively high speed of the vessel causes the water to swirl and eddy round the hull; under these conditions heat transfer is rapid. The deck, on the other hand, is more or less sheltered from the full effect of the breeze. These factors outweigh the effect of the higher air temperatures in summer, except, perhaps, in strong sunshine. Some vessels, moreover, spend a considerable time anchored in rapidly-flowing tideways, or in docks that can be as warm as 70°F, before discharging their fish. It should be noted that although peak air temperatures may occur in July and August, peak sea temperatures are usually in September. Consequently, although it is reasonable to insulate a steel deckhead even if the deck is wood-sheathed, it is more important to pay attention to ships' sides and bulkheads and to the bottom of the hold.

How a bulkhead is treated will depend partly upon the temperature of the space on the other side of it, but it must be remembered that the bulkhead is in any case connected to the ship's plating which in a steel vessel is practically at sea temperature. The aftermost part of the hold has always been regarded as a problem area, not only in coal burners with tunnels connecting the after fishroom and the stokehold, but even in modern motor trawlers. Usually, but not always, the first-caught fish is stowed in the aftermost pound, and it may be that criticisms of quality are sometimes merely a reflection of this fact. Since trouble is experienced in motor trawlers, it can only partly be due to heat leak from the engine room, since in some vessels the two spaces are separated both by fuel tanks and a coffer dam. It seems reasonable to assume that a more important cause is the connection of the bulkhead to the ship's plating.

Another possible source of excessive heat leak into the after end of a trawler fishroom is the slush well. The slush well is part of the structure of the ship and in a steel trawler is of necessity connected to the plating and to other parts of the structure. It is also connected to the engine room by substantial piping. Many modern trawlers are constructed with double-bottom tanks, and in some vessels the metal tank top forms the bottom of the fishroom. Measurements in one vessel showed that the temperature of this metal was within a degree or so of the sea temperature, and this was over 50°F for a number of days. The heat leak through the ship's plating and the structure of the double bottom to the tank top and thence to the interior of the fishroom must have been very considerable. Excessive melting of ice on the outward voyage due to heat from this source has also been experienced.

An entirely different source of heat leak is movement of air through the fishroom hatches. The extent of this varies greatly but will be of secondary importance to the inflow of heat from the sea. So, also, will be the heat actually produced within the fish itself as it spoils.

(2) *Mechanical refrigeration.* The heat entering the fishroom may be absorbed by a mechanically-refrigerated jacket system. The system usually employed on British trawlers, however, consists of a grid of plain pipes suspended from the deckhead and refrigerated by a reciprocating condensing unit usually driven electrically and controlled by hand. Such a system is limited in its usefulness and must be operated with care and intelligence. Certain refinements are possible which make it both more useful and more flexible and foolproof in operation.

A refrigerated pipe grid on the deckhead will not absorb heat entering the bottom of the fishroom and the ship's sides in the part of the hold where fish has been bulk-stowed or boxed. Here, the heat will either melt ice on the bottom or against the linings or, if insufficient ice remains, will warm up the fish, whether the linings, divisions and boards are of metal or not. Under these conditions the grid will continue to absorb only the heat coming through the deckhead and with any inleak of air. Where fish is shelved at about 9-inch intervals, with air spaces above the fish, part of the heat entering the ship's sides will be absorbed by ice, but once the ice against the linings has melted, some of the heat at least may be transferred by convection to the refrigerated pipe grid.

Overhead grids are quite useless for cooling fish to the temperature of melting ice and only intimate mixture of fish and ice will accomplish this. The importance of melt water in the cooling of fish and the preservation of appearance has already been mentioned. Taking all factors into account, it seems to be advisable to allow ice to melt and there is no case, therefore, for the use of refrigerated deckhead grids for bulked or boxed fish, in the climatic conditions in which British fishing vessels normally operate. If the hold is carefully insulated, and the shelf boards are of metal, ice alone might still be adequate to keep air temperatures in the low and middle thirties. In uninsulated, wood-lined Arctic trawlers, if insufficient ice is placed against the linings, the temperature of the air above the shelfed fish can, in summer, rise to 40°F or more. It is therefore, perhaps, advisable in this case to refrigerate the air as an insurance against the insufficient use of ice.

In some vessels, the bulked fish is carried below *staging level* (see Figure 2·3, *top*) and shelfed fish above; in such cases the usual single-circuit grid may prove adequate although temperature control may be difficult. The more usual stowage practice differs from this, possibly because of tradition, custom and training, and bulked fish is usually stowed right up to the deckhead aft and shelfed fish forward. Owing to the trim of the ship, warm air tends to accumulate under the deckhead in the forward pounds. In most trawlers, it is not practicable to stow bulked fish forward so that means have to be provided to refrigerate the air in the forward fishroom in the region of the shelfed fish. This can be done by a conventional single-circuit grid limited to the forward fishroom or, where the whole deckhead is refrigerated, the grid may be divided into two, or preferably four, circuits, so that the aftermost circuits may be turned off when fishing begins. If this is done, two fishroom thermometers should be provided and, if automatic control is used, one thermostat in the forward fishroom.

Four rather than two circuits are suggested in order to make easier the provision of rapid hot-gas defrosting. This allows operation of the grids until an hour or so before discharge, or for as long as power is available; without this method of defrosting, refrigeration is turned off perhaps 48 hours before the vessel docks in order to ensure that the grids are defrosted and dry before unloading actually begins. The already unpleasant job of unloading the fish may be rendered even worse by water dripping from the grids, if these are not allowed to dry before discharge begins.

Perhaps the principal effective use of refrigerated deckhead grids is the maintenance of relatively low fishroom air temperatures on the outward voyage. The advantages are that the ice is kept in a crisp condition in which it is more easily handled, the fishroom and its fittings are already cooled when stowage begins and the growth of bacterial slime on wood and other surfaces is considerably reduced. The refrigeration should be turned off, however, when fishing begins and full reliance placed upon ice except in those cases already indicated.

Control of fishroom refrigerating plant is usually by hand, but it is doubtful whether this is either wise or necessary. The maintenance of a correct temperature must of necessity be an item of low priority in the scale of values of either Skipper or Chief Engineer, preoccupied as they are with the safety and efficient running of the ship and machinery and with catching fish. Automatic thermostatic control overcomes these difficulties. A standard type of commercially available thermostat with automatic starter has been shown to be completely reliable over a period of months when installed in the forward end of a trawler's fishroom, with the starter in front of the bulkhead.

A more difficult problem is the siting of the sensitive elements of any thermostat and of distant-reading thermometers. The air space whose temperature is to be controlled is of very variable shape; it may vary from 30 feet wide by 50 feet long by 16 feet deep, through many intermediate shapes, to the same length and breadth but a depth of only 9 inches or even less. The trim of the ship and the amount of heat to be absorbed may also vary. It is furthermore necessary, if temperature is to be controlled within the desired limits or accurately measured, not to site the elements near fish, ice, bulkheads, stanchions, hatches and ship's sides, to shield the elements from the direct effects of the refrigerated pipes and also, on the outward voyage, from the effects of the close proximity of as much as 100 tons of ice. Thermometers placed on especially warm surfaces, such as after bulkheads, give readings that are difficult to interpret and which may mislead a Skipper or Mate unfamiliar with the vessel. The best guide to the positioning of the elements and interpretation of readings is experience, aided, if possible, by a thorough temperature survey of the hold. The following preliminary suggestions may be found useful. Two distant-reading thermometers should be fitted, one forward and one aft. If a thermostat is fitted the bulb should be placed beside the forward thermometer. The thermometers should be in the third pound from aft and the third pound from forward in a fishroom of 11 to 14 pounds. They should not be close to a hatch if possible and should be about half way between ship's side and the line of the hatch coaming and away from stanchions. They should also be below grid level and shielded from grids and ice by a box open at the ends and only perforated on the sides; care should be taken to ensure that the leads do not touch grids or grid hangers. The thermostat should be set to cut out at 33°F to 34°F and to cut in at 36°F to 38°F.

Fans should be used only in refrigerated jackets and not in open fish holds. Drying of the exposed surfaces of the shelfed fish is inevitable with mechanical refrigeration and fans will aggravate this without significantly assisting cooling.

It appears that in order to obtain completely trouble-free service with low maintenance costs, a refrigeration plant of somewhat more refined design than that normally fitted to fishing vessels may be advisable, and more care might be taken in the siting of the compressor unit. The size and horsepower that is desirable in any particular case can be estimated when the size of hold and type of insulation is known. To avoid leaks of refrigerant and loss of oil, all piping connections should, as seems appropriate, be welded or brazed, and hermetic compressors should be installed. This latter requirement would, where there is the usual d.c. electrical system, demand a magnetic gland; with an a.c. system, a standard hermetic unit could be fitted. An air-cooled condenser must be installed to avoid water circuits and pumps. Unless the condensing unit is in the engine room, care should be taken that the space is well ventilated. To avoid condensation in motor and starter when the unit is switched off, as it may be for several days at a time, the space should be heated or some heat provided locally near motor and starter. Finally, open driving belts may be avoided by using modern direct-driven compressors of moderately high speed.

It will be realized that there are several difficulties associated with the satisfactory operation of the usual type of refrigerating plant and it may be again emphasized that satisfactory results can be obtained by a trustworthy Mate using ice and insulation alone.

(3) *Fishroom fittings.* Apart from considerations of strength, corrosion and cost, the main points to be observed in the design and erection of a pound and stanchion structure in a fishroom are that it should be easy to clean thoroughly and that it should make stowage and discharge easier rather than more difficult.

Ease of thorough cleaning is best gained by making portable as much of the structure as possible. For example, pound divisions and rest angles should, wherever possible, be portable with a minimum of fixed rest angles, battens or cleats. Fixed wing-divisions should be fitted only where the shape of the ship would otherwise necessitate the use of portable boards of non-standard length. The fixed part should be neither higher nor further inboard than necessary to avoid the use of non-standard boards. This, incidentally, goes a long way towards giving a hold that is easily stowed and discharged. Stanchions, rest angles and cleats should be so designed that there are no sharp interior angles where fish slime and scales can collect, so making them difficult to clean.

Rest angles that bear on cleats or pegs directly attached to stanchions are preferable to those that are built into the vertical divisions. The advantage of such a design is that the breadth of the boards is then independent of the spacing of the shelves and also that, because the vertical divisions are free of vertical loads, the boards are less likely to jam in the stanchions. For shelfed stowage, the vertical spacing of the shelves is usually 9 inches and, except in the type of design recommended, this must mean a board 9 inches broad. Wooden boards are much less liable to split if they are only 6 inches broad. For bulked stowage, provision should be made for rest angles at vertical intervals of 18 inches.

Fishroom boards should be sufficiently strong and yet cheap and easy to clean. Some designs of metal boards are criticized on the grounds that they are likely to mark the fish; this is unjustifiable since the fish should always remain separated from boards, and everything else, by ice. Some designs of corrugated boards are also criticized because where fish is stowed *above the batten*, so that the next layer of shelf boards and the fish on it is initially supported by the fish below and not by rest angles, the shape of the board is such that it does not always come to rest on the battens but sometimes slips through. This might be dangerous, and in any case will harm the fish by continued excessive pressure from the layers above. The answer to this criticism is that in no circumstances should fish be stowed above the batten (see Figure 2·9).

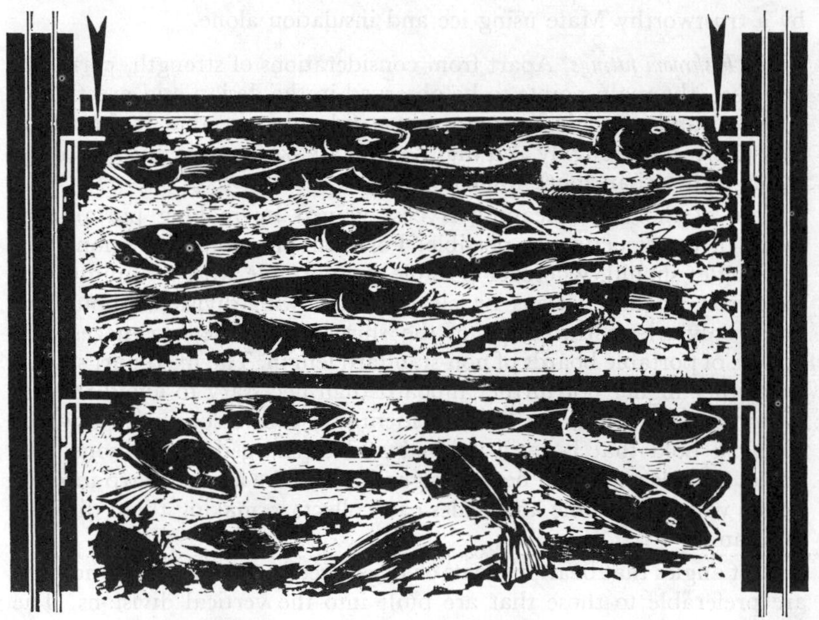

FIGURE 2·9 Stowage above the batten

The strength of metal boards is a somewhat controversial subject. Some companies are content with shelfing boards adequate to support only a single layer of fish on ice, with stronger boards for bulked stowage. Whether the boards, other than the centre staging boards, need to be designed so that they bend very little when one or even two men stand in the middle is not agreed. The argument is that the man is supported by the rest angles rather than by the boards. At the same time, if fish is stowed above the batten, the lowest shelves may have to bear a considerable load, and this is also true of the vertical divisions where the pound on one side is full and on the other is empty. The sensible criterion of strength appears to be that the board should be as strong in bending as the usual softwood board, after some use and when waterlogged, and, when the load is removed, should not remain bent but return to its original shape. A more severe criterion would be that for a given load the deflection of the metal board should be no greater than that of a wooden one.

There are certain other points that should be borne in mind when drawing up a specification for pound boards and linings. The surface should be hard, hygienic and easily scrubbed but the finish of boards should not be such as to make them slippery to stand on, especially at sea, or to make bundling and slinging of boards dangerous in dock. Boards laid fore and aft should be able to keep ice in place on a rolling ship. Boards and linings should, of course, contain nothing that must not come into contact with foodstuffs, and they must be able to withstand attack by sea water, weak ammonia and amines from fish, or fish spoilage bacteria, without corrosion or rotting. They should also be completely impermeable.

The strength in bending of boards has been discussed above. Boards must be able to withstand being dropped on to a hard surface from heights of up to 15 feet. They should also be able to stand up well to glancing blows from ice axes and should not be penetrated by blows from unloading hooks.

It should be possible to construct vertical divisions by sliding the ends of boards in channels in stanchions. The ends of the boards should be curved so that they can be easily introduced and withdrawn from the channels; they should be so constructed that they will rest upon each other edge to edge without jamming. It is important that the shape of boards should allow them to be used for constructing horizontal platforms by resting the boards on horizontal battens of about 1 inch effective breadth. The individual boards must be able to sit firmly without rocking, in spite of their curved ends.

Boards should be grooved or corrugated longitudinally to assist drainage; if this is impracticable, as, for instance, in the case of wooden boards, they should be perfectly smooth. They should preferably be made to interlock or overlap to prevent slime and melt water from pouring through shelves on to the fish below. The material used for construction of boards must be light as possible, compatible with the considerations of strength mentioned above. Pound boards should weigh 5 lb or less; waterlogged wooden boards may weigh 15 lb or more. There should be no great change in weight during use through, for example, waterlogging. This means in practice that wooden boards must be proofed with some impervious material.

Few fixtures are required for boxed stowage other than those already mentioned. A series of battens or beams, arranged to allow free fore-and-aft drainage and free flow of melt water to the slush well, should be provided at the bottom of the hold. The purpose of these is twofold; to support the bottom layer of boxes above the dirty melt water collecting at the bottom of the hold, and to create a space into which crushed ice can be placed to absorb the heat leak through the ship's bottom. The boxes should be carried 3 to 6 inches above the bottom of the hold, depending upon the degree of insulation and whether the ship is wooden or steel.

Stowage of boxed fish, owing to the trim of the ship, usually begins forward of a bulkhead which again should be provided with a series of vertical battens some 3 inches deep, and so disposed that they prevent boxes being stowed in close contact with the bulkhead, whilst allowing ice to be placed in the remaining cavities. The curvature of the sides of the hold may in itself be sufficient to create spaces at the ends of the tiers of boxes; ice can be put into these spaces. Where the sides of the hold are relatively straight, battens similar to those suggested for the bulkhead should be fitted unless the sides of the hold are directly refrigerated by jacket or otherwise.

Boxes should, for several reasons, be of light alloy or plastic. They are easier to clean than wooden ones and, once clean, bacteria and moulds do not begin to accumulate again. They are light and easy to handle and far less ice is needed to cool them than with wet wooden boxes. One possible disadvantage of light alloy boxes is that they are not suitable for long-distance transport either on open lorries or in rail vans, although they are used by firms with their own fleets of insulated road vehicles. Boxed iced fish should, in any case, be transported in closed insulated vehicles.

Light alloy and plastic boxes should be so designed that a stack of them is entirely stable. On the other hand, it should be impossible to build stable stacks when the boxes are over-filled to such an extent that the fish rather than the sides of the box bears the weight of the boxes above. There seems little point in providing lids for boxes to be transported in insulated vans. Drainage should for preference be towards the corners of the bottom, rather than in the centre; if possible a box should not drain into the box below. The box should be smooth, with no internal sharp corners that are difficult to clean. Some types of open box have been designed to save space in the stowage of empties. For example, two boxes may nest into each other in one position but when the top box is turned end for end, it is supported on the rim of the box below. A box that will nest, stack, and drain outside the area of the box below, but yet waste little storage space, would be ideal.

Finally, it should be noted that well distributed and adequate lighting, associated with light-coloured linings, boards and boxes, can do a great deal to ensure correct handling during stowage and discharge. There should preferably be lights in every pound.

3

The handling and distribution of fish on land

Here, as on a stone raft between the seas,
The fish-wives,
Armed with knives,
Called to each other over damp, slimy stalls
With thick briny laughter.
Here Mrs. Busk, a mountain in oilskin
With a creased tarpaulin face,
Bought her wares,
Squeezing, testing, prodding with appraising thumb and finger.

SIR OSBERT SITWELL
Mrs. Busk

THE fish landed at British ports is of two main types: white fish, such as cod, and fatty fish such as herring. Methods of fishing and subsequent handling and distribution in the two cases are so different that in this chapter they are dealt with separately.

THE LANDING OF WHITE FISH

Most British-caught white fish is obtained by trawlers and landed at the larger ports. Nearly all Hull's fish comes from distant waters and is mostly cod. Two-thirds of landings at Grimsby and a little of those at Fleetwood also come from distant waters. Each trawler usually lands between 100 and 200 tons of fish mixed with ice. Aberdeen is predominantly a near water and middle water trawling port; vessels are here more numerous and catches smaller, being about 40 tons from middle waters and 10 tons from near waters. North Shields and Milford Haven also fish the near waters and middle waters but on a much smaller scale. There is some long lining from Aberdeen for halibut, as far afield at times as Greenland. Vessels from Lowestoft, which fish mainly the near waters, make frequent but smaller landings of fresher fish. At small inshore ports there may be a range of size and type of vessel from seine netters, which can land from 5 to 10 tons, down to hand-lining craft each landing at the most only a few hundredweight.

Icelandic, German and Belgian trawlers occasionally come direct from the fishing grounds to land at the larger British ports. In addition, fish is imported by both passenger and cargo vessels; for example, boxed and iced North Sea fish is carried from Denmark to Harwich and from Norway to Tynemouth. It is then taken by road or rail to the main wholesale markets at the ports or inland.

Some British-caught fish is also landed frozen. For example, some is filleted and frozen at sea, and there is also a growing fleet of trawlers equipped to freeze their catches as whole fish. In addition, expensive species, such as salmon and halibut, are imported frozen from Canada and Japan through ports such as London and Liverpool, where there is available cold storage space.

At the distant water ports the market is usually notified several days in advance of the supplies that will probably be available from day to day. Landings are generally spaced out to meet the likely demand for fish at the ports. Extreme fluctuations in supply and demand are usually avoided, although numerous factors, including the weather, make the quantities of fish available predictable only within rather wide limits.

At all ports except the very small ones, the fresh fish, that is unfrozen and unprocessed fish landed from fishing vessels, is normally auctioned on the quayside at about 8 a.m. At the larger ports, vessels arriving during the day usually queue at the dockside waiting to be unloaded; this operation begins at different times from midnight to 4 a.m. in various ports.

The hatches are normally raised some time before unloading begins, so permitting ventilation of the fishroom an hour or so before men go into it. During a long trip it is possible for dangerously large amounts of carbon dioxide, produced by bacterial respiration and chemical changes in the flesh, to accumulate in the fishroom. These changes also use up oxygen and as little as 4% has been observed in fishroom air, whereas ordinary air contains 21% oxygen. The air returns to normal an hour or so after the fishroom is opened up, or even faster if a fan or pump is used. Although, therefore, there is in practice generally no danger, there is need for initial care, especially when trawlers arriving late for a market are unloaded as soon as they dock.

The large catches landed at the Humber ports make it necessary to employ special dock labourers, termed *bobbers* or *lumpers*. The trawler-owning companies employ foremen who have plans showing when and where the fish was stowed in the hold, and are in charge of the unloading operations. The crews of inshore and near water vessels may themselves help unload their catches and at small ports fishermen generally land their own catches.

At the smallest ports unloading is carried out entirely by hand. In most of the larger ports, however, fish is loaded by hand into wicker baskets with the help of hooks, which are supposed to be used only on the heads of the fish, but the flesh is also sometimes damaged. The filled baskets are then lifted out by electric winch. At Hull the baskets are swung on to the quayside where they are tipped into round aluminium alloy containers called *kits** and the net weight adjusted to 10 stones. The kits are then taken by hand barrow and set down on the market in lots, or squares, laid out more or less in order of catch, although there is inevitably some mixing of fish of different degrees of freshness within a square and even within a kit. This is also the case at other ports. At Grimsby, where the construction of the market is different, the baskets are landed on to movable boards that bridge

*At one time the smaller ports used various, curious local measures of weight and capacity. These terms have now practically all died out and today the stone of 14 lb is almost universal. Nevertheless, some shellfish are still sold by capacity measure, such as the gallon, and oysters and lobsters are still reckoned by number rather than weight. Examples of local measures are the *pad* of 5 stones at Brixham, the *wain* of 12 lb in N. Wales. The usual measure for sale of herring is still the *cran*, a volume measure. Formerly, owing to the method of filling the quarter-cran basket, the English cran was reckoned to be 3¾ cwt, whilst the Scots cran was 3½ cwt. It has now become standardized at the lower figure (28 stones, or 392 lb). The *long hundred*, of pilchards, herring and mackerel, varied from 120 to 133 individual fish in various places on the East and South Coast. The *last* was 100 long hundreds. The Yarmouth *swill*, which is still used, held about 131 lb, three making a cran. Eight *margarine* boxes also went to a cran. A *warp* of herring and mackerel on the East Coast consisted of four fish. A *maund* of sprats at Lowestoft was 33 lb, a *maze* was 1½ to 2½ cwt, or 600–720 herring, varying from place to place. The *wash* of whelks from the Thames estuary, once used as bait, amounted to about 41½ pints, and was about 40 lb.

the gap between ship and shore. They are then dragged to tables for sorting and weighing into aluminium boxes, called *kits* or *trunks*, which also hold 10 stones. At Aberdeen, the fish is laid out for sale without weighing in lots of about 8 stones, usually in wooden boxes. The weight of a small proportion of these lots is subsequently checked by employees of the trawler owners.

Granton trawlers stow white fish at sea in wooden boxes holding about 6 stones of fish. These boxes are sent mostly to Glasgow market, although some arrive by road for sale in Aberdeen, North Shields and even Hull. A few vessels land boxed fish at some of the other near water ports. Fish cannot easily be weighed at sea, so that the measure is only an approximate one.

Rough handling and bruising during unloading spoils the appearance and shortens the subsequent keeping time of the fish, although the effects are not immediately obvious. Fish boxed at sea is damaged less in unloading, and there is no danger of damage and disfigurement by hooks.

From time to time attempts have been made both here and abroad to reduce the labour required in unloading by installing various arrangements and machines to increase efficiency. In any such mechanical handling, damage to fish must be avoided, although, if the fish are to be canned or made into meal, damage to the appearance is of less importance. Pumps are used in some countries for unloading small fish like herring.

At Murmansk the task of unloading is said to be lightened by laying down nylon nets as the top layers of fish are stowed. The nets, each carrying about 6 cwt, are then pulled out. This avoids the delay caused by the initial digging down through the top layers of ice and fish into the fishroom before there is room for the full labour force.

Weighing, or estimating the weight, of the catch is an essential feature of unloading at all ports, except those where, as at Granton, the fish is boxed at sea. This weight is the basis for both the crew's pay and for sale. In order to weigh the fish at the market, all the ice, which up to this point has kept the fish chilled, must be removed.

CAN HANDLING OF WHITE FISH AT THE MARKET BE IMPROVED?

Immediately fish without ice is exposed on the market its temperature begins to approach that of the air. The freshness of fish is decided by the temperature at which it has been kept, and the time since it was caught. The higher the temperature during handling and distribution the more rapidly the fish goes bad. No matter whether the fish is fresh or already stale, the speed of *further* deterioration can be kept at a minimum only by thorough icing.

Despite the widespread use of ice in the fish industry, fish could nevertheless be kept fresher during distribution if still more ice were used and if it were more thoroughly mixed with the fish. This is the conclusion from an extensive survey of commercial practice carried out by DSIR.

The bulk of distant water fish is, and in fact has to be, iced sufficiently well at sea to keep its temperature at about 32°F. Although the average temperature of fish as it comes out of the trawler is about 33°F, some temperatures measured during the DSIR survey were as high as 40°F. In some instances fish aboard ship have been found to be at 50°F. Such high temperatures result from the use of insufficient ice, or from large amounts of heat leaking into the fishroom, through, for example, uninsulated double-bottomed tanks that form the bottom of the fishroom in some of the large modern trawlers.

Inshore fish, which is not always iced at sea, may often reach 60°F in summer by the time it is unloaded. Even where ice is available it is quite often not used because, it is said, the prompt sale of the fish makes icing unnecessary. Nevertheless, considerable deterioration can occur in a short time in fish that, for obvious reasons, ought to be very fresh.

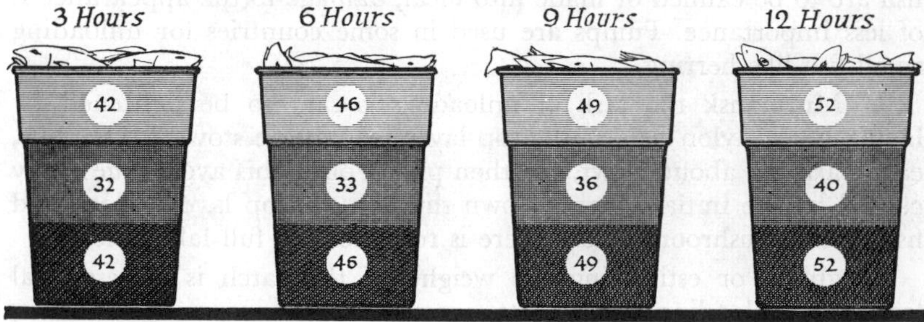

FIGURE 3·1 Temperature of fish in kits on Hull market in summer

Fish iced at sea normally begins to warm up after it is unloaded and while it is standing on the quay waiting to be auctioned. Figure 3·1 shows that the temperature of some fish at Hull can, in summer, become as high as 46°F by the time sales begin at 8 a.m. even though the average is considerably lower and some fish, probably those situated in the middle of a container, are still at 33°F. The average rise of temperature is a few degrees less at Grimsby, where the trunks of bulked distant water fish are piled four or five high, than it is at Hull where tub-shaped kits stand singly. North Sea and shelfed fish, however, which is laid out in single trunks, warms more quickly. Fish lying on the market floor either in piles or individually, such as halibut, for example, will warm even more rapidly. Although it would be preferable to put ice on top of all fish standing on the market, it is sometimes argued that fish merchants and Public Health Inspectors would not then be able to examine the fish. Some North Sea fish at Grimsby and Lowestoft has for a number of years, however, been top-iced in summer and this practice seems to be spreading. At Bremerhaven, which is the size of Hull, all boxes of fish standing on the market are top-iced even in winter. Fish, even if it is well iced, should always be shielded from direct sunlight.

As only the top layer of fish is affected by top-icing, this practice is more effective in a flat market container, as used at Grimsby and elsewhere, than it is with the taller kit used at Hull. Mixing ice with the fish would be much preferable to top-icing as it would distribute ice where it is needed instead of confining it in one place. This would, however, introduce handling and weighing difficulties.

The need for continued chilling still applies after the fish has been purchased and is taken away for filleting or other processing, whether

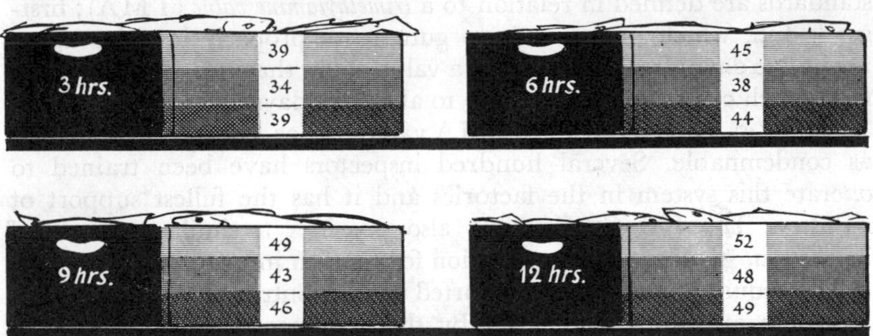

FIGURE 3·2 Temperature of fish in single trunks on Grimsby market in summer

on or off the market. Fish that is not to be dealt with in an hour or so should be iced. Generally speaking, ice is too sparingly used and not well enough distributed to keep fish as cool as it needs to be. By mixing fish with crushed ice it can be cooled much more quickly than it can be merely by placing kits or boxes in a chill room. Air is such a poor conductor of heat that to cool the centre of a 10-stone container of fish would take several days in a chill room running at about 32°F. A chill store is only of use in keeping cool fish that has already been thoroughly cooled to melting ice temperature.

CHECKING THE QUALITY OF WHITE FISH ON THE MARKET

Public Health Inspectors employed by the local Health Authority examine fish displayed for sale and condemn any that is 'unsound, unfit or unwholesome' for human food. This is then usually made into fish meal. An average of about 1% of the total landing at Hull and Grimsby is condemned.

In addition, there is a further 5% that, although not bad enough to be condemned, is not sold at the auction either because landings are greater than demand or because quality is poor, or the size not suitable. Much of this 5% is sold for canning for domestic pets, and most of the rest is made into fish meal.

To ensure that only the freshest fish goes into the normal fresh fish trade, the Hull trawler owners have introduced a scheme employing trained quality assessors, who recommend for withdrawal from the auction fish that is below a certain level of freshness.

In some fishing countries grading of the quality of fish is very thorough. In Denmark, for example, there is a rigid system of government inspection, particularly of fresh fish for export. In Canada, also, government inspectors examine white fish at landing and during processing and apply rather stringent standards of quality. The standards are defined in relation to a *trimethylamine value* (TMA); first-grade fish, which corresponds to gutted fish properly iced for up to about five days after catching has a value of less than 1·0. Second-grade fish, which corresponds to fish up to about 11 days in ice, are between 1·0 and 5·0. Any fish with a TMA value higher than 5·0 is regarded as condemnable. Several hundred inspectors have been trained to operate this system in the factories and it has the fullest support of industry. The US Government also provides a similar scheme of inspection, grading and certification for the fish industry.

The requirements for fish exported under contract from the United Kingdom are being influenced by the existence of these systems of inspection in other countries. There seems little doubt that the British fish industry will need eventually to adopt similar methods.

THE LANDING OF HERRING

The landing and marketing of herring differ from that of white fish in a number of ways. The fishing vessels, which are mainly drifters, are usually based on their home port. Nevertheless, they still follow to some extent, although less than formerly, the seasonal movement of herring shoals off the coasts. Thus, drifters from all the major herring ports fish off north east Scotland in the summer and East Anglia in the autumn, landing their catches side by side with the local vessels at the nearest port. At certain seasons herring can be caught by trawling. British vessels do not nowadays participate in this fishing.

In the past the drifters left port during the day, fished overnight, and returned in time for the morning sales. Nowadays drifting is carried out further afield and if, for example, the shoals are a hundred miles from port, vessels may dock at any time. If they arrive during the normal working day they are usually unloaded at once.

Herring are generally still sold by sample, because they can normally be expected to be of about the same standard of freshness, in contrast to white fish, which may be caught over a period of days. Bulk-stored fish are transferred from the hold by means of the quarter-cran (7-stone) basket which is emptied into a herring trunk on the dock side. Herring boxed at sea are removed several boxes at a time.

The most important single factor for the improvement of quality of the herring is for more ice to be used both at sea and on land. Boxing at sea, usually into boxes holding one-sixth of a cran, a method which in the post-war years has been widely adopted in preference to bulking, is useful for reducing the damage that results from the sheer weight on the fish, particularly when there is a large catch. The use of ice on the drifter, on the other hand, is still not common, and is often associated with longer trips, lasting two, three or even four days, so that the average freshness of the fish at landing is not necessarily much improved. There is some prejudice against iced herring in the trade, perhaps partly because in the past iced herrings have often been rather stale. Salt curers, for example, will not use them. Kipperers sometimes say that they do not like them so much because the scales which give a sheen to a kipper are often rubbed off by the ice. Some kipperers, however, intentionally remove scales from herring before splitting them and many of the herrings kippered today are therefore devoid of scales. Even freezers sometimes appear to be unwilling to buy iced herrings, perhaps because they cannot easily be sure of their freshness. It is also true that herring can be badly marked and crushed if stowed in bulk with ice.

There can, however, be no doubt that the best way to keep herring in good condition on the catching vessel is to mix them thoroughly with an adequate amount of ice and to stow them in shallow layers, or in boxes, immediately they are shaken out of the net. In fact, in the so-called 'icing at sea' as at present practised on British drifters, only a most inadequate sprinkling of ice is used. As a result, most of the 'iced' fish may still be as high as 50°F by the time they reach port. This is so even in winter because fish can come out of the sea, which may be around 50°F, and remain at this temperature even though the weather may be freezing cold. If more ice were used, and it were more intimately mixed with the fish, freshness would be much improved and wastage, for example due to torn bellies, greatly reduced.

There are, however, certain anomalies in the British herring trade which may render improvement difficult. For example, English drifters only very rarely land herring at Hull, which is the biggest kippering centre in Britain. Fish are therefore mostly sent there from the herring ports, chiefly by road. In the months August to October, however, a large quantity of iced, boxed herring caught by Swedish vessels in the northern North Sea on the Fladen ground is sometimes landed at Hull; the trips take five to six days. Also, until recently, herring caught between January and March off the coast of Norway were iced either in large cases of 16 stones, or in half-cases of 8 stones, and sent by carrier vessels to a number of British ports. In Hull, one of these ships, carrying about 200 tons, landed every few days during the season. The fish were unloaded and sold on the ordinary fish market. These herrings, reasonably well iced before leaving Norway, were usually only a few degrees above 32°F when they arrived.

In this connection, it should be mentioned that frozen Norwegian herring were also imported between March and June; these usually arrived in the regular passenger liner at Tynemouth. Their quality was often poor for a number of reasons. They were often rancid because they had been brine-frozen. They were also sometimes partially thawed when they reached the kippering port and they were then often placed, still in their large cases, in a cold store to re-freeze. This they did very slowly. The thawing of such a large block of herring was also difficult and quality often suffered as a result. The Norwegian winter herring fishery has failed over the last few years, but it is characteristic of pelagic fisheries that they may experience considerable fluctuations from year to year.

HANDLING BY THE FISH MERCHANT

The main function of the merchant handling fresh fish is to pack and transport it to one of the main centres of population. This operation frequently involves some additional treatment, such as washing, heading or filleting. Although the following is written with the fresh fish trade especially in mind, much of it equally applies to the preparation of fish for freezing and smoking.

Fish that are not gutted at sea, such as some inshore white fish, are normally gutted as a first procedure on shore, because the intestines rapidly decay and spoil the surrounding flesh. Redfish, which are not gutted at sea, are filleted without previous gutting. Herrings, pilchards, sprats and mackerel, however, are neither gutted at sea, nor usually dressed in any way before sale to the housewife. As a consequence, fish of this type are particularly perishable and are therefore in special need both of plenty of ice and rapid handling and despatch.

Gutted fish, such as cod and haddock, are usually sent off without further trimming only if they are fresh enough to look reasonably attractive when displayed in a shop. More often than not, however, the amount of spoilage between catching and landing has already had a disagreeable effect on the appearance of eyes, gills and belly walls. Much fish is therefore filleted, and this has the additional advantage of reducing transport costs. Offal can also be conveniently collected and made into fish meal at the ports.

Flat fish are often packed whole, after washing them to remove some of their surface slime. Although it is considered by some merchants that fish keep better if the natural slime is retained, this may be due to the fact that too vigorous scrubbing bruises the fish and allows bacteria to enter, so that washed fish may not keep as well as unwashed. The chief varieties of flat fish are relatively valuable, so that the

FIGURE 3·3 Hand filleting of fish

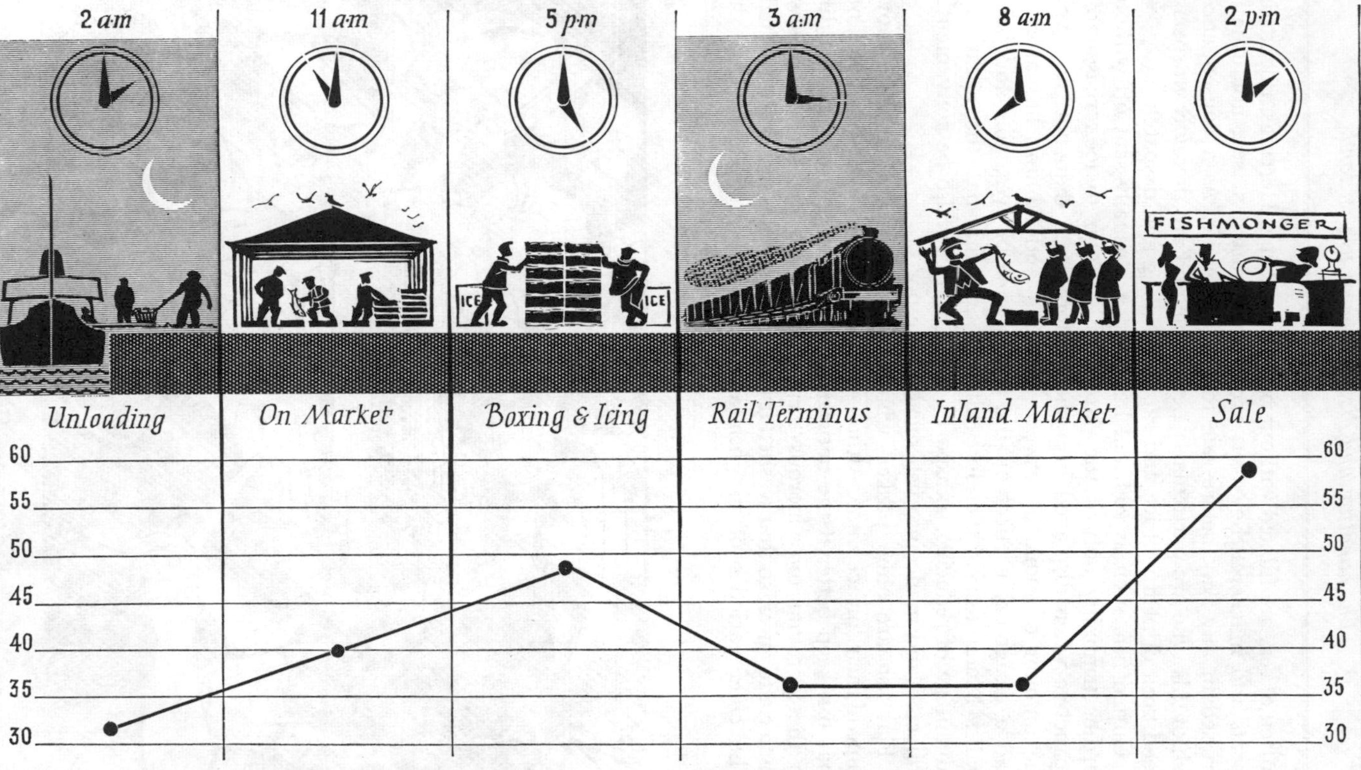

FIGURE 3·4 Temperature of fish during distribution

additional cost of transporting head and bone is proportionately less than with cheaper fish. Furthermore, if the expensive fish is transported whole there can be no doubt of its identity.

Many fish fillets are skinned, often by machine. Some species, such as haddock, are usually despatched with skin on, both because the flesh may be so soft that it tears if the fillet is skinned and also because it cannot then be confused with cod fillets. Most white fish filleting is still done by hand. Machines are in general use, however, for filleting herring and for splitting them for kippering. Machines, mainly of German manufacture, are also available for heading and filleting the commoner species of white fish, such as cod and haddock, of a number of size grades. Although these machines save some labour, their yield of flesh is lower than that obtained by hand filleting and this does not encourage their general use in Great Britain. Furthermore, fish that is soft is sometimes damaged by the machine. They are, however, used on factory freezing ships, where the fish is much firmer and the yield is less important than saving labour. Attempts are being made to develop a block filleting machine for small whitings and similar fish. Even where yield is as good as with hand filleting, machines may still not be used if their depreciation is greater than the cost of the labour available.

THE NEED FOR LOWER TEMPERATURES

Surveys of the temperatures of fish during commercial distribution have led to the conclusion that, from the time that fish is purchased until it is processed, the speed with which it becomes staler could be reduced by more effective cooling. At the large ports, considerable quantities of fish can still be seen late in the afternoon standing without ice in the market containers, waiting, perhaps, until the following day, to be filleted. As seen from Figure 3·4, the temperature of the topmost fish on a normal summer's day can rise above 50°F even although in the middle of a pile of kits individual fish may still be at 32°F. The average temperature of all fish left standing on the quay or at the factory was found to rise steadily throughout the day at the rate of between 1 deg F and 2 deg F an hour.

More cod and haddock are filleted than fish of any other species. They are usually first tipped into a tank of water to remove slime. The temperature of tap water is in many places about 50°F even in winter and, as a result, the temperature of fillets immediately before they are packed is usually fairly close to 50°F. If the washing water is cooled by means of crushed ice, the temperature of fillets can be kept below 40°F. If this is done it will reduce the amount of ice melted when the fillets are packed. Fish or fillets should never be stacked near radiators or other sources of heat.

Two criticisms can be levelled at existing methods of packing fish. In the first place, often not enough ice is used. Secondly, the ice is often put in the wrong places. These two points will now be discussed.

WHY ICE NEEDS TO BE CLOSELY MIXED WITH FISH

The ice packed in a box of fish is intended to do two things:

First, cool the fish to 32°F.

Secondly, keep it cool, ideally at 32°F, despite the heat entering the box from its surroundings.

Wet fish is a rather poor conductor of heat, that is to say, heat takes a long time to pass through fish. It takes much longer to cool a box of fillets from 50°F to 32°F, for example, than might be thought. It is common practice to pack fillets about 4 inches deep in 2-stone boxes with a layer of ice roughly an inch thick on top. It takes about 18 hours to cool to about 32°F all the fillets packed in this way. The time it takes a fish or fillet to cool depends on its distance from the ice layer. Cooling through the last few degrees from 35°F to 32°F is very slow, especially if the layers of fish are thick. Although the top fillets which are close to the ice layer cool quickly, those at the bottom of the box cool very slowly indeed. The box may well arrive at its destination with plenty of ice left and yet contain fillets which are at 40°F or above. In some places 1-stone boxes of fillets are iced only at the ends. It would take several days to cool fish in the centre of such a box, if they ever cool at all.

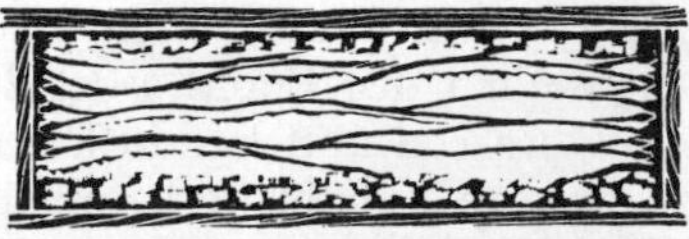

FIGURE 3·5 (*left*) Incorrect and (*right*) correct icing of a box of fillets

Much better practice for 1-stone and 2-stone boxes of fillets is to place a layer of ice at both top and bottom of the box. Provided enough ice is added in the first place, all the fish then reach their destination at temperatures between 32°F and 34°F. The fish in the middle are in this case the slowest to cool. The thicker the layer of fish the slower it cools at the centre. Table 3·1 shows that after 18 hours with a layer of ice only on top, a thick layer of fish at 50°F has only cooled to 40°F at a depth of 3 inches and has not even started to cool at a depth of 13 inches. A 2-stone box of fillets standing on the floor, however well it is iced on top, will take well over a day to reach 35°F in the middle. No temperature change whatsoever is detectable after as long as 18 hours at a depth of 13 inches from the ice.

TABLE 3·1

Time to cool layers of fish of various thicknesses, iced only on top

Thickness of layer of fish	*Time to cool from 50°F to:*	
inches	*40°F*	*35°F*
½	less than 1 hour	4 hours
1	2 hours	18 hours
2	8 hours	more than 24 hours
3	18 hours	more than 24 hours

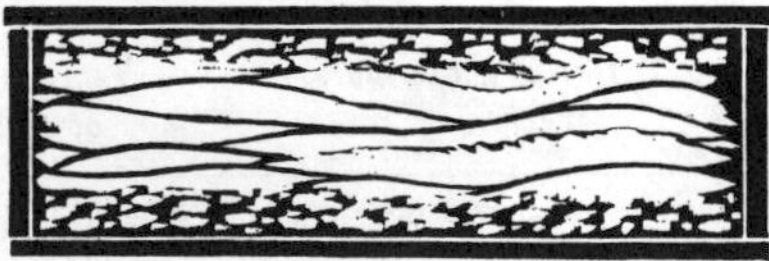

Fish in the middle of the deeper box take much longer to cool

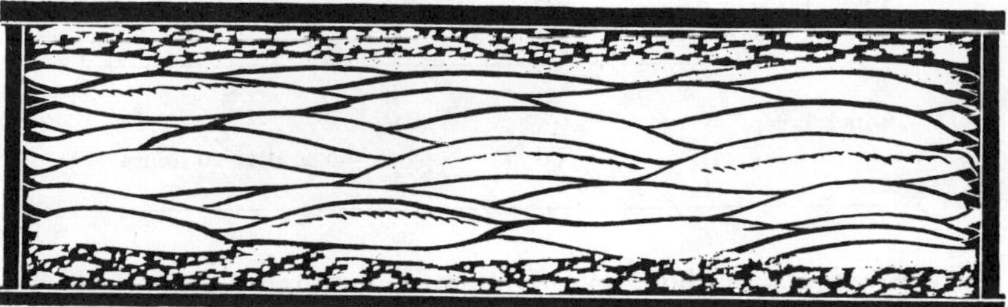

FIGURE 3·6 Comparison between a 1-stone box and a 6-stone box of fillets

TABLE 3·2

Time to cool layers of fish of various thicknesses, well iced top and bottom

Thickness of fish inches	*Time to cool from 50°F to 35°F at the centre*
3	2 hours
4	4 hours
5	6½ hours
6	9 hours
10	24 hours
24	5 days

A typical 6-stone box is between 5 inches and 6 inches deep. Even if it is iced at both top and bottom, fish in the middle take a long time to cool. If the 6-inch layer is split into two 3-inch layers with ice sandwiched in between, it cools more than four times as quickly.

Although the temperature of the fish when ice is first put on to it does have some effect on how long the ice takes to cool it to near 32°F, it is the distribution of ice in the box that, far more than anything else, decides this (see Table 3·3).

TABLE 3·3

Number of hours to cool a box of fillets, iced top and bottom to 35°F

	Initial temperature at centre of box	*Time to reach 35°F at centre of box*
	°F	hours
3-inch layer	40	1½
	50	2
	60	2¾
6-inch layer	40	6
	50	9
	60	36°F after 10 hours

These weights are the absolute minimum, assuming that all the fish in a box remains cooled to 32°F, until the last speck of ice has melted. Actually, some of the fish will have begun to be warmed again by the

TABLE 3·4

Weight of ice to cool a 1-stone box of fish to 32°F

Starting temperature of fish (°F)	*Minimum weight of ice* (lb)
65	3¼
60	2¾
55	2¼
50	1¾
45	1¼
40	¾
35	¼

heat from the surroundings long before the ice has all disappeared. Obviously to distribute fish at 32°F more ice always has to be used than that required simply for cooling it down. Table 3·4 makes no allowance at all for any ice melted by the heat of the surroundings.

There are obvious advantages in cooling fillets *before* they are packed, for example, by keeping the washing water cool, as suggested earlier.

It is not so easy to estimate how much extra ice is needed to keep fish cool during transport. This depends not only on the duration of the journey and the state of the weather, but on the insulation or lack of it in the vehicle in which the box is travelling and on the position of the box in relation to others. For example, a box on the floor of a railway fish van needs more ice than one that is sheltered by standing on another box of fish. A single box travelling in the guard's van of a passenger train requires most ice.

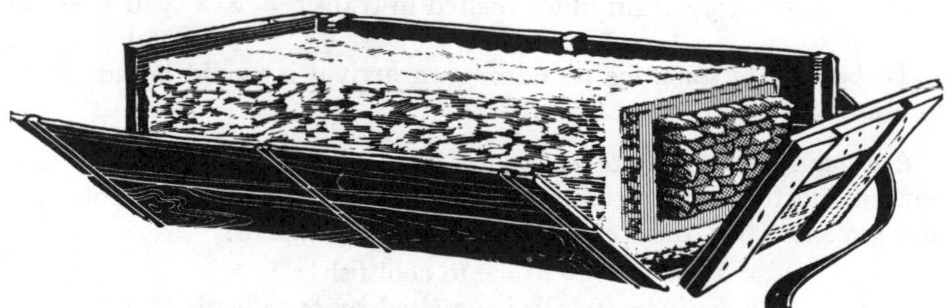

FIGURE 3·7 The ice required in a box of fish does two jobs; the outer layer absorbs the heat coming in from outside, the inner layer cools the fish

In many cases more ice is melted by the heat that enters into the box from the surroundings than is required for cooling the fish. Table 3·5 gives approximate figures for the rate at which ice melts from this cause in ordinary wooden fish boxes *standing alone*, in air at different temperatures.

TABLE 3·5

Meltage of ice in a single wooden box by heat from surrounding air

Outside air temperature (°F)	*Weight of ice melted in 12 hours* (lb)	
	2-stone box	6-stone box
80	10	20
70	8	16
60	6	11½
50	4	7½
40	2	3½

If the box is part of a stack, it is protected by other boxes and the weight of ice melting may be only a half or even a quarter of these quantities.

From Tables 3·4 and 3·5 it is easy to make a rough calculation of the quantity of ice required in a given box of fish.

Example 1. Imagine a single 2-stone box of fillets packed at a starting temperature of 40°F and making a 24-hour journey at an average outside air temperature of 50°F.

From Table 3·4 weight of ice required to cool fish
= 2 × ¾ lb = 1½ lb
From Table 3·5 weight of ice melted in transit = 2 × 4 lb = 8 lb
Total = 9½ lb

To be sure of some ice remaining on arrival it would be advisable to use about 11 lb.

Example 2. If, as is more likely, the fillets are at 50°F before they are packed and the temperature of the surrounding air is 60°F, Tables 3·4 and 3·5 show that the ice required would amount to:

2 × 1¾ lb = 3½ lb ice to cool fish
2 × 6 lb = 12 lb ice melted by outside air
Total = 15½ lb
or 17 lb, so that a little would be left on arrival.

These quantities are much greater than are normally employed. The size of the 2-stone box commonly used can hold only somewhere between 4 and 7 lb of crushed ice in addition to the 2 stones of fish.

The most usual size of box for despatch of whole fish is packed with 6 stones of fish with about 1 stone of ice in winter and 5 stones of fish and about 2 stones of ice in summer.

Example 3. In this case, with fish at 50°F and air at 60°F, the amount of ice required (from Tables 3·4 and 3·5) would be:

$$5 \times 1\tfrac{3}{4} \text{ lb} = 8\tfrac{3}{4} \text{ lb ice to cool fish}$$
$$2 \times 11\tfrac{1}{2} \text{ lb} = 23 \text{ lb ice melted in transit}$$
$$\text{Total} = 31\tfrac{3}{4} \text{ lb or about 32 lb.}$$

The ice normally allowed in summer therefore seems scarcely adequate, even for normal, let alone exceptional, conditions. It is, of course, particularly necessary in a deep box to have layers of ice if possible between every 2 or 3 inches of fish.

The most uncertain factor in this calculation is the outside air temperature which varies throughout the day as the fish passes through the fish merchant's premises, into the railway van, through the inland market, and into the fishmonger's van to the shop. For part of the time the box may be in a stack and for part of it may be standing alone completely exposed.

It is nevertheless possible, as shown in the calculations given above, to estimate the approximate quantity of ice required for a particular journey. In most cases it will be found that more ice should be used than is general in commercial practice. This is borne out by the fact that there is usually very little ice remaining on fish by the time it reaches inland wholesale markets in summer, and that fish temperatures can be as high as 60°F.

It would be sensible to vary the amount of ice used for a certain weight of fish according to outside temperature conditions. This would, however, mean using more than one size of box for a fixed net weight of fish, and packing and transport costs would consequently be greater in summer.

It is noteworthy that fish imported from Denmark arrives at the wholesale market, in spite of the road journey from Harwich, considerably cooler than fish from British ports. The reason seems to be simply that more ice is used in Denmark, 22 lb of ice to 55 lb of fish in winter and 33 lb of ice to 44 lb of fish in summer. It has nothing to do with the type of ice used since all kinds of ice, however they are made, are equally effective, weight for weight.

FISH BOXES

Fish, whether whole or filleted, has to be packed for transport in a suitably shaped container capable of standing up to handling. Fish boxes, which are usually made of wood, may hold anything from ½ stone for smoked fish to 12 stones for a large halibut. The commonest sizes are, however, 1 stone for smoked fish, 1 and 2 stones for white fish fillets, and 3, 4, 5 and 6 stones for whole fish. Typical dimensions for such boxes are given in Table 3·6. The larger boxes are usually returned to the port for use again, while the 1-stone and 2-stone sizes are now almost all non-returnable. Although returnable boxes are stronger and therefore more expensive, packaging costs for a stone of fish have in the past been lower because the initial cost is spread over a number of journeys. The situation is changing, however, and returnable boxes are falling into disuse.

TABLE 3·6

Approximate dimensions of fish boxes

Capacity of box	*Size (inches)*			*Contents*
	Length	*Width*	*Depth*	
1 stone	15	9¾	4½	Kippers
	16	11½	3½	Finnans
	15	9½	4	Wet fish
2 stones	17¾	11½	5	Wet fish
3 stones	24	14	5	Wet fish
	21	13½	6½	Wet fish
4 stones	22	13½	6	Wet fish
5 or 6 stones	28	15	7½	Wet fish
	28	15	7	Wet fish

There are a number of disadvantages in using returnable boxes. There is a charge, which has recently risen steeply, for returning an empty box by rail. In fact, it is difficult to ensure that the box will be returned, and impracticable to enforce payment of a deposit charge. In addition, boxes get progressively dirtier every time they are used. The use of an aluminium sheet liner was not an improvement from the point of view of hygiene because the space between liner and crate was rarely cleaned and in summer maggots could develop there. This type of container has now fallen out of favour. On average a returnable box needs a big repair after only a few journeys, and this may cost, perhaps, 2s. 6d.; after a few more journeys it is beyond repair. There is therefore an increasing trend towards the use of non-returnable boxes, even of the 5-stone to 6-stone size. On the other hand, a non-returnable box is not, of course, as strong as a new returnable box. As a result, there is some wastage of fish in non-returnable boxes damaged by rough handling and stacking in transit. Often fillets in particular are spoiled by being squeezed through the slits in the box.

Attempts have been made to replace the wooden non-returnable box by one made of metal or plastic. Although promising results have been obtained with a box made from paper pulp incorporating a synthetic resin, subsequent developments have proved disappointing. More promising is the recent introduction of a fibreboard box that appears capable of withstanding the rigours of the distribution system when filled with fish and ice. Boxes made of synthetic material such as hardened polythene, expanded polyvinyl chloride and polyester, still cost more than wooden ones, so that they would probably have to be returnable. As already pointed out, returnable boxes are in any event going out of favour.

The most successful competitor to wood for returnable boxes has hitherto been aluminium. This has been widely adopted by firms that control their own road transport and can rely on regular collection of empty boxes.

Metal boxes cannot, however, be used on the railway. Special racking to prevent boxes sliding about is not provided in the vans, although it is used in the privately operated road vehicles. Moreover, ice melts faster in a metal box than in a wooden one, because metal is a better conductor of heat than wood. Obviously, the better the contents of a box are insulated against heat entering from outside, the less quickly the ice will melt and the fish warm up. The differences in use, however, are much less than might have been expected. Thus a thin non-returnable wooden box lets in heat at practically the same rate as a thick returnable one, and it does not matter much whether the box is dry or wet. A light alloy box conducts only about twice as much heat as a normal wooden box. At the other extreme a highly insulated experimental container lined with aluminium was found to be only between three and four times as good an insulator as the standard wooden box. Although in theory the differences might have been expected to be much greater, in practice thin layers of non-conducting still air at the surfaces have the effect of reducing them.

UP-ENDING OF BOXES

It is practically impossible for one man to handle a 6-stone box unaided without lifting it up on one end. Stacking or transporting boxes on end is, however, undesirable, particularly with fillets and small fish, because the contents become disarranged as the ice melts, especially in warm weather. This also occasionally happens even in boxes stacked flat as a result of fierce braking or of shunting. Such disarrangement damages the fish by compressing and creasing it, as a result of which flakes of fish can gape and tear and the surface becomes dull.

In addition, up-ending for prolonged periods results in an increased loss of weight of the fish itself; pressure squeezes more fluid out of the fillets than would normally be lost if the boxes were stacked flat. In some experiments it was found that boxes of fish up-ended on an all-night journey lost about 2% by weight in cold weather and 4% in warm weather, whereas the losses in fish from boxes stowed flat were about 0·5% and 1% respectively. Up-ending for short periods before and after the journey did not, in the few tests made, appear to have much effect on the fish in the boxes transported flat.

TRANSPORT OF FISH

The fish trade as it is known today has developed largely because for many years it has relied upon the railways for distribution. More recently road transport of fish has become more important and nowadays less than half the fish leaving the ports travels by rail.

RAIL TRANSPORT

In most large fishing ports there are railway sidings adjoining the fish dock premises and in many cases fish can be graded or filleted, packed into boxes and put straight into a railway van.

Several thousand fish vans are used by British Railways. They vary somewhat in size and in details of construction, but a typical, modern fish van is internally 20 feet 6 inches long, 8 feet wide and 7 feet 9 inches high at the centre and 7 feet at the eaves. Sliding doors on either side of the van are 4 feet 10 inches wide.

The inside of the van is lined with light alloy mounted on plywood. Between this lining and the outer shell there is a 2-inch space filled with glass fibre or expanded ebonite insulation. This is, however, pierced in a number of places by bolts and strengthening pieces, which transmit heat from outside and so reduce the effectiveness of the insulation. At present only the walls, doors and roof of the van are insulated; the floor is of wood covered with asphalt, and has a number of drain holes. Future vans will, however, have one inch of insulation beneath the floor.

FIGURE 3·8 Interior of modern rail fish van, showing shelf for smoked fish

In addition to these modern vans, there are a number of vans of the older type still in service. These have merely double wooden walls with no insulation between them and there is no sheet metal lining. A number of vans of this older type have been partly or wholly modernised. All fish vans are fitted with vacuum brakes, and this enables fish trains to run at much greater speeds than ordinary goods traffic.

Measurements of the effectiveness of the insulation show that under steady conditions, the heat flow into a modern fish van, when stationary, is between 160 and 180 Btu an hour for every degree Fahrenheit difference between inside and outside temperature. Under the same conditions, the heat flow into the older van, which is also slightly smaller, is about 300 Btu an hour for every degree F difference. These values may, however, be as much as four times as great when the van is travelling. Although in theory it would be possible to cool a van before use, this is often impracticable under commercial conditions. At some large ports, trains stand two or three abreast by the side of the fish dock, and the vans have to be kept open to provide access until the last moment before the train draws out. Vans, therefore, can begin to cool only when they start their journey; some additional ice, either inside the boxes or sprinkled over the top of them, must therefore be allowed to absorb the heat from the inside structure of insulated vans.

Fish vans are seldom filled to capacity. Fish boxes are commonly stacked three, or at the most four, high and a van is considered to be full when the floor is covered with boxes to a depth of about three feet. Four tons of wet fish with ½ ton to 1 ton of ice is reckoned to be a load above average. The demands of distribution are, however, such that many vans are much more lightly laden. In the more modern vans a shelf is provided at each end which can together carry up to 1 ton of smoked fish.

The advantages and limitations of dry ice (solid carbon dioxide) are not generally understood. Often it is wastefully employed in the transport of wet fish where ordinary ice would be equally, if not more, effective at much less cost. Nevertheless, at all the larger ports depots of solid carbon dioxide are maintained for this purpose by British Railways and hooks are fitted in the newer type of fish vans to hang up to six 25 lb blocks of this material. The object, no doubt, is to lower the air temperature in the van and thus indirectly keep the fish in the boxes cooler by reducing the rate of melting of ice. The temperature of fish packed in boxes is decided primarily by how much ice is used in the box and the way it is mixed with the fish. Outside air temperature can have only a subsidiary effect. Furthermore, although dry ice is very cold, minus 140°F, compared with the 32°F of melting ice, the total cooling effect of a given weight of solid carbon dioxide is only about twice that of the same weight of ice. Six 25-lb blocks, therefore, have about the same effect as 3 cwt of ice but cost about 20 times as much. The carbon dioxide gas produced has no perceptible preservative effect when used with wet fish in this way. There is, in addition, the risk of actually freezing fish close to the intensely cold blocks.

Much the same observations apply to a newer process, known as *blast chilling,* in which liquid carbon dioxide under pressure is used. When pressure is released through a nozzle, a mixture of cold gas and powdered solid carbon dioxide in roughly equal proportions is forced over the load. The cooling effect of a certain quantity of carbon dioxide is in this way concentrated much more at the beginning of the journey. Inevitably much of the cold gas produced must escape from the vehicle, although some of it also blows out and replaces the warm air. The solid is distributed as a thin layer over the load, from which it evaporates and escapes as gas more quickly than it would if the same weight of solid dry ice in blocks were used because the blocks have a smaller surface area from which to evaporate. There is, again, the risk of freezing wet fish.

Solid carbon dioxide, whether in blocks or as 'snow' from blast chilling, may, however, be advantageous for the transport of smoked fish on its own, which cannot be iced, and also for frozen fish.

A safer, cheaper and more effective method of absorbing the heat entering a rail van is to cover the whole stack of boxes with a layer of crushed ice. Portable machines are available which are supplied with ordinary crushed ice to produce a spray of coarse snow that can be used to blanket a load. Such a procedure certainly assists in preserving the ice in the fish boxes; the ice can then cool the fish, instead of being melted by the heat of the surroundings. British Railways generally make no charge for the carriage of this extra ice. In the modern fish van, 6 cwt of ice used in this way is ample for a 10-hour journey in hot summer weather. Naturally, only large firms sending complete van loads to one destination can reap the full advantage of such a measure. It would, of course, be preferable for this extra ice to be used inside the boxes instead of outside them but this would only be feasible if transport charges were computed on the basis of net weight of contents rather than gross weight of fish, ice and package.

ROAD TRANSPORT

Part of the journey from the quayside to the fishmonger always involves some road transport. Even if the journey from the port to the inland market is made by rail, fish goes from the station to the inland market, and from there to the retailer, by road. In many cases, too, fish goes by road from the quay to the wholesaler's premises and then back to the dock to be put on a train.

Once the fish has left the dock there are obvious attractions for the wholesale merchant in sending it by road to his customers inland. For short journeys of, for example, less than 50 miles, the ease of handling may make road transport quicker than rail, although a train is clearly much faster than a lorry for a long journey, for instance from Aberdeen to London.

One group of long journeys for which lorries are commonly used, however, is the transport of herrings for kippering from the West Coast of Scotland to Hull. In this case, boxes of herrings with little or no ice are sometimes merely covered with a tarpaulin and carried on open lorries. Fish temperatures in the region of 50°F are commonly found when the lorries arrive in Hull. When well-iced boxes of fish are sent on short journeys, for example, Hull to Leeds, there seems little objection purely on grounds of temperature to the use of open lorries, although hygiene needs to be considered, particularly when, as with herring, the boxes usually have no lids.

Some large firms have their own transport organization, and use specially designed insulated vehicles. Such vehicles are also used by some transport contractors. The thickness of insulation is usually greater than in rail vans, 3 inches to 5 inches being common, and some vehicles are designed for the dual purpose of carrying either wet or frozen fish. When insulated road vehicles are used it is common to pack the wet fish in returnable boxes of light alloy, which are collected from the fishmonger when the next consignment is delivered. These boxes may be carried on shelves or racks fitted inside the vehicle to make easier the unloading of a number of small orders, but in this case they lose much of the protective effect of a stack of boxes in keeping down the temperature of the bulk of the fish. Heat flow into a light alloy box will also be somewhat faster than into a similar wooden box. These points are, however, relatively unimportant in an insulated vehicle in which the air temperature is low.

It has been demonstrated that fish sent commercially from Aberdeen to the south of England by road in the summer can arrive at temperatures of 32°F to 34°F provided that each box is adequately and correctly iced before it is put into the insulated vehicles. The use of insulated vehicles without proper and adequate icing of the fish they contain can, however, be worse than useless because it may lead users or purchasers to believe that everything possible is being done to keep the fish cool. Thus, it has been shown that fillets in 1-stone boxes iced only at the ends and transported in insulated containers, can arrive at their destination considerably warmer than fish properly iced and carried on an open lorry. In fact, in winter, when sea temperatures may be higher than air temperatures on the land, inadequately iced fish initially at 50°F, for example, can arrive at their destination warmer when transported in an insulated van than similar fish carried on an open lorry. The previous remarks about the use of solid carbon dioxide in rail vans are equally applicable to road vehicles.

The value of dual purpose insulated vehicles to carry both fresh and frozen fish seems doubtful. When wet fish is carried there is always the risk that the insulation of the floor may become soaked with water and its efficiency thereby impaired. It is not easy to make an inner lining that will remain watertight in a moving vehicle. Deterioration of the insulation is progressive and a vehicle that gave satisfactory results for frozen fish when new may have a poor performance when it is a few years old. Finally, thicker insulation is needed to restrict the amount of heat entering a load of frozen fish than is necessary for iced, wet fish, so that this extra insulation often has to be transported needlessly. The problems of the road transport of frozen fish, including the advantages or otherwise of refrigeration and solid carbon dioxide are discussed in Chapter 7.

WRAPPERS AND PRE-PACKAGING

The only wrapping material in general use in the wet fish trade is the sheet of parchment paper put over fillets when packed in a box. One object of this is to prevent deposition on the fish of dirty scum left by melting ice made from hard water. It also prevents the ice and water from coming in direct contact with fillets, and so softening the flesh, washing out the flavour and damaging the surface appearance.

In the past few years, however, a change has appeared in the pattern of retail buying habits of the British consumer. Self-service stores, similar to the North American supermarkets, have appeared in many places and many foodstuffs, meat and bacon among them, can be readily and attractively packaged before display for sale. Customers can handle and select what they require without delay and carry away their choice along with other food commodities without risk of contamination. It is therefore only natural that the fish industry should give serious thought to this development, and various attempts have been made to pre-pack fish.

Wet fish can be pre-packed either by the processor at the port or by the retailer immediately prior to display in his shop. Pre-packaging by the processor can be of two kinds; the fish can be wrapped in an impermeable material and the package then evacuated and sealed, or the fish can be given a simpler wrapping that will make the product easier to handle, more attractive to look at and less prone to contamination, but will not necessarily be airtight or watertight.

The advantage of vacuum packaging is that storage life of wet fish can be extended by a few days; the less permeable the film to oxygen and other gases, the longer the shelf life of the product. Disadvantages are that there is often a build-up of odour trapped in the package and unsightly fluid can accumulate in the bag. The odour usually disperses quickly when the package is opened, but the appearance of a transparent pack can be unattractive if the amount of fluid present is considerable. However, treatment of the wet fillets by dipping them in a solution of polyphosphate before packing can successfully prevent large accumulations of drip fluid in the bag.

Another disadvantage is that certain food-poisoning bacteria, if present on the raw material before it is packed, can flourish in the absence of oxygen at temperatures above 40°F. Cooking of the contents will destroy any such bacteria present, but if the product is one that is usually eaten without being cooked, for example smoked salmon, and the packages have not been kept chilled throughout the distribution chain, there is a possibility that cases of food-poisoning could occur from contaminated vacuum-packaged fish.

Among the technical problems of vacuum packaging wet fish must be mentioned the difficulty of satisfactorily heat sealing the lip of a pouch that is moist. Even if such problems are satisfactorily overcome and the method becomes a practicable one, it must be remembered that the wrapped product is still a very perishable commodity which spoils in a manner similar to unwrapped fish. Retention of the product at, or close to, a temperature of 32°F throughout distribution is essential if quality is to be maintained.

Pre-packaging, but without deliberate exclusion of air, either by the port processor or by the retailer, is a much simpler process, and has fewer disadvantages than vacuum packaging. Shelf life of the wet fish will not be extended by wrapping in this way, but attractive presentation and convenience of handling can make such packaging well worth while. An absorbent pulpboard tray can be used with a transparent film overwrap; the tray will absorb much of the unsightly fluid that forms, and will also help to prevent deposit of moisture on the inside of the transparent film.

Ice can be an inconvenient means of cooling wrapped wet fish during transit from the port, since the outside of the packages become unpleasantly wet. The packages should preferably be cooled to about 32°F before dispatch; they can then be maintained at chill temperature in a pre-cooled insulated vehicle by judicious use of mechanical refrigeration or dry ice. Some commercial packs of this kind are, however, transported satisfactorily in crushed ice.

New plastic films, foils and laminates are continually becoming available for packaging, so that recommendations about types most suitable for the wrapping of wet fish rapidly become obsolete; apart from the obvious requirements of mechanical strength, the absence of adverse odours and flavours that can be imparted to the fish, and inert towards the oil and other constituents of fish flesh, the materials most suitable for vacuum packaging are those that offer the greatest resistance to the passage of oxygen and other gases. The best wrappers at present available are usually laminates of two or more materials, and are not particularly cheap. Packaging of wet fish, without evacuation, for short periods during retail display can be carried out quite successfully with less costly films and pulpboards.

HANDLING AT INLAND MARKETS

Fish arrives at wholesale markets in the large cities by road and rail from most of the ports. Although each inland market has its special features, their general pattern is similar. Boxes of fish arrive in the early hours, possibly after a short journey by road from a railway station, and are put up for sale at about 6 a.m.

At Billingsgate, London, through which about 100 000 tons of fish pass every year, only a small sample is actually brought in to the markct itself to show what is available on the lorries parked outside. Fish porters then carry fish to purchasers' vehicles.

Normally, most of the fish is cleared by about 9 a.m., but there is nearly always some fish held over until next day. These boxes are usually merely stacked on the floor and a layer of crushed ice shovelled on top. In some markets boxes remaining unsold are placed in a chill room. Cooling is then very slow, taking many hours or even days because of the slow removal of heat by cold, still air. If temperatures are to be kept down, all boxes should be examined and any fish needing it iced again; this may not by any means be easy.

It is true, of course, that if the port wholesale merchant iced the fish adequately and if it was transported under good conditions, further icing at the market would normally be unnecessary except for fish held overnight. For this there is no alternative but to open up the boxes and add more ice. It is, of course, preferable to repack the box with ice at both top and bottom.

The total quantity of ice required will depend on the temperature of the fish. If this is roughly known, the weight of ice can be calculated accurately enough as shown earlier in this chapter. After adequate and proper icing it is useful to place the boxes of fish in a chill room. This prevents the cool fish warming again on a warm day as a result of all the ice melting, although if sufficient ice is applied, the chill room should be unnecessary.

CLEAN HANDLING OF FISH

By paying attention to hygiene in handling, fish can be prevented from becoming unduly contaminated with the types of microbe that make it go bad. The same care can greatly reduce the risk of contamination with food-poisoning bacteria.

The clean handling of food is nowadays essential not only for the production and display of attractive-looking and tasty articles but also for the prevention of wastage and disease.

In keeping a factory hygienic, two different types of chemicals are used, known as *detergents* and *sterilizers*, which act in different ways. A detergent loosens and helps to remove dirt, such as stale fish slime or infected pieces of waste fish, whilst a sterilizer* kills bacteria.

FIGURE 3·9 Cleaning by applying a detergent . . .

It is important to use detergents and sterilizers in the correct manner. Manufacturers are usually prepared to advise on the amount to use and the best temperature range for the purpose required. These instructions must be followed if satisfactory results are to be obtained. If the temperature and dilution are too low the treatment may be

* This term here means a chemical sterilizing agent. In other connections the term sterilizer may mean a vessel in which bacteria are killed by heat.

ineffective; if too high, corrosion of equipment or contamination of the food product may result. It is especially important to follow the directions to rinse off the chemical with water after treatment for the correct time. In a cleaning operation it is generally good practice to apply the detergent first to loosen the dirt, then to remove the bulk of the dirt mechanically, as for example, by scrubbing, and finally to apply the sterilizer to destroy any bacteria still remaining. If the sterilizer were applied at the outset it might be largely used up and inactivated by the relatively large amounts of dirt present.

Fish is a foodstuff in which bacteria causing spoilage multiply very rapidly, particularly if it is not kept thoroughly chilled. Storage time in fresh condition can be markedly affected by the large numbers of fish spoilage bacteria that may be picked up from unclean surfaces. Normally fish is cooked before it is eaten, and this is usually sufficient to destroy most bacteria of the types liable to cause food poisoning,

FIGURE 3·10 . . . and then sterilizing

which in any event do not multiply rapidly in fish that is thoroughly chilled. In factories producing cooked food products, such as fishcakes and fish sticks, there is the danger of contamination with food-poisoning bacteria because of unhygienic handling after cooking. Subsequent freezing will not destroy all potentially dangerous bacteria which can begin to multiply again once the product is thawed. The keeping quality of these products in palatable condition is also adversely affected unless scrupulous attention is paid to thorough cleaning of the mixing and other machinery at frequent and appropriate intervals.

There are a number of cleaning products available that are suitable for use in the fish industry. These all have various advantages and disadvantages and the most expensive product is not necessarily the best.

The principles underlying proper cleaning procedure have been laid down in a previous publication, and the conclusions are summarized there in a collection of Do's and Don'ts. In an official publication actual trade names cannot be mentioned although the various types of product are described. In the dairy industry a British Standards Institution specification has been formulated governing the composition and effectiveness expected of products used for certain cleaning purposes. It is hoped that a similar specification for use in the fish industry may soon become available. All that a prospective user will then need to do is ascertain that a particular product comes within the relevant BSI specification.

HYGIENE AND SANITATION AT THE MARKET

There are various problems of hygiene and sanitation at the port fish market that concern the fishing vessels, the containers used to hold the fish for display and sale, and the market itself.

The market containers, used to hold the fish for display and sale, are at most of the larger ports now made of aluminium and are washed mechanically. This is satisfactory provided that the boxes are not allowed to dry before washing, for then brown slime is likely to accumulate. This slime is removed only with difficulty and, although it has probably little effect on the fish in the containers, is unsightly. Wooden containers are extremely difficult to clean unless they are coated with a plastic material that is in good condition; the coating requires frequent renewal if it is to remain effective. Fish in contact with dirty wooden surfaces may pick up unpleasant odours. Even if there is no obvious effect on the flavour of the fish, the practice of using dirty wooden containers is an unsavoury one.

The fish market floor should be frequently cleaned. High pressure hoses may be used for this purpose. Quite apart from the desirability of handling food in hygienic surroundings there is the constant danger of fishworkers contracting Weil's disease, spirochaetal jaundice, as a result of rat excreta infecting pools of water on the market floor. For this reason, the surface of floors should be kept in good condition. The use of chlorinated water for washing the market floors would lessen this risk. As the market floor cannot be kept clean all the time, it is obviously undesirable that fish should be laid out directly on the floor.

HYGIENE AND SANITATION DURING PROCESSING OF WET FISH

When fish is landed at the ports, it has considerable numbers of bacteria on its surface and even in the flesh. Typical distant water fish have several million bacteria spread out over every square inch of skin and also in every ounce of flesh. The precise numbers vary considerably depending upon how fresh the fish is. These bacteria are almost entirely types that cause fish to go bad even at melting ice temperature.

The standards of hygiene and sanitation at the filleting stage vary enormously. Much of the fish is filleted on the market under conditions that are far from ideal, although an increasing amount is processed in modern factories where conditions are a good deal better.

There is no doubt that contamination with large numbers of bacteria during the production of fillets results in a reduced keeping time even in ice. It is important that fish be thoroughly washed before filleting, to remove the surface slime that contains large numbers of bacteria. Almost 99% of these surface bacteria can be removed in laboratory experiments by careful and thorough washing in running water. Suitable mechanical washing gives results approaching this figure. The current practice of casually rinsing fish in troughs of water already heavily laden with bacteria is unsatisfactory. Not much of the slime is removed and in addition the water in the trough may contaminate the working surfaces and the fillets. Where filleting troughs have to be used, some system of spraying the fish with fresh water seems to be the only satisfactory answer. In any case, the fish should not be kept in washing water for longer than is necessary because of the risk of undue softening.

The use of easily cleaned filleting tables swept continuously with running water, in some cases chlorinated to around five parts per million (p.p.m.) of chlorine, has also been shown to reduce the contamination of the fillet.

For washing equipment at the end of a shift, higher concentrations of 25 or 50 p.p.m. of chlorine can be used, provided that the equipment is rinsed after the appropriate interval to avoid corrosion and tainting of food. It is important that the cut surface of the fillet, during filleting and skinning, should not come into contact with working surfaces, such as filleting boards and conveyor belts, on which contamination may build up very rapidly. It is possible to develop a filleting and skinning technique that prevents this. The use of mechanical skinners instead of hand skinning also reduces fillet contamination.

The total effect of steps taken to improve hygiene and sanitation can only be measured adequately by bacteriological tests on the products. Realistic bacteriological standards can then be formulated.

4

The smoking of fish

And he opened the bottomless pit; and there arose a smoke out of the pit, as the smoke of a great furnace; and the sun and the air were darkened by reason of the smoke of the pit.

The Revelation, 9: 2

THE PURPOSE OF SMOKING

FISH is smoked nowadays in order to give it a pleasant taste rather than to preserve it. It was probably prehistoric man who discovered that flesh could be preserved for long periods by heavy smoking and salting. During the Middle Ages a variety of traditional foodstuffs were developed and one of the most important was the red herring, made by smoking heavily salted herring for some weeks. The strong tarry and salty flavour and characteristic tough texture of the red herring and similar traditional products is not widely relished in Britain today, although there is still a considerable export of Yarmouth red herring to the Mediterranean. Modern products are therefore salted and smoked mainly to give them a mild, savoury flavour, and although the combined effects of salting, smoking and drying alter their texture, they will not keep in edible condition for more than a week or so at ordinary temperatures.

Typical British dishes such as the kipper and the finnan haddock have been widely known only for the last hundred years or so; it is unlikely that they would ever have become popular had the coming of the railways from 1840 onwards not made rapid and cheap distribution possible.

What preservative effect there is, is due to the combined effect of drying and of bacteria-killing chemicals in the smoke. The 2% or 3% salt in most present day products has only a slight effect on the bacteria that spoil smoked fish. Traditional products owed their long storage life to much higher salt concentration, of perhaps 15%, and to much more smoking and drying.

HOW SMOKING IS CARRIED OUT

Two main types of smoking processes are used. In *cold smoking,* which is used for most British cures, the temperature of the smoke ought not to rise above about 85°F or the fish will begin to cook. In *hot smoking,* however, the intention is to cook the fish as well as smoke it. The smoke reaches 250°F or so and the centre of the fish may be at 140°F. Most continental products are hot smoked; in Britain the method is used only for the relatively small quantities of sprats, eels, trout, *buckling* made from herring, and *Arbroath smokies* made from small haddock.

The equipment used for smoking, whether hot or cold, is either a traditional or a mechanical kiln.

The *traditional kiln* is simply a chimney in which fish are hung over a fire of smouldering sawdust. Although experienced smokers can usually produce good results from them, traditional kilns nevertheless have a number of serious disadvantages. Control of the amount of heat and smoke produced by the burning sawdust is difficult. One moment the warm smoke will follow one path and the next a quite different one. Occasionally the smouldering fire will burst into flames and the lower rails of fish will be cooked. Uniform drying of the fish is impossible, because the smoke is practically saturated with water after it has passed over the first few rails and cannot therefore dry any fish higher up the kiln. These difficulties are to some extent overcome by the laborious and unpleasant operation of shifting fish in the kiln during smoking. When those fish nearest the fires are judged, by their colour and feel, to be ready, they are removed and replaced by those from higher up. This is known as *stripping the kiln*.

On a warm, humid night it may be impossible to use a traditional kiln at all. The speed of the draught and the capacity to dry fish depend largely upon how much the air entering the kiln is warmed by the fires. If very cold air enters and is warmed to 85°F, then it will be very much lighter than the cold air outside and will rapidly travel up the kiln. If already warm air enters, on the other hand, it still cannot be warmed above 85°F or the fish will cook and drop into the fires, and it may, therefore, be impossible to warm it sufficiently to create draught at all.

The drying capacity of the air is also at a minimum on warm, humid nights. The amount of water vapour that air can hold depends upon the temperature; if warm air is cooled, it can hold less water vapour and the excess condenses, for example as mist or dew, or as a film on a bathroom window. If cold air is warmed, on the other hand, it can then hold more water vapour and its drying capacity is therefore increased. On a warm, humid night the air cannot be warmed very much further without the risk of cooking the fish and therefore the drying capacity of the incoming saturated air cannot be greatly increased. Indeed, fewer fires than usual are lighted on warm nights and smoking time is consequently longer.

A much greater draught of this nearly saturated air on a warm night than of the nearly dry air on a cold night would be required to dry the fish to the same extent. Consequently, it is when most draught is required that least is produced.

The wastage and financial loss from *droppers,* that is cooked fish that falls on to the fires, may be very considerable. Even if this fish can be salvaged, it is difficult to pack because it is so squashy and it looks unattractive because it is without the usual gloss.

Usually only one batch of cold-smoked fish can be produced every 24 hours. Red herrings are smoked for a much longer period and the fires lighted only on alternate nights; nowadays they are usually hung in the kiln for a week or two. Although hot smoking is a quicker operation, and several batches can be produced every day, control is still difficult and it is necessary frequently to shift frames of fish up or down the kiln. In Denmark, the fires themselves are sometimes raked to and fro across the floor or moved about on trolleys.

FIGURE 4·1 The traditional kiln

There are other disadvantages quite apart from the serious ones mentioned above, namely, the difficulty of producing a uniform product and the occasional disastrous effects of the weather. A great deal of labour is required to hang and strip a kiln, and the process is not a pleasant one to carry out. There is, indeed, an increasing difficulty in obtaining skilled fish smokers. Although the skilled smoker can turn out a good product it nevertheless may be covered with smuts, dust and ashes. In some firms these are carefully removed with a soft brush, so adding to production cost. The actual time involved in the smoking process itself is also considerable.

Mechanical kilns of numerous designs have been developed in an attempt to improve the process. Unfortunately, some are even less satisfactory than traditional ones. The mechanical kiln that has now become established in Britain, almost to the exclusion of other types, is the *Torry kiln* (see Figure 4·2) first designed in 1939.

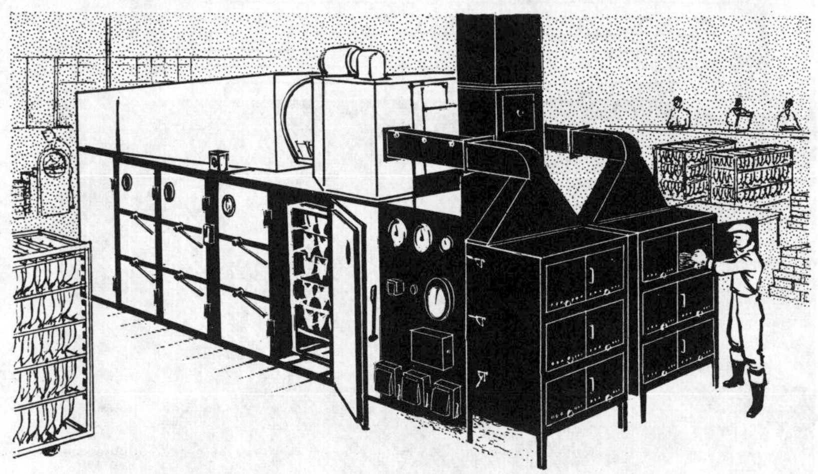

FIGURE 4·2 The Torry kiln

Smoke is obtained from fires laid in special hearths outside the kiln (see Figure 4·3). The fires are usually of hardwood *dust* on top of white wood shavings or *mush*. The smoke is led into the kiln by ducts and mixed with air. Temperature is maintained by either electric or steam heaters which are thermostatically controlled. The humidity of the warm smoke can be controlled by altering the amount of fresh air entering the kiln.

The warm smoke is blown by a fan at an even speed over trolleys of fish in a horizontal tunnel (see Figures 4·2, 4·3). A proportion of it then passes up a chimney but most is circulated again; on its way round it is mixed with fresh smoke and more air. Half way through the process, the trolleys of fish are interchanged because those nearest the incoming stream of smoke dry more rapidly than the rest.

Although in Britain most smoked fish is still made in traditional kilns, the Torry kiln is being rapidly adopted. The control it gives (see Table 4·1) means that a uniform and cleaner product can be made with less waste, more quickly and with less labour, and to a considerable extent independently of the weather. Of recent years it has been adopted commercially with very successful results for producing many products including finnans, kippers, sprats for canning, and smoked salmon.

Kilns of various capacities are available. Large-scale production is best based on big kilns that can smoke 100–150 stones every four hours or so. Smaller kilns, for example, with 80 stones capacity, are more suitable for some firms while, for the production of luxury and delicatessen products, such as buckling and smoked salmon, there is even a 10-stone size.

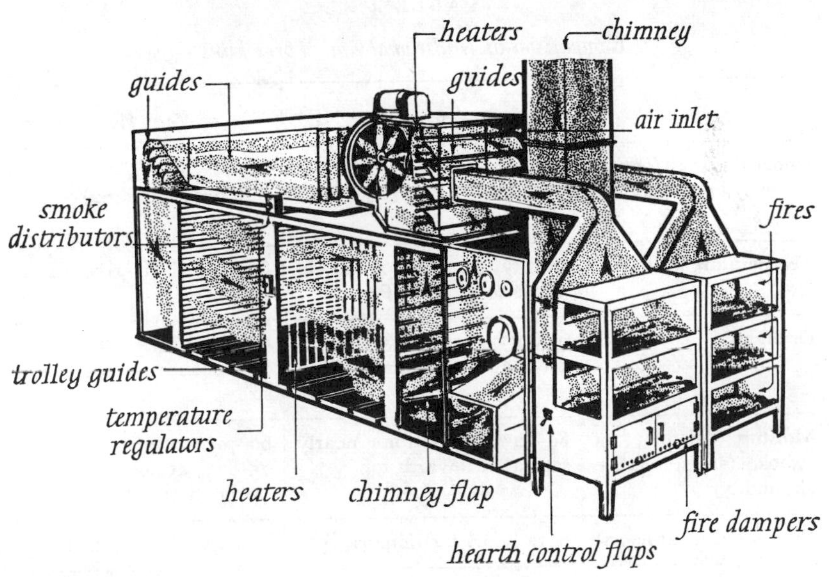

FIGURE 4·3 Cutaway diagram of Torry kiln

It is necessary, however, to add a word of caution. Use of a mechanical kiln, Torry or otherwise, does not guarantee satisfactory results. Mechanical kilns are machines and, like cars, need proper control and maintenance. The product from a Torry kiln, for example, will be quite uniform although it may be uniformly dried but without any smoke, or uniformly smoked but still almost as wet as when it was put in. Tar and dust may accumulate in some of the ducts and these need to be regularly cleaned or breakdowns and fires may occur, perhaps causing damage and certainly interfering with production.

The need for process and production control is not appreciated as widely in the fish smoking industries as in some other food industries. Quality control in fish smoking is only possible if the smoking process itself can be controlled, and control is possible only in a mechanical kiln. Unless the mechanical kiln is operated together with a proper quality control scheme, the product may be less satisfactory than that from traditional kilns, since the limitations of these are well known to expert smokers.

There is no doubt that any firm wishing to modernise its production would be well advised to consider replanning factory operations round 100–150 stone Torry kilns. Continuous shift operation, with its

TABLE 4·1

Comparison of traditional and Torry kilns

	Traditional kiln	*Torry kiln*
Smoke thickness (optical density per foot of smoke)	0–1·0 (irregular)	0 to 0·5 (as desired)
Temperature of smoke	Mostly 65–75°F rising in bursts to 95°–140°F	85°F ± 2°F (thermostat as desired)
Draught	0–5 ft/sec, in gusts; sometimes a down draught	6–8 ft/sec for 50 stones size 10–14 ft/sec for 100 stones size
Moisture content of smoke (as % relative humidity)	60–80% at bottom, nearly saturated towards top	60–70% at inlet end 75–80% at outlet end (reversed at half-time)
Time in kiln (in hours)	6–12 hours for finnans	4 hours (as necessary)
Drying; loss of weight	10–20%	14–18% (as required)

economies of both time and labour, is then possible. The cost of putting up brick kilns is nowadays greater than that of installing Torry kilns of equivalent capacity. Total costs of operating Torry kilns are also rather lower. This is in spite of the fact that, since mechanical kilns are regarded as machinery, accountants customarily allow an annual 10% reduction in their initial value; brick kilns, on the other hand, are regarded as part of the building and are depreciated by 5% every year.

THE SCIENCE OF SMOKING

Wood smoke contains both vapours and droplets. It is made up of millions upon millions of tarry droplets so small that a hundred thousand of them placed side by side would only make an inch. The vapours, however, are invisible, although they can be smelt. It is the vapours given off from the walls of an empty kiln which make it smell smoky.

Although the same chemicals are present in the vapours as in the droplets, their relative proportions are different. The substances that evaporate most easily are present mainly in vapours; those that have to be heated to make them readily evaporate are mainly in the droplets.

By a special process known as *electrostatic precipitation* it is possible to remove all the droplets without affecting the vapours. It has been shown at Torry that it is mainly the vapours that are taken up by fish during smoking; the droplets are not essential. The substances in the vapours dissolve in the liquid on the surface of the fish. The moister the fish, and the faster the smoke flows over its surface, the more rapidly chemicals from the vapours are absorbed.

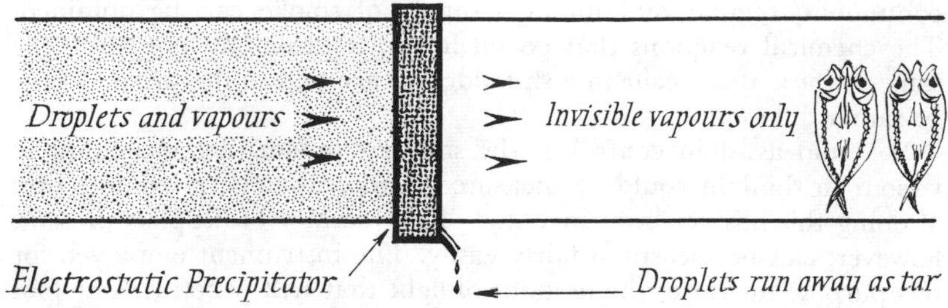

FIGURE 4·4 Electrostatic precipitation of smoke droplets

It has been suggested from time to time that electrostatic precipitation could be used to speed up the smoking process. Smoke droplets when treated electrically can be made to settle on objects almost instantly. This method is used industrially for removing troublesome fine dusts produced in various manufacturing processes. It has also been employed in Germany, USA and USSR for giving a smoky flavour to sardines for canning. As mentioned above, however, it is the vapours and not the droplets that give smoked fish its characteristic flavour and keeping qualities. Even if smoking were speeded up, drying would not be and it still would take as long to perform this essential operation.

It has also been suggested that smoking could be speeded up if the fish were dipped into a solution of smoke chemicals, perhaps added to brine. It would, of course, still be necessary to dry the product. Attempts to make liquid dips by distilling wood have not been very successful because the flavour is not the same as that of ordinary wood smoke. The discovery that the vapours are responsible for the smoky flavour provides a clue to this difference and has made possible a dip treatment that under laboratory conditions gives something very similar to the product of normal smoking.

The precise chemical composition of smoke no doubt depends upon many factors. In an ordinary sawdust fire there is a whole range of processes going on side by side. At one extreme there is complete burning of sawdust to give carbon dioxide and water, through a whole host of complicated reactions between the wood and oxygen in the air to, at the other extreme, a simple warming during which some of the sawdust components are evaporated.

Both evaporation and oxidation of sawdust without burning can be carried out in a *fluidizer*. In this apparatus a steady stream of hot air is blown through a layer of sawdust, which consequently behaves like a fluid. The sawdust does not burst into flames, but gives off smoke at a steady rate and, in doing so, becomes charred. By continuously adding sawdust, and removing that which has become charred, a continuous, regular and uniform supply of smoke can be obtained. The chemical reactions that go on in the fluidizer are not quite the same as those that occur in a smouldering fire and so the product may be rather different.

It would assist in control of the smoking process if the amount of vapour in the kiln could be measured. Unfortunately, no simple way of doing this has yet been invented. The quantity of droplets present, however, can be measured fairly easily. The instrument employed for this purpose measures the amount of light that will penetrate one foot of smoke. If only 1/10th of the light penetrates the smoke, the *optical density* is said to be 1·0. If the smoke is thicker and only 1/100th of the light penetrates, then the optical density is 2·0; if only 1/1000th part penetrates, it is 3·0 and so on. Such a scale is said to be *logarithmic*. Intermediate values can be calculated.

The smoke treatment of a batch of fish can be calculated if the optical density throughout the process is known. If, for example, a batch of fish has been in smoke of an optical density of 0·1 for one hour, 0·3 for two hours, and 0·5 for one hour, the total smoke treatment is:

0·1 + 0·3 + 0·3 + 0·5 or 1·2 units of smoke.

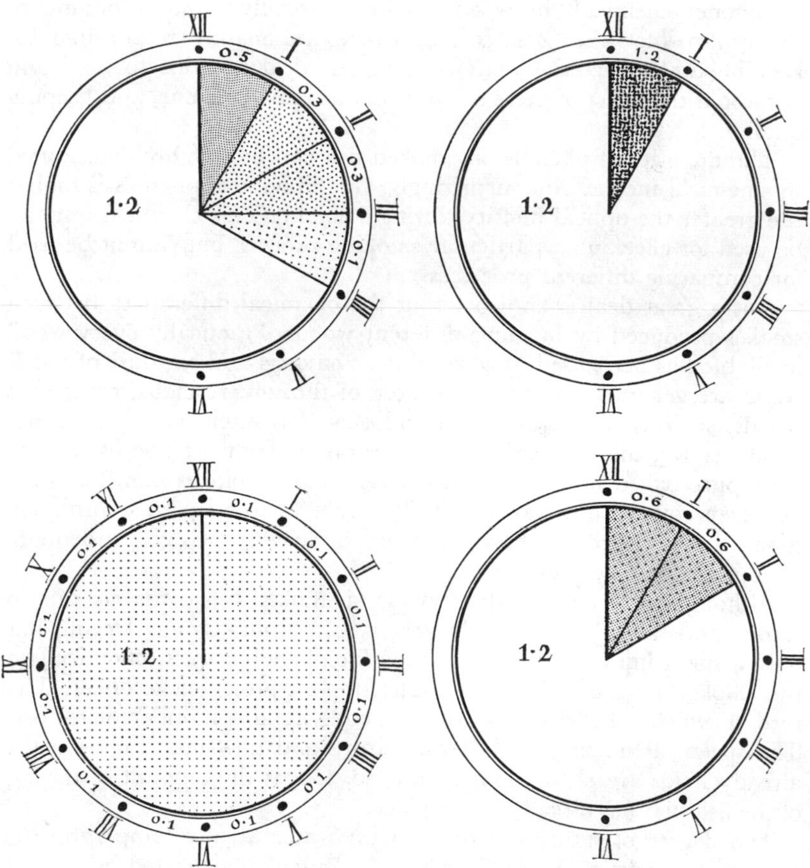

FIGURE 4·5 Optical density of smoke can be used as a measure of the smoke treatment of fish. In every instance shown above, the smoke treatment is the same, although the time taken differs

Although, as already indicated, the optical density is a measure only of the density of the droplets and not of the vapours, it nevertheless can be a valuable aid to the control of the smoking process. It has been found in the laboratory that a dense smoke is usually rich in vapours and, in fact, the quantity present of one is under most commercial conditions a good guide to the amount of the other.

The chemistry of the processes that go on when wood burns is by no means understood. It has long been known that burning wood produces a considerable amount of water; in fact, on a warm, humid night a traditional kiln may become filled with a mist which cannot be dispersed. Scientists at Torry have recently identified many of the commoner chemicals in wood smoke, especially those belonging to the group known as *phenols* and which are mainly responsible for keeping down bacterial activity. A great deal remains to be found out about the effect of these substances on colour, flavour, and keeping quality.

The quantity of phenols on smoked fish can be roughly determined by chemical means. Not surprisingly, the amount of phenols is higher the greater the optical density during smoking. This chemical test can be used for checking a particular smoking process, but cannot be used for comparing different processes.

Not a great deal is known about the chemical differences between smokes produced by burning different woods. Practically every wood available has been used at some time. Shavings and sawdust of hardwood are generally preferred to those of the soft, resinous, coniferous woods, such as pine, but in the industry it is often necessary to use whatever is available. Although oak is often claimed to be used, very little pure oak dust is used nowadays; what is sold as oak dust may not even contain any oak at all. The type of wood used is probably much less important than smoking sufficiently to give adequate flavour and preservation.

Although the characteristic flavour of smoked fish is mainly due to smoke and salt, the texture largely results from drying. Drying has only a slight influence on the storage life of the product, chemicals in the smoke being mainly responsible for any preservative effect. The rate at which fish dries in smoke depends upon a number of factors. The speed of the smoke, its temperature and how much moisture it already holds are clearly important; so, too, is the amount of drying of the fish that has already taken place.

The degree of saturation of air with water at any temperature is termed the *relative humidity* (r.h.). Completely saturated air is at 100% r.h. and air that has only half as much as it could hold at any temperature is at 50% r.h. The best range of r.h. for cold smoking at about 85°F is between 60% and 70%. If it is above 70% drying is too slow and if much below 60% the outside of the fish dries too rapidly.

Recent study at Torry has shown that there are two quite distinct stages in the drying of fish. While the surface of the fish is wet, moisture is lost regularly at a speed depending only upon the r.h. or drying capacity of the air or smoke and not upon the temperature. The surface is cooled by the evaporation of moisture and it may at first nearly reach the wet bulb temperature.

In the second stage, when the surface has dried somewhat, its temperature approaches ever closer to the temperature of the air or smoke. The rate of drying becomes progressively slower and slower as layers beneath the surface become drier and moisture moves out ever more slowly from the middle of the fish.

It is unfortunately not possible to calculate from the water content of the final product how much drying has occurred in the kiln. Cod, for example, contains about 80% water and 20% of other constituents. If it loses 20% of its initial weight in the kiln, and this would be a greater amount of drying than is usual, the water content of the final product will still be 75%. Some samples of fresh cod may have water contents approaching 75%; different parts of the same smoked fish could give quite different results.

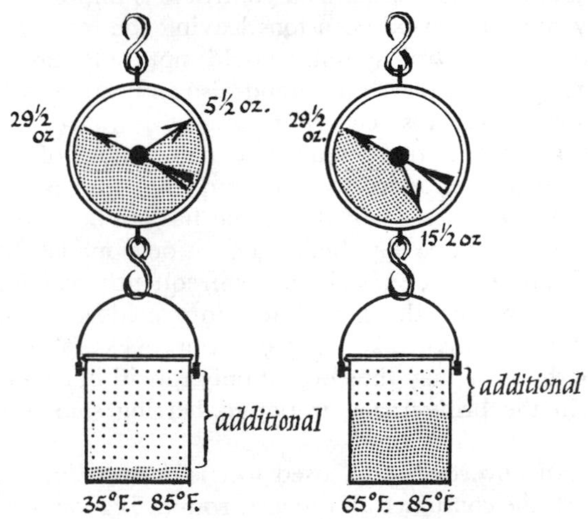

FIGURE 4·6 The drying power of a kiln depends upon the humidity of the air entering it; on a warm, humid night the potential drying power of the incoming air is much less than in cold dry weather

SMOKED PRODUCTS

GENERAL

Good smoked products cannot be made from stale fish. Smoking cannot be used to rescue what is already too stale to sell as wet fish. The quality of the product reaching the purchaser depends upon the freshness of the fish before it is smoked and the care in handling it, the smoking process itself, and subsequent history. Of these three, the smoking process is least important; most poor quality smoked fish is due to bad handling before and after smoking or to using inferior fish in the first place. Adequate smoking is nevertheless essential for a first class article which will keep well. These facts are not as widely appreciated in the fish smoking industry as they should be.

During a DSIR survey of the kippers produced in one of the large centres, it was found that only one pair out of every four leaving the producers' premises was of good quality. More than one pair out of four were so poor that they could not be eaten with any enjoyment whatsoever and some fish were so bad that they could not be eaten at all. It has since been shown that there is a further serious deterioration in quality by the time kippers are sold to the public.

Without some properly organized control it is impossible to prevent poor quality smoked fish on occasions leaving the factory. A properly organized quality control system should not only prevent a poor product leaving the factory but should also indicate at what stage in the process deterioration is occurring.

Use of fish that is initially stale is a major cause of poor quality. It is still not general practice to put ice on fish that is awaiting processing. In some factories, smoked fish that has not been sold is allowed to remain for two or three days before a decision is made about whether or not to freeze it. Freezing is not infrequently carried out very inefficiently by putting the boxed fish into cold store; under these conditions, freezing may take days or even weeks. What eventually comes out of the store may thus be not only rancid but stale. Rancidity also occurs in the brine-frozen imported herring sometimes used for kippering.

Freezing can nevertheless be used to ensure that first-class smoked products reach the consumer all the year round. If good quality herring are properly frozen at the height of one of the main seasons and stored at the right temperature, they can be used for kippering when supplies of suitable fresh herring are unobtainable. This procedure has a number of advantages. Herring will remain in good condition in cold store for a longer period than kippers; staff can furthermore be employed regularly throughout the year so that kippering need no longer be an industry with seasonal bursts of activity. It cannot be assumed, however, that frozen fish is always of good quality. There are a number of points during manufacture and distribution where faulty handling may seriously damage the product.

Although the producers of smoked fish bear a considerable measure of responsibility for poor quality, distributors and retailers are by no means blameless. Supplies are sometimes held at inland markets and in shops for much longer than they should be. Smoked fish is a perishable commodity, although some retailers, and many housewives, too, still apparently believe that it will keep for weeks without deterioration. Temperatures of smoked fish during distribution and especially in shops can be so high that even the freshest products cannot remain in edible condition for longer than a day or so. Ice cannot, of course, be applied directly to keep smoked fish cool; some firms add a few lumps of solid carbon dioxide, dry ice, to boxes of smoked fish distributed in summer and this is, no doubt, better than nothing. It is an expensive method of cooling and, furthermore, cannot be used by retailers. The practice of chilling the product to about 32°F before despatch does reduce temperature during distribution. The additional shelf life so obtained may, however, be taken up entirely in cooling the product with no advantage to either producer or consumer. If good quality smoked fish is to reach the consumer, then, unless it is to be frozen, there is no alternative to rapid distribution and sale.

THE PROCESS

The production of smoked fish involves, apart from the smoking process itself, a number of preparatory operations, many of them vitally important for the manufacture of good saleable articles. The industrial operations are splitting and cleaning, salting, hanging, smoking and packing.

(1) *Splitting and cleaning*. The precise treatment depends on the product; care must always be taken not to bruise or tear the fish. All pieces of gut, gill and kidney must be removed, because these go bad quickly and may spoil otherwise good fish.

(2) *Salting*. This is nowadays almost always done by soaking the fish in a strong brine. One of the permitted dyestuffs is added if it is wished to colour the fish; this practice is usual in making kippers and smoked fillets of various kinds, but not for finnans or bloaters. The duration of the brine dip depends on brine strength and on the size and the fattiness of the fish, although if the flesh is not very thick most

of the salt probably enters during the first three or four minutes. Brining time also depends upon the amount of stirring of the fish in the brine.

A 70–80% saturated brine is usually employed for all the common types of smoked fish. If a fully saturated, or 100%, brine is used, the appearance of the finished product may be marred by fine powdery crystals of salt on the gill covers or skin. In a 50% brine, fish swell slightly and although they will still take up 2–3% salt, they also gain 2–3% in weight. This additional water has to be evaporated during smoking. In a 90–100% brine, there is a 2–3% weight loss.

Brine becomes weaker during use. Water on the surface of the fish dilutes the brine and the fish also absorb salt from it. Brine strength is usually maintained by adding solid salt, which often settles on the bottom of the tank in a layer several inches thick. This can be prevented by efficient stirring, for example, with a suitable electrically driven propeller.

It has sometimes been suggested that it would be easier to maintain the strength of brining baths by adding saturated brine rather than by adding solid salt. Unfortunately such a procedure is quite unpractical. It would, for example, require a volume of saturated brine equal to that of the brining bath itself to make up 60% brine to 80%. Table 4·2 shows how brines of various strengths can be made up.

TABLE 4·2

*How to make up brine**

Salinometer degrees (% of saturation)	*Number of pounds of salt to be added to container and made up to 1 gallon with water*		*Number of pounds of salt to be added to 1 gallon of water*	
	lb	oz	lb	oz
10	0	4¼	0	4½
20	0	8¾	0	9
30	0	13½	0	14
40	1	2½	1	3¼
50	1	7½	1	8¾
60	1	12¾	1	14½
70	2	2¼	2	4¾
80	2	7¾	2	11¼
90	2	13½	3	2½
100	3	3½	3	10¼

*The solubility of salt in hot water is greater than in cold. These figures are for a brine made up with water at 68° F but the quantities required for lower temperatures are not very different from these.

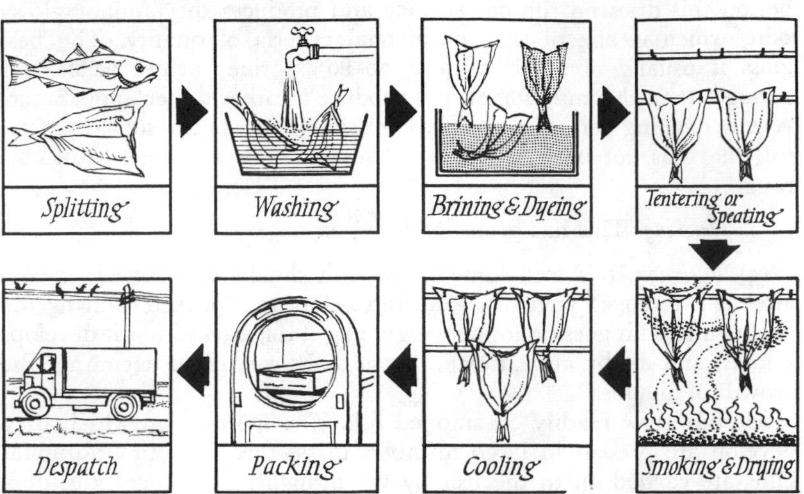

FIGURE 4·7 Sequence of the production of smoked fish

Greater care could be exercised in the control of the strength of brine. At present, in many factories, salt is added to brines haphazardly. The result is that one tank may have a layer of salt on the bottom while another may be only 50% saturated.

The most convenient instrument for measuring brine strength is the salinometer. Table 4·2 gives the number of pounds of salt to a gallon of water for every 10% of saturation.

Normal brining procedure does not produce a uniform salt content, even in fish of uniform size. Better results are obtained if the brine is stirred during the dip. A mechanical process would certainly improve matters as well as allowing continuous filtration and control of brine strength. Continuous mechanical briners, in which the fish are conveyed at a controlled speed through a long trough of saturated brine by means of a motor-driven chain of paddles, are used by a number of the larger processing firms.

Brines should be changed at least once a day. Frequent removal of sawdust, scales and pieces of gut is necessary. Fish can be contaminated by stale brine and they do not then keep as well as they would otherwise. In addition, some of the fish protein that at first dissolves in the brine subsequently forms clots which settle on the fish and make it look unpleasantly blotchy.

(3) *Hanging*. Fish is hung to drip either on racks or in the kiln. Protein dissolves in brine to give a sticky solution; during the dripping

period this dries on the cut surface and produces the familiar glossy skin, which is one of the commercial criteria of quality. The best gloss is usually obtained with a 70–80% brine; brines weaker or stronger than this may sometimes produce a rather duller appearance. A long hanging period of up to 18 hours gives the best results. Smoked fish that has not been salted at all has no gloss and looks dull and rough.

(4) *Smoking*. This has been described on page 71.

(5) *Packing*. It is well known that fish should be allowed to cool after it is removed from the kiln and before it is packed. During the cooling period it goes on losing weight and if it is packed hot it develops a moist and flabby appearance. These moist conditions encourage the growth of moulds.

Moulds grow readily on smoked fish. The spores from which they develop are present in large numbers in sawdust, and in traditional kilns are carried on to the fish by the draught. The Torry kiln does

TABLE 4·3

Fish smoking in a Torry kiln

	Time in kiln (hours)	*Temperature (°F)*	*No. of fires*	*Desired weight loss (%)*
Kippers	4	85	All	14–18
Kippers for canning (mild cure)	2½	85	All	12–14
Kipper fillets	2½– 3	85	All	10–12
Cod or haddock fillets	3 – 5	80	2/3 to All	10–12
Finnans	4 – 6	80	2/3	12–14
Pale cure	2 – 3	80	1/3 to 2/3	8–12
Golden cutlets	2 – 3	80	2/3	10–12
Bloaters	3	90	1/3	14–16
Red herring	36*	85	All	20–25
Silver cure	8	85	1/3	5
Salmon	8 –12	85	1/2	10
Cod roe	8	100	1/2	20–25
Buckling	2 – 3	80–160	All	10–12
Sprats for canning	½– ¾	80–160	1/3	10–12
Kielersprotten	1 – 1½	80–160	All	15–20
Smokies	2	80–160	All	30
Trout	3	80–160	All	10–12
Eels	2	80–160	All	10–15

*Intermittent

not have this defect because the fires are not immediately beneath the hanging fish. Because smoked fish so easily picks up mould spores, sawdust stores should never be close to brining tanks, horses or packing areas.

SPECIFIC PRODUCTS

Kippers and finnan haddocks account for over 90% of all smoked fish produced in Britain. These will therefore be considered in greatest detail. Table 4·3 gives a summary of the treatment necessary to make all the types of smoked product usually encountered commercially.

(1) *Kippers and kipper fillets.* Only fresh, or good quality frozen, fatty herring will make first-class kippers. Most landings of ring-netted herring are of fairly uniform freshness, because the vessels are usually away from port only overnight. The voyages of herring drifters may be of four or five days' duration, however, and there is therefore considerable variation in the freshness of the iced herring they land. Herring should, if possible, be smoked the same day as they are

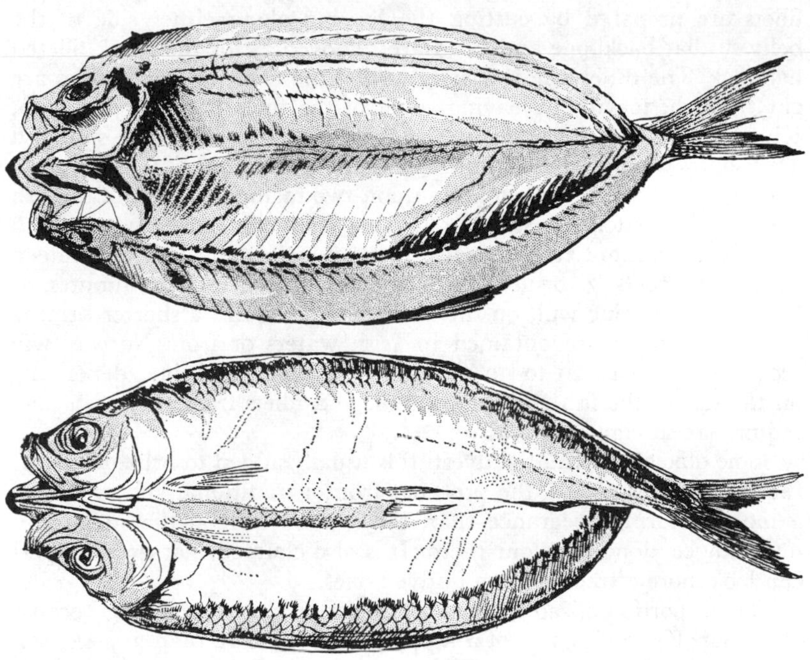

FIGURE 4·8 Kipper

caught; after three days in ice they are certainly past their best and should not be used.

There is need for the greater use of ice in the transport of herring from port to kippering centre, which may be several hundred miles away. Fish may in summer be as warm as 65°F by the time it reaches the curing house. Even in February, the temperature of herring arriving in England after a road journey on a frosty night from the west coast of Scotland has been found to be 50°F. The reason for this is that very little ice is used for transporting herring in cold weather because it is not generally realized that sea temperatures, which decide the fish temperature in the first place, may yet be quite high even in winter.

Herring caught in early spring and after spawning may have only a little fat in them. Although these spent and lean fish are generally regarded in the trade as inferior, they are certainly preferable to stale or rancid fatty fish. Provided they are fresh, they can make a pleasant, although not particularly succulent, dish.

The whole herring are first washed to remove any loose scales. Kippers are prepared by splitting the fish down the back and removing guts and roes, an operation which is nowadays, with rare exceptions, done by a machine although it was formerly done by hand. Kipper fillets are prepared by cutting the herring along either side of the belly so that backbone and head are removed as for the block filleted haddock. The difference is that in a fillet the chest or *barrel* bones are cut through, inevitably leaving some of the finer bones in the flesh. Kipper fillets also are nowadays prepared by machine. The split fish are usually washed again before brining. The brining time for kippers and kipper fillets depends mainly upon two factors, the size of the fish and the fat content. A summer herring from the north east Scottish coast is a medium-sized fish with up to 20% fat, and will take fifteen minutes in 70–80% brine. The fillets take three to four minutes. A lean winter herring will, on the other hand, require a shorter brining time. The larger fish obtained in Irish waters or from Norway will require as long as 20 to 30 minutes in a 70–80% brine, depending on the size of the fish and fat content, and fillets from these fish may require six to eight minutes.

Some officially permitted dyestuff is usually added to brine to colour the fish. By this means, the products from a traditional kiln are given a more uniform appearance than they would have if they depended upon smoke alone to colour them. It is also claimed that dyed kippers can look more attractive than undyed ones.

The proportion of salt in the flesh increases during smoking because of drying. If the thin side of a kipper contains more than 3·5–4% salt it is too salty for most tastes. The salt content of the thick side, which is always less than that of the thin side, should not be below about

1·5% or it will not taste salty enough. Cooking by grilling or frying, since this involves further drying, makes the fish taste more salty. It is sometimes claimed that freezing and cold storage also make fish taste more salty but it has not been possible to confirm this by experiment.

The brined fish for kippering are hung on tenter sticks. The fillets can be hung in the same way as block filleted haddock on a special tenter stick. A dripping time of an hour or so should be allowed before smoking commences. The dripping time of herring products in general does not need to be as long as for white fish products, since the gloss depends to a considerable extent on the oil which comes to the surface during smoking.

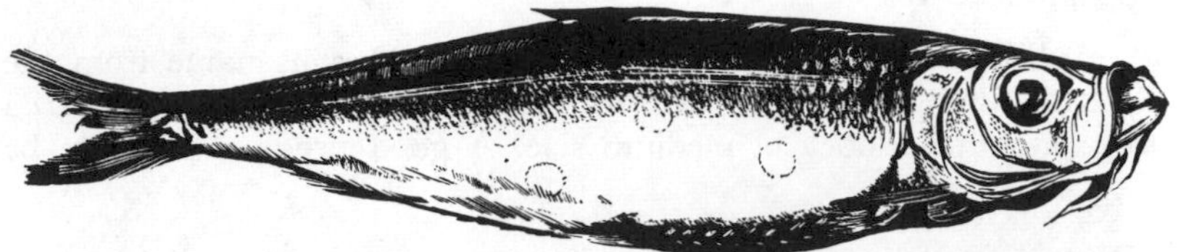

FIGURE 4·9 Bloater

During the smoking process itself, the temperature should not exceed 85°F or cooking will occur. For an average, well-cured, dyed kipper, about 4 hours' smoking is required with all the fires lighted in a properly operated Torry kiln. During this period it should lose 14–15% of its dripped weight. If good undyed kippers are to be produced, about six hours' smoking and weight loss of about 20% is necessary. During the last half-hour or so the temperature may be raised to 95°F in order to develop a deeper colour and to bring oil to the surface. Very oily summer fish become too soft to handle, however, if warmed to this temperature. Kippers for canning should be subjected to less smoke and need to lose only about 12–14% of the original weight (see Table 4·3) and, like kipper fillets, can be made in a Torry kiln in under three hours.

Keeping quality of the final product is directly related to the amount of smoking and drying. Fresh herring smoked in a Torry kiln to lose about 5% of their original weight keep reasonably fresh for about two days at a summer temperature of 60°F; with a weight loss of 13% the keeping time is about four days and for 15% weight loss, about six days. Kipper fillets are usually not smoked as much as kippers and consequently do not keep so well.

(2) *Bloaters*. Bloaters are made from whole, ungutted and slighted salted herring. They owe their characteristic flavour, which is sometimes described as gamey, to the enzymes or ferments from the gut.

The fish are normally salted by mixing them with solid salt and leaving them overnight. They lose roughly 6% of their weight during this process. After salting the herring are washed to remove the surface salt.

The fish are then threaded on speats, either through the gills and mouth or through the eyes. The trolley-loads of washed fish need not be given time to drip and can be run into the kiln whilst still wet, since there is no question of developing a protein gloss.

Smoking and drying of bloaters takes about four hours at a temperature of 85–90°F. During the first three hours the fish will only be dried and in the last hour two fires (or one in the smaller kilns) should be lighted. In consequence, the product has a faint flavour of smoke but very little, if any, colour, and the herring retains a bright silver appearance.

(3) *Finnan haddocks.* The finnan haddock takes its name from the Scottish fishing village of Findon, south of Aberdeen. It is prepared from a split haddock of medium size. A good product can only be

FIGURE 4·10 Splitting a haddock to make a finnan

made from a plump fresh fish, since in stale fish the lugs, or belly flaps, and the flesh along the backbone are discoloured and the flavour is inferior. Haddocks are often thin and in poor condition at spawning time in the early months of the year. A fungal parasite that sometimes occurs in the flesh gives the product a white blotchy appearance and an unpleasant taste. These infected fish are called *greasers* in the trade, and seem to be commonest in inshore and near waters.

The haddocks are prepared by beheading them and opening out the abdominal cavity to the vent. The gut cavity is cleaned to the extent of scraping or brushing off the black skin (peritoneum) and removing the blood and kidney lying beneath the backbone. This brushing and scraping must be done carefully since haddock flesh tends to be softer than cod flesh and is more easily torn. The fish are then split by cutting them open all the way down the backbone from neck to tail fin (see Figure 4·10). Most finnan curers prefer the so-called London cut where the backbone lies on the left side of the product (see Figure 4·12). The thick part of the flesh is cut in such a manner along the side of the backbone that the flesh stands up and gives the fish a plump appearance. This method of cutting also opens out the thick part of the flesh and exposes a greater surface for drying and smoke penetration. Aberdeen curers, however, frequently cut their fish in a slightly different manner (see Figure 4·13).

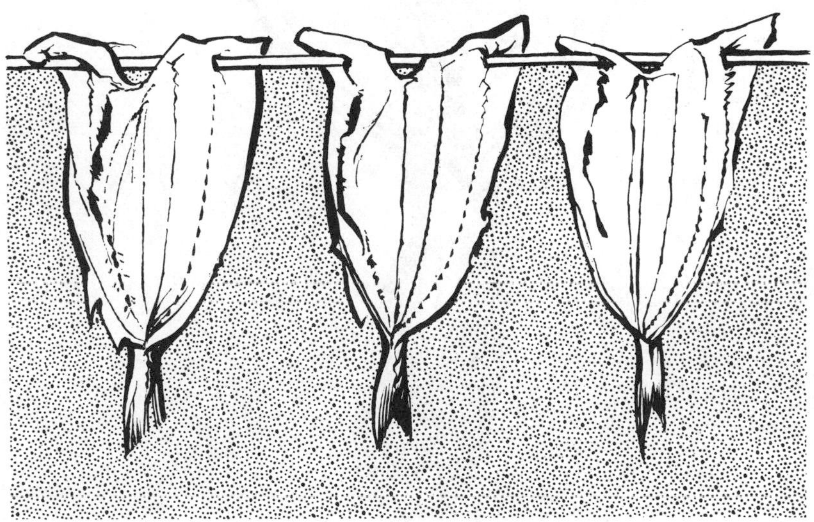

FIGURE 4·11 Split haddocks threaded on speat for smoking

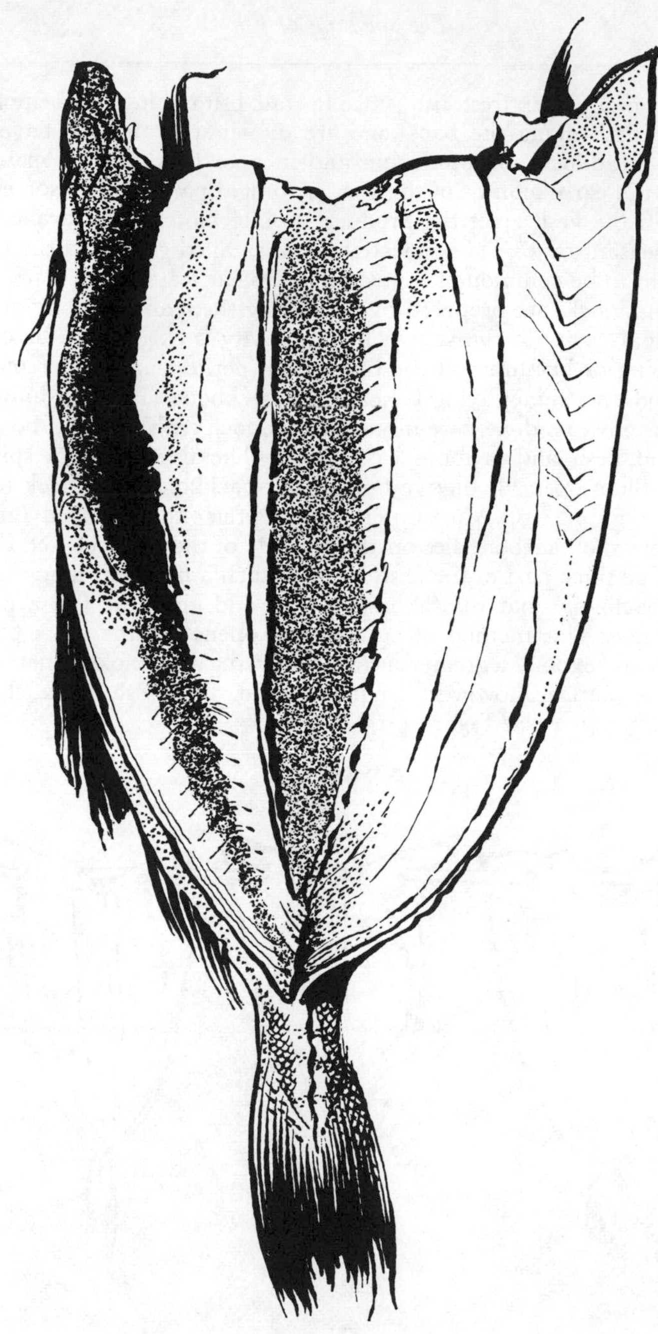

FIGURE 4·12 London cut finnan haddock

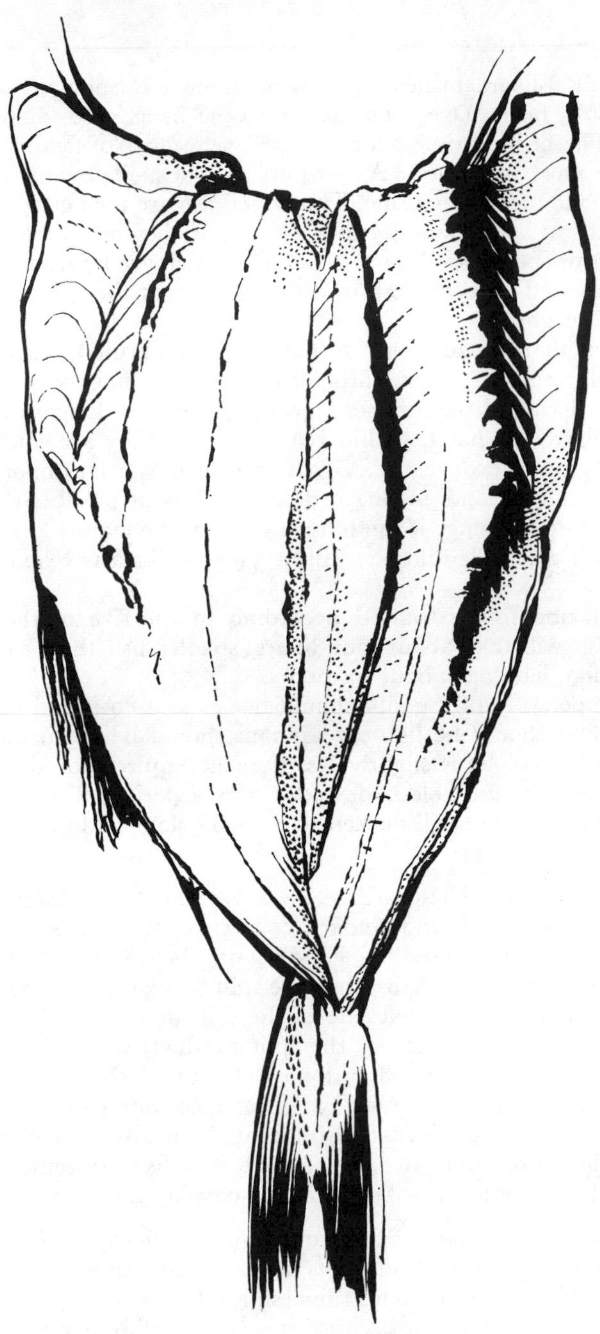

FIGURE 4·13 Aberdeen cut finnan haddock

The split fish are brined by soaking them for from 7–15 minutes in a 70–80% brine. Dye is not added to the brine used for preparing finnans. The brining time depends upon a number of factors, the size being the most important. A 1–1¼ lb haddock, for example, would require 10 minutes in the brine. The brined fish are then either tentered like kippers, as is done mostly in Scotland, or they are threaded on to metal rods or speats (see Figure 4·11). If the fish are large, small hooks are also inserted into the flesh in order to prevent them tearing under their own weight.

The gloss that develops on fish that have been brined, is mainly due to the swelling of the protein of the fish under the influence of the brine, and to the drying of the surface. In consequence, the longer the brined fish are allowed to hang, within reason, before they are smoked, the better the gloss that develops. A good gloss is of special importance in the case of finnans and as long a dripping time as possible should be allowed before smoking; if the temperature in the factory is high, it is best to push the trolley into a chill at 35–40°F for a few hours before smoking.

The smoking time is varied according to the size of the fish. A 1–1¼ lb fish will take about four hours, smaller fish three hours and large, jumbo, haddock about six hours.

The temperature of the kiln should not exceed 80°F and in a large kiln four fires should be lighted. Finnans should be taken out of the kiln when the colour is slightly less than is required for the finished product since the full colour develops over a period of time. A pale, straw-coloured finnan will darken to a good colour within a few hours of cooling.

(4) *Glasgow pales.* These are really a variant of the finnan, made from a small haddock and smoked to a pale straw-colour. The fish usually do not weigh more than ¾ lb and after beheading and cleaning, in the same way as for finnans, they are split along the belly so that the backbone remains on the left side of the fish, that is, the backbone is seen on the right hand side of the split product, which has a much flatter appearance than the London-cut finnan. Fish are brined in a 70–80% brine for 4–5 minutes. Dye is not used. Smoking time is 2–2½ hours with, in a large kiln, four fires lighted. In a smaller kiln, which has only three fires, only two are lighted. The fish are removed from the kiln when they are just beginning to develop a colour.

(5) *Cod and haddock fillets.* Cod and large haddock are filleted and most of the lug removed. Cod fillets are normally skinned but the skin is left on haddock, both because the flesh is softer and tends to tear and gape more easily, and also because it serves to differentiate between the species.

The fillets are brined in a 70–80% brine for from 4–10 minutes depending upon size. Dye is usually added to the brine bath. The brined fillets are then laid over banjoes. The brined fish should be allowed to stand and drip on the trolleys for at least two hours before smoking. Smoking time varies from two to five hours according to size. Four fires are lighted (two in smaller kilns) and the temperature is controlled at 80°F.

Since white fish fillets are generally dyed, the smoking process should be aimed at giving flavour and texture and not colour. In consequence, if the fillets are too heavily smoked, they will develop a very dark colour on cooking and may also have a tough skin on the cut surface.

(6) *Block fillets or golden cutlets.* Small haddocks or whiting are used for the preparation of this product. A cut is made behind the head and is continued down either side of the backbone so that head and backbone are removed, leaving the two fillets joined down the back. The skin is left on the double fillet.

These fillets are very thin and four minutes' brining is usually sufficient. After brining the block fillets can either be laid over banjoes or hung by the tail on special tenter sticks. The brined fillets are then treated in the same way as is described for small fillets above.

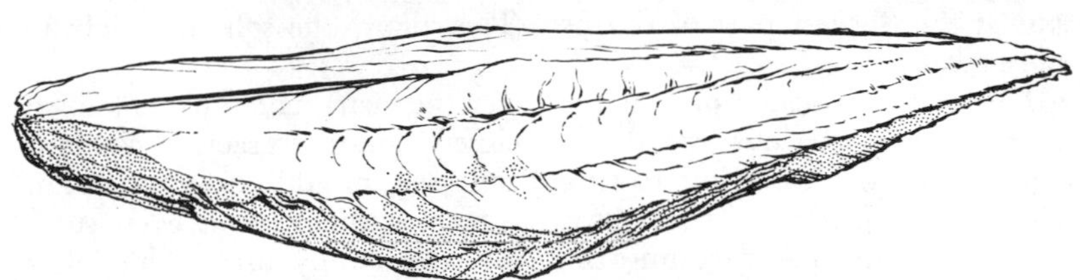

FIGURE 4·14 Cod fillet

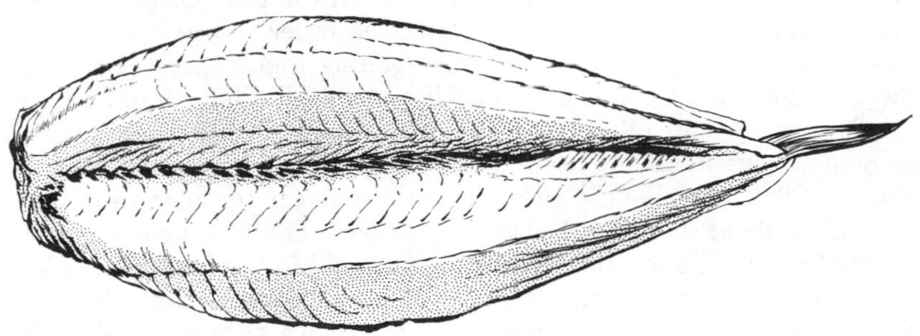

FIGURE 4·15 Block fillet or golden cutlet

(7) *Salmon.* The whole salmon are gutted and the gut cavity scraped and all blood removed. The fish are then beheaded and filleted, although the lug bones are left on the fillets. To give the finished product a better appearance, the ribs, which run along the inner surface of the lug flap, can be removed at this stage by trimming off a thin layer of flesh.

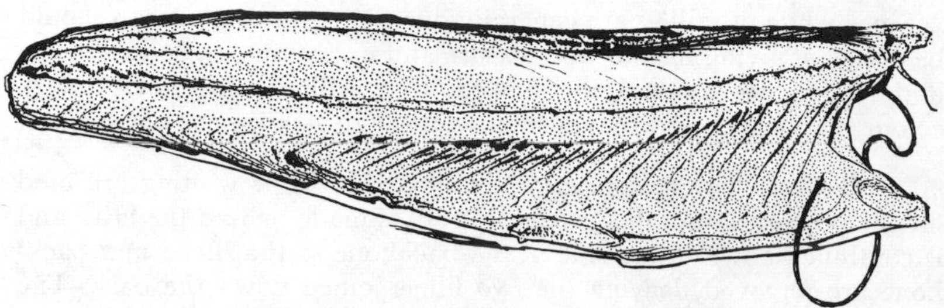

FIGURE 4·16 Side of smoked salmon

String should be threaded through the shoulder of each fillet under the lug bones to form a small loop. This facilitates the handling of the fillet during its subsequent processing. The skin can be scored either with a very sharp knife or with a razor blade set in a wooden handle. This operation must be done very carefully to cut just through the skin at the thickest part of the fish. This allows the salt to penetrate more evenly.

The next part of the process has, over the years, come to be looked upon as a mysterious ritual and closely guarded secret. Various ingredients have from time to time been used, in addition to common salt. These include molasses, brown sugar, saltpetre, and even rum. Nowadays only fine vacuum-dried salt is generally used. The fillets are laid down on a layer of salt about 1 inch thick. The cut surface of the fillet is then covered with another layer of salt about ½ inch deep at the thick end tapering down to the tail. At the thinnest part of the tail the salt is only lightly sprinkled over, in order to prevent the tail end from becoming too salty and dry. Salting time depends both on the size and the fat content of the fillet. A rough guide would be 12 hours for a 1½–2 lb fillet, 16–20 hours for a 3–4 lb fillet, and about 24 hours for a 5 lb fillet. The weight loss after salting should be 9–10% and the fillet, when lightly pressed with the finger tip at the thickest part of the flesh, should feel firm and springy. The salted surface will feel slightly tough but if the underlying flesh still feels soft and jelly-like a longer period in the salt is required. After salting, the sides are thoroughly washed in clean cold water to remove all surface salt and are then hung up to drip.

There are two methods generally used at the next stage of curing. One is to hang fillets for up to 24 hours in a drying room maintained at a temperature of 70°F and then to smoke them for 6–7 hours in very dense smoke. The other method is to smoke the fish slowly at 80°F in a light smoke for up to 12 hours depending on size. Both methods produce acceptable results.

In the Torry kiln, the fillets from the first method, smoked for 6–7 hours, would require all the fires to be lighted. For the second method, where a longer smoking period is used, two fires (one in the smaller kiln) should be lighted and allowed to burn out. This will take about four hours and a further two (one) fires can then be lighted and so on. The fish should be examined periodically and when the desired colour has been obtained the smoke should be turned off. Larger sides of salmon may require a still longer period in the kiln in order to dry them completely. A well-cured side should lose 9–10% of its weight during the actual smoking and drying process. Thus, including the loss in weight brought about by salting, a fresh fillet loses in all about 20% of its weight during processing.

(8) *Cod roes.* Cod roes must be handled very carefully before salting to avoid bursting their delicate skins. The roes are washed in cold water and are then covered with dry salt. They are usually salted in layers in boxes up to about two feet deep and are left for about six to eight hours.

After salting, the roes are placed in large-mesh baskets and thoroughly washed in cold water. The baskets of washed roes are then dipped for a minute or two into very hot or even boiling water. This makes the roes swell and gives them a plump appearance. Dye is sometimes added to the hot water in order to produce a more uniform colour. They are then either hung over speats of wood or laid on wire mesh trays and smoked at 90–100°F for six to eight hours with all the fires alight.

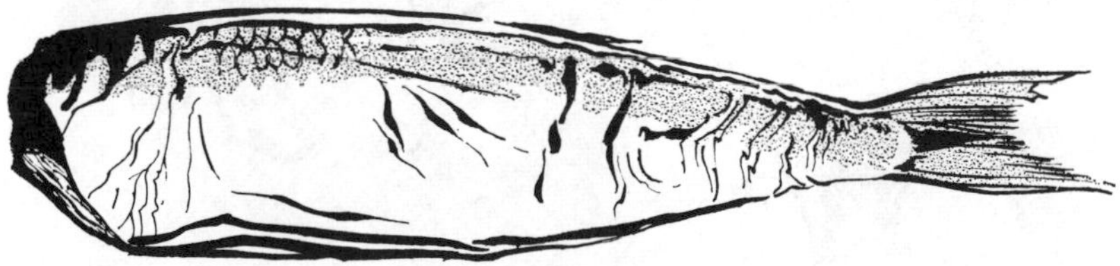

FIGURE 4·17 Buckling

(9) *Buckling.* These are hot smoked herrings. This means that the temperature of the smoke is such that not only is smoke deposited

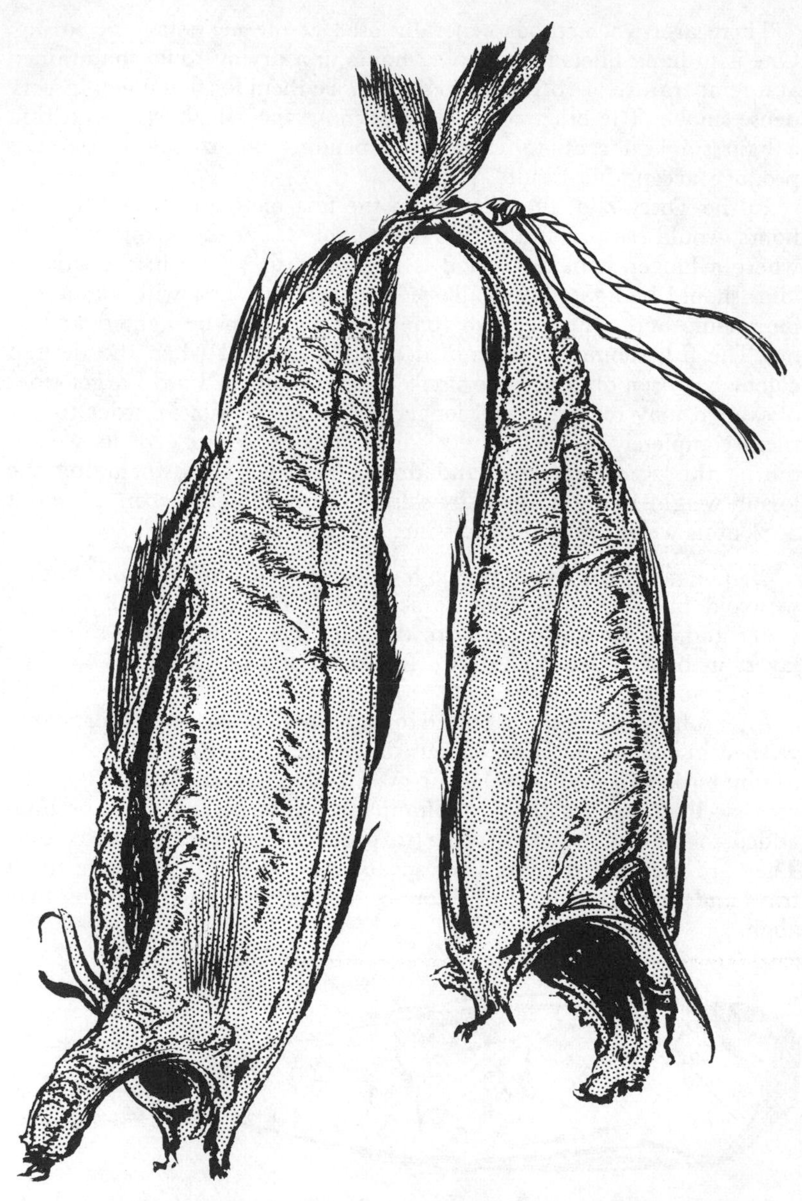

FIGURE 4·18 Smokies

during the smoking process but also the flesh is cooked. Buckling is originally a German product, very popular on the Continent, but less popular than it deserves to be in Britain. Most of those eaten in Britain are either produced in London or are imported.

The initial preparation of the herring varies. In some cases the fish are beheaded but in others the head is left on. Also, the gut may or may not be removed. British curers, however, favour nobbing the herring, that is, after washing, the whole herring are beheaded by hand and the gut pulled out, leaving any roe or milt in position. The nobbed herring are immersed in a 70–80% brine for one hour and then hung on speats. The point of the speat is pushed through the thick part of the flesh near the 'shoulder'.

In order to obtain a good golden brown colour on the final product, it is better to start smoking the brined herring while they are still wet with brine. The first stage should be carried out at 80–90°F. With all the fires lighted and with the air inlet of the kiln practically closed, the heat from the fires should bring the air temperature to approximately 90°F. The fish are then smoked for one hour at this temperature. The skins will dry off and this will prevent them from bursting when the herring are cooked. The second hour of smoking should be carried out at 110–120°F and the temperature is then raised to 160–170°F for a further hour, when the fish should be cooked. If the fish are not quite cooked in this time a further half-hour at 170°F may be necessary.

(10) *Smokies.* Smokies were first made in Arbroath, south of Aberdeen, in Scotland. They are hot-smoked, round, beheaded haddocks or whiting.

The traditional Arbroath smokie has a very dark, tarry appearance and was made in a barrel over a fire. It is not possible to produce this tarry appearance in a Torry kiln although a fair imitation can be made as follows. Gutted haddocks weighing $\frac{1}{2}$–$\frac{3}{4}$ lb are used. The fish are beheaded by hand and the gut cavity cleaned out as for finnan haddocks. The fish are then tied together by the tail in pairs and brined for $\frac{1}{2}$–$\frac{3}{4}$ of an hour in a 70–80% brine. The brined haddocks are hung over wooden sticks and are smoked as for buckling.

(11) *Trout.* Small rainbow trout are gutted and the gut cavity cleaned out, care being taken to remove all black skin and blood along the backbone. The fish are brined in an 80% brine for one hour. The brined fish are then pierced through the eyes with metal speats. Small strips of wood, like matchsticks, are placed between the lug flap to keep the gut cavity open and to allow drying and smoke penetration. Trolleys of brined fish are put into the kiln whilst they are still wet with brine. All the fires are lighted and the temperature of the kiln is allowed

to rise to about 90°F. After $\frac{1}{2}$–$\frac{3}{4}$ of an hour when the skins have dried off, the temperature is raised to 180°F and the trout are cooked. This takes about two hours. The trout have a lower fat content than herring, and in consequence they should be dried less than buckling or they will tend to be somewhat dry and hard and will lose their soft texture.

(12) *Sprats.* Sprats for canning are first washed and then soaked in a 70–80% brine for 15 minutes. The fish are then threaded through the eyes on thin speats which are fitted into frames. Smoking takes about one hour with two fires (one in the smaller kilns). The finished product should be cooked and have a silvery appearance. After smoking, the sprats are beheaded before they are packed in the cans.

The continental hot-smoked sprats, or *Kielersprotten,* are washed and brined as for canning but they are then speated through the gill and mouth. Smoking takes 1$\frac{1}{4}$–1$\frac{1}{2}$ hours with all fires lighted. The wet sprats are smoked at 90°F for half an hour and the temperature is then raised to 180°F until the fish are a golden brown colour and are cooked.

(13) *Eels.* Freshwater eels up to $\frac{3}{4}$ lb are gutted and cleaned and are then either soaked in strong brine for about an hour or are dry salted for up to 12 hours. If dry salted, the fish lose about 20% of their original weight, but if brined they lose no weight at this stage.

The dry-salted eels are washed thoroughly before smoking and hung on hooks which pierce them from the throat to the back of the neck. When they are hung on the hooks, the eels are dipped for a few seconds in a vat of boiling water or brine until the sides of the body open out. The eels are then smoked at 140°F with all fires lighted for 2–4 hours depending on size. The smoked eel is another delicacy that is much appreciated on the Continent.

(14) *Seelachs.* This delicacy was developed in Germany during the First World War when salmon was unobtainable. Although it does not compare with good quality smoked salmon either in texture or flavour it is nevertheless a pleasant delicatessen product that has sufficient merit to be worthy of greater attention than it receives in Britain. It is very popular on the continent of Europe, especially in Germany.

Although the best product is made from coalfish, other species of the cod family, including cod and ling, can be used. Coalfish are usually at their best in the early part of the year and in Germany many firms salt the sides or fillets at this time and use them throughout the rest of the year.

The first stage of the process is to fillet the fish, leaving on the lug bone and skin. If it is desired to use the fish immediately, then it can be salted in solid vacuum salt for about 24 hours, or less if the sides are small, as for salmon. The liquor that is produced should be allowed to flow away. If the sides are to be stored for use at a later date the fillets should be salted in solid salt in a vat, in which the liquor is allowed to collect. Provided that the sides are weighted so that they remain well below the surface, they will keep in good condition for many months if the room temperature is not too high.

If the fillets have been kept in brine for a long time, then they must first be desalted in water or dilute brine. Fillets salted overnight can be used at once after a preliminary wash to remove adhering salt. At this stage the fillets should be firm yet springy and it should be possible to cut them into thin slices. The salt content in the flesh should not be below 9%; this is important for subsequent keeping quality.

The fillets are then cut into pieces suitable for slicing, and sliced either by hand or, preferably, by machine. The slices, which should be thin, are then immersed in a dye bath containing 5% saturated brine, 0·2% acetic acid, and a suitable dye. Dye manufacturers can make a mixture of edible dyes suitable for this purpose. The slices are then laid out individually on oiled wire-mesh trays and smoked in a dense smoke for about 30 minutes with just enough smoke passing up the chimney to keep the fires burning.

The smoked slices should contain no more than 60% water. They are then packed in a pure vegetable oil. Olive oil makes a very good product but is costly. Arachis or peanut oil, provided it is of good quality and is not rancid, makes a perfectly acceptable product.

The shelf life at chill temperatures, 32°F, is at least four weeks. At higher temperatures it is rather less and it should therefore be kept chilled if it is to remain in first class condition until eaten.

(15) *Oysters*. A considerable proportion of the total production of oysters in Britain consists of misshapen native oysters or of Portuguese oysters that do not command the high price of top quality natives. In America these second-grade shellfish are frequently smoked and packed in oil to give a popular delicatessen product.

The oysters, which should first be cleansed to ensure that there is no risk of food poisoning from them, are steamed in a cooker for 20 to 30 minutes. The meats are then picked and brined in a 50% brine for 5 minutes. They are quickly dipped in oil and spread on wire-mesh trays. They are smoked for 30 minutes in a dense smoke at 180°F, the kiln being raised to this temperature before smoking begins. During the smoking, the oysters should be turned over, to ensure uniform smoking.

The smoked meats are packed in small jars, covered with good quality edible oil, and the jars sterilized at 15 lb pressure for 15 minutes.

5

Salt curing

The Herring is a lucky fish,
From all disease inured.
Should he be ill when caught at sea;
Immediately — he's cured.

SPIKE MILLIGAN

COMMON salt, or sodium chloride, if present in sufficient strength, will slow down or prevent bacterial spoilage of fish. This property of salt is used in salt curing to make products that will keep in good condition at ordinary temperatures for a long time. Salt curing is a traditional process which is now used less in western European countries than it was formerly, although total world production of salt fish is still increasing. The decrease in popularity of salt fish in western Europe follows a general decline in the taste for heavily salted products especially in countries where the corresponding fresh and frozen commodities are available. Salt-cured products are still eaten in large quantities in some parts of the world, however, and there is a fair-sized export trade in salt fish products from Britain and a number of other European countries.

There are two main types of curing. *Dry salting* is carried out by burying split fish in salt and allowing the brine liquor to escape. After a period in salt, the product is usually dried. Dry salting is suitable for white fish such as cod and ling but cannot be used for fatty fish such as herring. *Pickle curing,* in which the fish is preserved in airtight barrels in a strong pickle formed by salt dissolving in the body fluids, is used for fatty fish.

PRINCIPLES OF SALT CURING

The main features of salting are the removal of some of the water from fish flesh and its partial replacement by salt. In dry salting of cod, for example, there is a rapid loss of weight over the first four to five days of about 25% of the original weight of the dressed fish and a subsequently slower decrease to just over 30% loss. The salt content conversely rises to about 18% in the wet tissue in six to eight days, and the further uptake, to a maximum of about 20%, occurs at a much slower rate. When the amount of salt in the flesh rises above about 9%, certain irreversible changes occur in the muscle proteins, which become denatured.

In pickle curing the weight changes are most complicated. During the first few days of curing herring, there is a marked loss in weight of about 20% accompanied by a rapid intake of salt into the flesh of up to 18%. After this the fish gradually gain in weight again until by about the tenth day the cured fish weigh about the same as the original fresh herring. Many rather complicated theories have been suggested to explain these phenomena.

Salt uptake and water loss are influenced by the fattiness of the fish, the thickness of the flesh, freshness, temperature, chemical purity of the curing salt, and probably a number of other factors. The fat in the herring acts as a barrier both to the entry of salt and withdrawal of moisture; experiment has shown that both salt uptake and moisture loss become progressively slower with increasing fat content.

The skin, at least in fish such as cod, does not prevent diffusion of salt into the underlying tissue although it has been stated that it does slow it down, particularly in fatty fish. This effect may possibly be due to the fatty layer immediately beneath the skin.

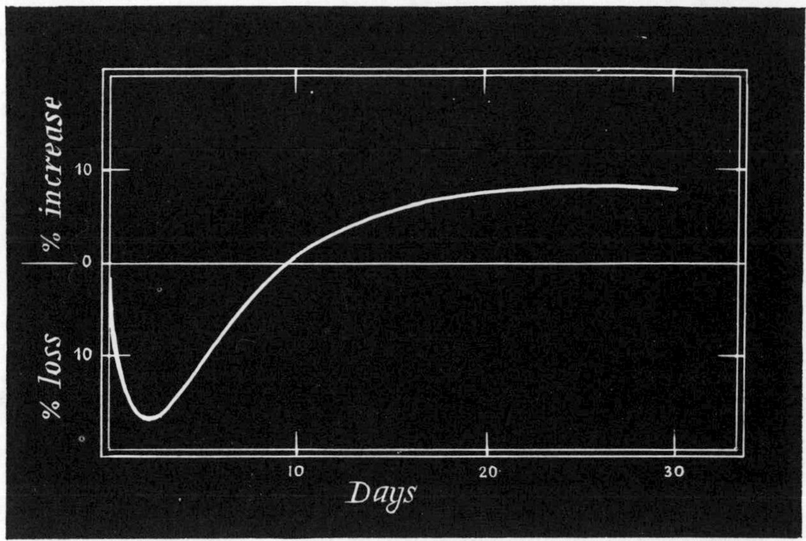

FIGURE 5·1 Change in weight of herring during pickle curing

The thickness of flesh also has a pronounced effect. The concentration of salt at the centre of a fillet one inch thick can be as high as 10% after 24 hours; in a fillet two inches thick, three days are required to reach the same concentration.

It has been found at Torry that the staler the fish, the quicker the salt uptake and the greater the weight loss. Salt uptake is also faster at higher temperatures. This table shows the effect of temperature:

TABLE 5·1

Percentage salt in cod fillets 1 inch thick at different times and temperatures

Temperature	*1 day*	*2 days*	*5 days*
40°F	1·1	4·0	14·1
80°F	4·8	8·6	15·2

Unfortunately, the higher the temperature the faster the spoilage bacteria multiply, so that a point is reached where putrefaction is proceeding more quickly than the increasing preservative effect due to entry of salt. This is most likely to occur in the deeper parts of large thick fish and can give rise to the so-called *putty fish* in which the centre of the flesh looks and feels like putty. It is therefore usually safer to keep the fish cool during salting.

The purity of the salt used for curing is largely responsible for the physical character of the final cure and this is of particular importance in dry salting. Pure common salt consists only of the chemical sodium chloride; curing salt may contain traces of certain impurities, some of which are important because of their effects on the cure. The main impurities in ordinary curing salts, calcium and magnesium chlorides and sulphates, even though present in small amounts, slow down the penetration of salt into the flesh. Salt fish, such as salt cod, containing these impurities spoils more quickly than products made with pure salt. Some of these compounds, particularly magnesium chloride, readily absorb moisture from the air and become first moist and then wet. If, therefore, salt fish contains a sufficient quantity of these impurities, it may become damp again after it has been dried. Damp fish will grow bacteria and moulds: *pink* and *dun* almost inevitably occur.

The calcium and magnesium salts give, on the other hand, a whiter cure. Pure salt results in a softer and yellower product which is more tasty, however, because calcium and magnesium compounds have a strong bitter flavour. In spite of this, it is generally thought advisable there should be some calcium and magnesium compounds in the salt used for dry salting, in order to produce the attractive white cure which the customer demands. French workers suggest that the total amount of calcium and magnesium in the salt should be between 0·3% and 0·6% and Russian scientists have stated that for dry salting, calcium and magnesium should not exceed 0·5% and sulphate 1·0%

Another impurity in the salt which affects the appearance of the cut surface of dry-salted fish is copper. Even in the presence of copper compounds to the extent of only a few parts in 10 million of salt, the white surface of the fish can become brown. This effect can occur even in very fresh fish that otherwise is just as good in eating quality as the normal cure. To the ordinary buyer, however, it looks very like the discolouration of stale or spoiling poor-quality salt fish and its market value is severely reduced.

Three types of salt are generally used in the trade. *Rock* or *mined* salt occurs in many parts of the world including Britain. *Manufactured* salt is obtained by the evaporation of naturally-occurring brines or brines made from naturally-occurring salt beds. *Solar* salt is made in the warmer parts of the world from sea water or inland salt lakes wherever weather is suitable. Most of the rock salt used for curing in Britain is imported from Germany and is known as *Hamburg* salt. The purification of salt is an important activity of the British chemical industry. European supplies of solar salt come mainly from the Mediterranean and North Africa; Caglari, Setubal and Sfax salts are well known in the trade.

The preservative effect of salt is often said to be due to the removal of moisture from the flesh, because fish spoilage bacteria cannot grow in the absence of sufficient moisture. Another factor is also, however, apparently involved. Salt itself reduces or prevents the activity of these bacteria if more than about 6% is present.

METHODS OF SALT CURING

It is not necessary to discuss here the many varieties of product that are made by dry salting and pickle curing. Some typical examples of particular interest in Britain will, however, be mentioned.

DRY SALTING OF COD

(1) *Heavy or hard cure.* The fish are headed, split along the belly, and the backbone removed, except for the tail portion which gives rigidity to the split fish and makes handling easier. The Norwegian and Icelandic regulations for the manufacture of salt cod, which show how to make good quality products, require that the fish be bled immediately after catching, to obtain white flesh and a better keeping quality in the final product.

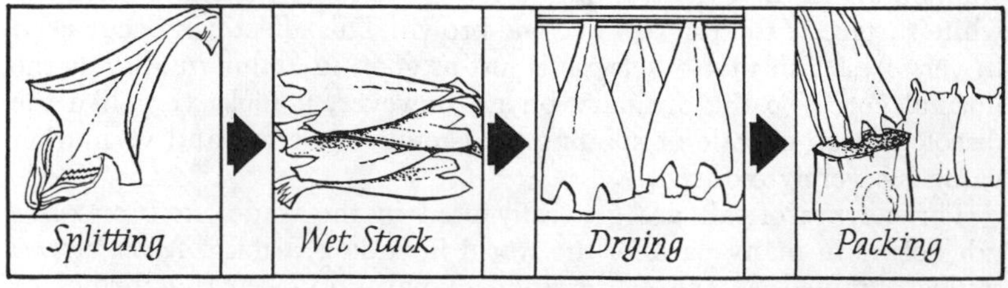

FIGURE 5·2 Sequence of operations in dry salting of cod

The fish are then laid in piles with layers of salt in between and the juices, withdrawn from the fish by the salt, are allowed to run away. Within 15 days or so the salt has penetrated or *struck through* the flesh, so saturating the juices that remain. The *green* cure so obtained is now said to be lying in *wet stack*. At frequent intervals the fish are re-stacked, the top layers going to the bottom and the bottom ones to the top. This produces an even cure, because the pressure of the pile squeezes out a certain amount of juice; the lower they are in the pile, the more heavily the fish are pressed.

Although most wet-stacked fish produced in Britain is made on shore, this part of the process is carried out by the French, Portuguese and some other nations on board their vessels while they are actually on the fishing grounds. The fish may remain in wet stack for months before it is finally dried. Its water content is at this stage between 53% and 58%.

The drying operation is always carried out on shore. In Norway and Iceland some fish is still dried by exposing it to the wind and sun in the open air but in Canada and Britain some kind of drying chamber

FIGURE 5·3 Heading and splitting cod for dry salting

FIGURE 5·4 Split salt cod lying in wet stack

is employed. British production is normally exported and has consequently to be packed in a suitable container. At one time metal-lined cases were used to prevent the absorption of moisture, and hence spoilage, during shipment to their destinations which were often warm and humid countries. This method has now been abandoned in favour of the much less expensive and almost as effective one, in which the fish are packed in one-hundredweight bundles in hessian and then wrapped in stout brown paper.

The moisture content of the final product varies widely depending on the extent of drying and the thickness of the fish and may range from 10% to 30%. Consequently the salt content is also quite variable, ranging from 25% to 35%. It has been calculated that every 100 lb of dressed fish require about 30 lb of salt for a heavy cure.

(2) *Gaspé or light cure.* For many years now there has been an increasing demand in the major importing countries, such as Spain, Portugal and Italy, for less heavily salted cures. Producers in Britain and elsewhere have been faced with increasing competition from the *Gaspé*

cure, prepared in the Gaspé peninsula of Quebec Province and elsewhere in Canada. This cure is still made in much the same way as it was when first invented by the early Canadian settlers over three hundred years ago.

The fish are gutted, headed and split in the same way as for the heavy cure. The dressed cod are then laid cut side uppermost, in tubs approximately 36 inches in diameter. Fresh fish up to about 48 hours after catching must be used if successful results are to be obtained. Salt is then added in the proportion 7–9 lb to 100 lb of split fish, depending upon the weather, more salt being used if it is warm. Alternate layers of fish and salt are put in until the tub is full. After 24 hours sufficient brine is formed to float the fish which is then kept under with flat stones or other heavy weights. After 48–72 hours, the fish are removed, washed in their own brine, stacked 5–6 hours to press out some of the liquid and then dried in the sun or in driers.

FIGURE 5·5 Salt cod being hung in a drying chamber

The Gaspé cure is yellowish when it is removed from the brine and the dried product, which tastes rather like cheese, has a characteristic amber-coloured, translucent flesh. This is in contrast to the heavy cure which is usually greyish-white before it is dried and, after drying, is almost snow white. This appearance is due to a thin surface layer of salt, although the flesh underneath is straw-yellow and opaque.

PICKLE CURING OF HERRING

As already mentioned, pickle curing is used for fatty fish, such as herring, mackerel and salmon. Pickled herring is by far the most important British product of this type. The herring, as they are landed, are sprinkled with about 14 lb salt to every cran. This salt both helps to prevent deterioriation during transport and while awaiting further processing at the curing yard but also, by coagulating the slime on the fish, helps to make subsequent handling easier. When they arrive at the curing yard, the boxes are emptied into *farlins* or long wooden troughs and in these the fish are *roused,* that is, well mixed with more salt and gutted, *gipped* or *gibbed* by women.

A knife is inserted at the gills and by a very quick and dexterous movement, the gills, long gut and stomach are removed. The milt or roe and some of the pyloric caeca remain. The pyloric caeca are necessary if the desired cured flavour is to be obtained. As the women

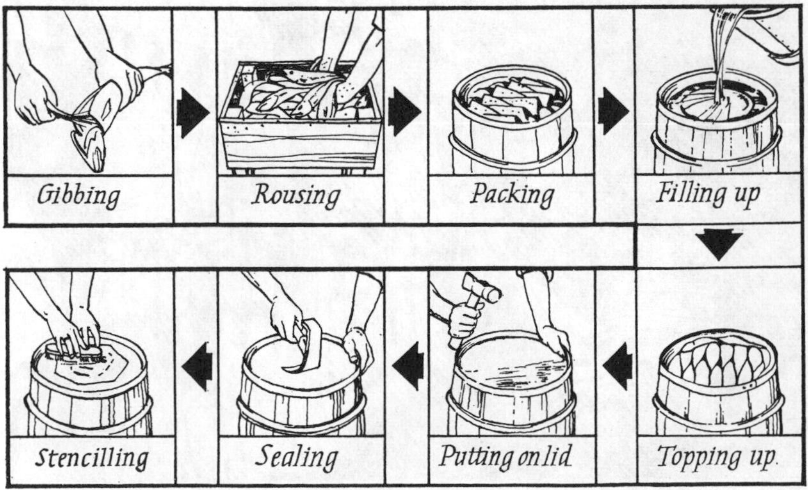

FIGURE 5·6 Sequence of events in pickle curing of herring

gut the fish they grade them, mainly for size but sometimes also for the condition of the roe or milt. An experienced crew of three, two gutters and one packer, can gib, grade and pack at the rate of 2–3 cran an hour. Women usually stand at the farlin and as they gut the herring, throw them into a series of tubs or baskets arranged in a semicircle

FIGURE 5·7 Herring being gibbed at the farlin

behind them. Owing to the decline in the curing industry a fall in the number of skilled workers has occurred over the past decade. As a result, gibbing machines have now been introduced which require comparatively unskilled labour for their operation. The Herring Industry Board, which has been instrumental in introducing these machines into this country, state that a crew of six people can gib, grade and pack 6–7 cran an hour.

After gibbing the herrings are then taken to the packer, who lays the fish in the barrel in the traditional manner, which consists of first putting a layer of salt on the bottom, then a layer of fish belly uppermost and head to tail until the layer is complete. More salt is placed on top and a new layer of fish laid at right angles to the layer beneath. These alternate layers of salt and fish are continued until the barrel is full. On average, a crew of two women gutting and one packing can cure 20 cran of herring in an eight-hour day.

After a day or two the herring have shrunk appreciably. Fish from the same day's cure and grade are used to fill up the barrel which is then headed, that is, the lid is put on, and it is placed on its side for a

FIGURE 5·8 Packing pickle-cured herring into barrels

further eight to ten days. It is now known as a *sea stick*. At the end of this period, it is up-ended again, the lid removed, and a bung hole bored in the middle of a stave. *Blood pickle*, so called mainly because of its reddish-brown colour from the herring blood, is drained off from the top half of the barrel. More fish of the same cure are now added to make up the loss of brine until the barrel is tightly packed up to the *croze*, that is, up to the groove at the top of the barrel into which the wooden lid fits. The lid is then replaced. The barrel is again placed on its side, blood pickle is run in to fill up spaces between the fish and the bung inserted. The barrel is now ready for storage.

Fresh saturated pickle is never used for the final packing because, it is said, this causes rancidity. So far as is known, there is no scientific evidence to support this view, although it may well be that some of the natural antioxidants in the herring are removed by the brine. Freshly made brine would not, of course, contain them.

In Britain, all the herring curing is done on shore but the Dutch, Poles and Russians cure on board ship. In many cases the fish are merely *rough packed,* or salted in barrels without gibbing. Such a procedure can only be carried out with certain classes of herring, for example, those with little food in the gut, for otherwise spoilage, and particularly perforation of the belly, quickly occurs. Even so, rough-packed herring are normally used for further processing on shore, such as the production of half-conserves, for example, the Scandinavian tidbits or gaffelbitter.

There has been an increasing tendency to produce lighter herring cures. The most important is the so-called *matje cure.* Here the brine is no longer fully saturated, but may be anything from 80–90% saturated.

In addition to the lower salt content, it is believed in the trade that herring with the true matje flavour can be made only from fish caught early in the season, May to June in Britain, before the fat has been properly assimilated into the tissue. The matje cure is soft in texture and, like the Gaspé cure, has its own peculiar flavour, said to be like cheese. Owing to its lower salt content, it readily spoils at ordinary temperatures and has to be kept chilled.

As might be expected of a traditional technique such as herring curing, it has acquired over the centuries a number of practices and rules, some based on fact but many on mere prejudices.

It has, for example, long been accepted as a fact that fatty fish make the best cures and, indeed, fish with less than about 10% fat are never cured. There are many rules that appear to have no substance in fact regarding the best type of salt to use. It was, for example, believed that one type of solar salt was better than another. All that appears to be established is that the salt should not be of too fine a grain. Crystals should be sufficiently large to keep the fish apart and allow the brine to diffuse into the fish but yet sufficiently fine to dissolve easily and form pickle. So far as is known, however, there has been no definite recommendation as to the proportion of fine to coarse grains to give the best cure. Rock salt is not usually recommended apparently because the sand that is unavoidably present makes the cure gritty.

Many curers formerly believed that the development of a properly flavoured cure could take place only in a wooden barrel, preferably of spruce or fir. Experiment has shown, however, that cures, indistinguishable even by experts from those made in wooden barrels, could be produced by using concrete or enamel tubs and by finally packing in lacquered tins.

SPOILAGE OF SALT FISH

Salt stops the growth of spoilage bacteria. There are, unfortunately, many types of bacteria found on fish, and they are not all equally affected by salt; some micro-organisms can tolerate much greater quantities of salt in their environment than can others.

Micro-organisms may be divided into three groups according to their sensitivity to salt. Group 1: those that are held in check and even killed by a few per cent of salt. This group includes most of the bacteria that commonly make things go bad. Group 2: those that can tolerate large amounts of salt, even up to completely saturated brine, but which find it increasingly difficult to grow as the percentage of salt rises. Group 3: those that cannot grow without salt. These are the so-called *halophiles* or salt-lovers.

When fish such as herring are heavily salted at, for example, 50–60°F, there is an increase in the number of bacteria over the first few days before the salt has had time to act effectively. There is then a steady decrease in numbers in the brine and a suppression of the putrefactive types. After about three months, the numbers may be only 10% of those present after the first days of the cure. If storage is continued at these temperatures, the surviving bacteria gradually begin to grow. After a period of about six to eight months their numbers may have increased considerably. It is of interest to note, however, that although the brine always contains some micro-organisms, the flesh of the cured herring itself is, after about three months, practically sterile. This is in marked contrast to dry salting where the flesh always contains some micro-organisms, often in considerable numbers. Indeed, in dry salting, the evidence is that although the fish-spoiling types of bacteria in the flesh are almost completely suppressed, the salt-tolerant and halophilic types continue to grow right through the curing process up to the end of the wet stack stage. The probable explanation for this situation is that the salt-tolerant and halophilic types all require oxygen for growth and this is excluded in pickle curing.

With the less heavily salted products such as the Gaspé, and possibly also the matje, cure the peculiar flavours of the products are in fact due to the activity of some of the surviving micro-organisms, just as the flavour of many varieties of cheese is also the result of controlled microbial growth. With herring, however, it has long been held and recently scientifically confirmed that part of the true cured flavour, particularly in the matje cure, is the result of the activity of the enzymes present in the pyloric caeca which remain in the herring after gibbing.

As might be expected, the harder the cure or the greater the salt content of the cure, the longer it will keep. Even so, given favourable temperatures and sufficient time, some of the micro-organisms, including those producing the desired flavour, increase to such an extent that they cause putrefaction. At temperatures of 60–70°F such spoilage with heavy cures may become evident within two or three months. It is correspondingly quicker with lighter cures. At the low temperatures of 32–40°F, no bacterial spoilage in the hard cures may be evident for years, although the product may be so soft and rancid that it is inedible. With the light cures, storage at chill temperatures is, of course, essential. It is well known in the trade that the conditions of production and storage of light cures must be rigidly controlled if success is to be certain.

PINK SPOILAGE AND ITS CONTROL

It was mentioned earlier that heavily cured white fish, such as cod, often lie in wet stack at ordinary temperatures of 50–70°F for considerable periods before drying. This may be due to a variety of factors, such as lack of drying facilities, or to the fact that different markets require products dried to differing degrees and, until a firm order has been received, it is not, therefore, desirable to commence drying operations.

During this storage period in wet stack not only are the salt-tolerant micro-organisms slowly increasing in numbers but so, also, are the halophilic ones. Among the latter is one group that has a distinctive pink to rose-red colour, and which is responsible for one of the most troublesome types of spoilage encountered in the dry salting trade; it is known as *pink*. In the early stages of pink spoilage, the surfaces of the wet stacked fish are covered with small separate pink patches which may gradually run together and cover considerable areas of the flesh and skin of the fish. At this stage, the fish are usually little damaged, and the pink patches can be removed by scrubbing and washing the surfaces in running fresh water. This treatment not only removes the pink but, in the process, destroys the bacterial cells; when they are placed in fresh water, or even in salt solutions containing less than 10% salt, they break up completely. Once pink has been found in a pile of fish it is good commercial practice to scrub and wash them. If no precautions are taken at this stage, however, more serious changes occur. Usually another group of pink bacteria, and also some colourless types, begin to grow, and vigorously attack the flesh with the production of the characteristic off and putrid odours, which have been variously described as stale, sweaty, or like over-ripe cheese. The flesh becomes muddy brown and very soft. Indeed, it may become so soft that it falls to pieces when the fish is held up by the tail.

It is of interest to note that these halophiles fail to grow unless at least 10% salt is present and consequently pink never occurs in the lighter Gaspé cure. It should also be added that, although they cause putrefaction of salted fish, these pink halophiles are quite harmless. Although pink salt fish has been incriminated as a source of food poisoning, the evidence is that this was due to the presence of a well known type of food poisoning organism which can tolerate high concentrations of salt.

All the available evidence seems to show that the curing salt is the major source of the pink micro-organism. As a general rule it may be stated that rock salts are usually free from pink bacteria whilst solar salts often contain very large numbers. Many manufactured salts are also relatively free from pink bacteria but occasionally even they can contain several thousands in a gram.

If a curing salt is heavily infected it is almost certain that pink spoilage will occur during the wet stack stage if the fish remains long enough at a suitable temperature. It need hardly be added, therefore, that the trade should avoid using salts that are heavily infected with pink bacteria. Since the numbers of these halophiles varies markedly from salt to salt, and even from one year's production to the next, the only way to test whether or not a salt is suitable for curing is to find out how many pink micro-organisms are present in a known weight of salt. Unfortunately this is a skilled job that cannot be done by the curer himself. Some Public Analysts will, however, carry out the necessary tests.* It has been found, in general, that freshly prepared solar salts are the most heavily infected; the older the salt the fewer the numbers of pink halophiles it contains. Hence, it is generally advisable not to cure with salt less than a few years old.

Although a salt may be free from pink, however, it does not follow that it will not occur in the cure. The reason for this is that in most curing yards and premises pink has occurred at one time or another and the walls, floors and ceilings are so heavily infected that they continue to contaminate subsequent cures. It is very important in commercial

*Numbers of bacteria greater than 1000 in a gram of salt indicate that the sample is not suitable for curing (28·35 grams go to an ounce).

practice to ensure, therefore, that once premises have become infected they should be thoroughly cleaned. For this purpose, all walls, floors, ceilings, tubs, and other equipment should be thoroughly scrubbed and washed in fresh running water. Canadian scientists state that after scrubbing and washing plant and equipment it is an advantage also to disinfect it. Hypochlorite is one of the most suitable disinfectants; it should be diluted to give a solution containing 500 to 1000 parts per million of free chlorine. Commercial solutions may be suitably diluted, but, since the strengths of these solutions vary widely, it is better to calculate the necessary dilution from the figures given by the manufacturer, or to ask his advice. The solution is sprayed on to the surfaces and the treatment repeated after a period of 20–30 minutes.

An interesting feature of these pink bacteria is that they grow best at about 104°F, very slowly at 50°F, and not at all below 40°F. For this reason many modern plants, particularly in France, store the wet stacked fish in a chill room. Under these conditions, not only is pink spoilage prevented but the growth of the other putrefactive types is also held in check and hence a better product can be obtained. Unfortunately, in this country and elsewhere, chill storage is either not available for this purpose or is too expensive to instal. Consequently a variety of chemical inhibitors, including antibiotics, have been tested with a view to preventing or restraining the growth of pink. None has yet proved to be completely successful in practice, although the search for an inhibitor still continues.

DUN SPOILAGE AND ITS CONTROL

As already mentioned, light cures, such as the Gaspé cure, are seldom attacked by pink. There is, however, another type of spoilage that is particularly noticeable in the light cures, although it can also occur in the heavy ones, and is known in the trade as *dun*. The fish surfaces become covered with numerous small tufts or spots of a black, brown or fawn material making the fish look as if it has been peppered. The 'pepper spots' are not caused by bacteria but by moulds. Fortunately these moulds do not damage the flesh of the fish as pink does and, although the discoloration is unsightly, are harmless. Not unexpectedly, however, fish covered with dun fetch lower prices. Another important difference between pink and dun is that the moulds causing dun can grow, although only very slowly, when the moisture content of the flesh is well below that at which pink spoilage is checked. Dun can consequently occur on the surfaces of dried fish, particularly if they have picked up some moisture from a humid atmosphere, for example.

Canadian experience has shown that dun flourishes in old factories built on low ground, and grows well on damp, salt-soaked boards, tubs and soil. Although these moulds do not require such high concentrations of salt for growth as do the pink bacteria, they are halophilic, and require at least 5% salt.

It is still not very clear where dun comes from. It seems almost certain that some moulds are present in the solar salts. It also seems probable that since moulds are universally present in soils, dun may have arisen from the types normally present there that have become adapted to the conditions in salt-soaked soil, surrounding the older and less clean plants.

The best method of preventing dun is to have a system of control in the curing yard and storage premises. These moulds grow best in damp, dark corners, on dirt and rotting wood; consequently proper ventilation and cleanliness would do much to eliminate such sources of contamination. Special paints that prevent the growth of moulds are now available and these would also prevent wooden surfaces from becoming salt soaked and hence an ideal breeding ground for moulds.

Whilst much can be done to prevent infection, dun can, nevertheless, sometimes occur on the fish. Recent experience has shown that this can be prevented or controlled by dipping the fish into a 0·1% solution of sorbic acid.

DRYING OF SALT FISH

Although some of the dry-salted cured products are consumed locally, most of them are exported, usually to warm, humid climates, and it may take months before they finally reach the consumer. Light cures, like the Gaspé one, keep in an edible condition for only a few days at the wet stack stage, particularly at relatively high temperatures of 60–80°F; even the most heavily salted fish begin to spoil within a few weeks at these temperatures. It is consequently necessary to dry the fish to prevent further spoilage.

KLONDYKING

Mention should be made here of another method of preservation of ungutted herring involving the use of salt, although it is not strictly speaking a method of curing. This is *Klondyking*, possibly so called because the method was developed at about the same time as the famous Gold Rush of 1897. It was formerly employed on a fairly considerable scale for preserving herring for export, mainly to Germany, where it was chiefly used for marinades and similar products.

Typically, two quarter-cran baskets of herring were sprinkled with coarse fishery salt at the rate of about 22 lb salt to a cran, as they were tipped into a half-cran wooden box. Ice was then put on top of the mixture of fish and salt at the rate of 2 to 2½ cwt of ice to a cran.

If fishing was light and loading of freighters therefore slow, rather larger proportions of ice and salt were used. Herring treated in this way could probably be kept in edible condition for at least a week.

6

Drying

'Ahem!' said the Mouse with an important air. 'Are you all ready? This is the driest thing I know.'

LEWIS CARROLL

WATER is essential for life, and micro-organisms are no exception in requiring plenty of water for their growth and multiplication. Lack of water, or loss of it, can bring to a standstill the activities of the bacteria and moulds that spoil foodstuffs and hence drying can be used as a means of preservation.

Primitive methods of drying fish merely by hanging them, either in the sun and wind, or over wood fires, have been used by man for thousands of years. The process is slow and may take up to several weeks to complete, depending on the size and thickness of the piece of fish, and on the climate. Some traditional dried products are salted as well as dried. They mostly have a strong flavour.

Drying is involved in various ways in processes used by the fish trade. Unintentional drying occurs frequently, for example when wet fish stands in a breeze on the market or when fish is cold stored, and various measures can be taken, such as suitable packaging or covering, to reduce or prevent unwanted drying. This chapter, however, is concerned with those processes in which drying is an essential part of the operation. For example, the evaporation of moisture from the product is a complementary part of the smoking process, and is the main object in the production of dried salted cod, stockfish and fish meal.

The term drying usually implies the removal of water vapour by evaporation, but water can also be removed from fish by pressure, or by the use of absorbent pads or by salt.

In the British fish industry, drying is involved mainly in the manufacture of smoked fish, fish meal and salt cod. Hard-dried fish, such as stockfish, is an important commodity in some other countries. Various attempts have been made in Britain and elsewhere to improve by dehydration the character and palatability of dried fish products.

STOCKFISH

Stockfish has been produced in Iceland and Norway for many centuries. It is made by hanging up headless cod in the open air until it is dry and hard. Larger fish is usually split and may be hung in two pieces. Drying takes up to six weeks, and during this time the water content falls from about 80% to about 15%; this means removing over 95% of the water.* This final water content is about the minimum at which moulds can grow; bacterial activity ceases when the commodity contains less than about 25% of water.

The product, which remains edible for several years, was an invaluable stand-by in Britain in the Middle Ages, and until this

*Care must be taken in calculating percentage loss of weight during drying from percentage water content before and after drying. It might be thought, for example, that when half of the water present in wet fish initially containing 80% of water had been removed, the half-dried product would contain 40% of water. In fact, calculation shows that the water content would then be almost 67%. Even if dried until the gross weight is only half what it was originally, the water content is still 60%. This apparent inconsistency is because the percentage water content of an article during drying is expressed on a changing basis.

In the drying of stockfish it may be assumed for a simple calculation that the initial water content is 80%. The dry weight if all water were removed would therefore be 20% of the original. There is therefore 4 lb of water for every one pound of absolutely dry material.

At a final water content of 15%, the dry weight is 85% of the product. It therefore contains $\frac{15}{85} = 0{\cdot}175$ lb of water to every lb of dry material.

Assuming that there has been no loss of dry material, 4 lb − 0·175 lb = 3·825 lb of water has been removed for every pound of dry material, and the weight of the article has been reduced from 5 lb to 1·175 lb. The percentage of water lost is therefore $\frac{3{\cdot}825}{4} \times 100 = 95{\cdot}5\%$.

The percentage loss of weight in drying is $\frac{3{\cdot}825}{5} \times 100 = 76{\cdot}5\%$.

century on sea voyages, but most of present-day production goes to the Mediterranean countries and Africa. Its flavour is characteristic and it is extremely tough. The Scandinavians, who still greatly relish it, soften it by soaking in alkali to produce a special dish, known as *lutefisk*. No use seems to have been made of mechanical dryers for the commercial production of stockfish, although products have been made experimentally from surplus British fish.

FIGURE 6·1 Stockfish drying in the open air

DRIED SALTED COD

The salting of white fish, principally cod, is normally followed by drying; the water content of the wet-salted split fish must be reduced from about 58% or 59% to between 35% and 43% depending on the product and the particular market. The drying of cod and similar fish in this way is, apart from the manufacture of fish meal, by far the most important application of drying to fish preservation throughout the world.

The traditional method of drying is to expose the fish to sun and wind, by laying them either on suitable rocks near the curing sheds, or on frames or hurdles, known as *flakes*. These are long wooden slatted or wire-meshed benches raised off the ground to allow free circulation of air. Not only is the method slow but the fish are also at the mercy of the weather; if the sun is too hot and there is no wind, they can suffer sunburn—that is, become cooked and hence soft and mushy. On the other hand, as soon as the weather becomes damp or it rains, the fish have to be collected into piles and covered. The high labour costs involved have consequently forced curers, particularly in Canada and Britain, to replace outdoor natural drying by artificial drying.

The simplest form of dryer is a room in which the fish are hung on racks, or *horses,* over fires, sometimes of coke. These create a draught and also heat the air so that it can hold more water vapour. Although still used, this process is slow and inefficient, and specially designed mechanical dryers are now more usual. Many of the dryers in use have had to be designed by engineers without the necessary knowledge of the best conditions for drying salted fish. In Canada, however, where the salting and drying of cod is an important industry, the Government research workers have made the following recommendations and comments. These have not been tested under British conditions.

(1) The most suitable air speed over the fish is between 200 and 300 feet a minute. Drying is slower at lower air speeds, and is not much faster at higher air speeds, at which power costs become heavy.

(2) The most suitable air temperature in the dryer is 75°F and it should never be allowed to go outside the range of 60°F to 80°F. At 90°F drying is very slightly faster but the fish is cooked. At lower temperatures drying is slower.

(3) The relative humidity of the air in the dryer should be 50% to 55%. Some modern Canadian dryers are fitted with automatic humidity control.

(4) Drying that is too slow at the beginning of the process causes rough, pebbly crystals of salt to form on the fish.

(5) Drying becomes progressively slower during the first six to ten hours and by this time, a salt crust has formed on the surface. The fish is then taken from the dryer and piled for a while in stacks. The pressure of the fish above squeezes water from the centre of the fish below and wets the crust so that the subsequent drying is faster until the crust becomes dry again. Even large, thick fish can be dried in about 40 hours if piled two or even three times.

(6) At atmospheric relative humidities above 76%, solid salt, and therefore dried salt fish, absorbs water. Under these conditions dried salted fish goes bad.

WARTIME WORK ON DRYING

Traditional dried products are found unpalatable by many people accustomed to fresh fish. These products are not only tough and often very salty, but have characteristic strong flavours.

In 1940 attempts were first made at Torry to produce improved products by controlled drying, at that time usually referred to as *dehydration*. It was soon found that drying of whole uncooked fish or fillets, with or without salting, gave unacceptable products. Various standard types of dryer, including steam jacketed dryers, spray dryers and roller dryers were tried without much success. Ultimately a process was developed of drying cooked, minced fish in hot air. For a time small quantities of dehydrated cod and herring products were made for the Forces.

The product was made in the following way. The fish was first filleted. The fillets, preferably skinned, were cooked on perforated trays in steam retorts, under a slight pressure of 2 lb/in^2 to 6 lb/in^2. About 40% of the initial weight of fillet was lost as liquor. The cooked fillets were cooled and then minced. The mince was spread evenly over wire mesh trays at the rate of 2 lb to a square foot. The trays were placed on trucks in a hot air dryer based on the design of the Torry mechanical kiln. In about four hours the moisture content was reduced from 70% to less than 10%. The dried fish was sealed in cans in which the air was replaced by nitrogen to prevent oxidation.

Careful bacteriological control of production was required in order to ensure a low bacterial count in the product and freedom from food-poisoning bacteria.

The product could be made into acceptable fish cakes and similar made-up dishes. It kept in reasonably good condition for up to a year at ordinary temperatures. A product that can be used only as minced fish in fish cakes, however, has a very limited appeal.

VACUUM CONTACT DRYING

Attempts were made to devise a process that would give acceptable dried fish fillets. Workers at the Ministry of Agriculture, Fisheries and Food Experimental Factory at Aberdeen, which is now closed, experimented for some years with a method originally developed in Denmark of drying fish by contact with heated plates in a vacuum.

This process, known originally as *Vacuum Contact Drying* (VCD), has two theoretical advantages:

There is no oxygen to produce undesirable oxidation in the fish during drying and the evaporation of moisture in a vacuum is so rapid that the fish is kept very cool; this slows down bacterial and enzymic deterioration.

Hot water is pumped through the plates which are housed in an evacuated cabinet. The fish is loaded to a regular depth of about ¾-inch on aluminium trays. Each tray is placed between the horizontal, water-heated plates, and these are brought together hydraulically.

The original plant would dry 1 ton of fish fillets to 15% water content in about seven hours. In this process the dried fish was then compressed in laminated blocks and dried further to a final water content of 5%. The final product was known in Scandinavia as *pressfisk*.

FREEZE DRYING

If the pressure in the vacuum cabinet mentioned above is kept low enough, evaporation occurs so rapidly that the fish freezes and remains frozen whilst it dries. The process then becomes *freeze drying*. For freeze drying to be possible, the water vapour pressure in the cabinet must be maintained below the saturation water vapour pressure at the freezing point of fish, about 30°F. Although in theory this is no more than about 0·15 inches of mercury, in practice the water vapour pressure should be in the region of 0·06 to 0·08 inches of mercury for satisfactory freeze drying.

Freeze drying has two advantages. First, the low temperature and virtual absence of oxygen prevent oxidation during drying. Secondly, there is no shrinkage when material is dried in the frozen state, because it cannot contract. The open structure resulting from freeze drying encourages *reconstitution*. Freeze-dried fish was, in fact, produced experimentally in the laboratory before 1939.

In the process as now developed the fish is frozen before it is put into the cabinet. The fastest possible rate of drying is obtained by balancing the heat input against the rate of evaporative cooling.

Furthermore, to obtain uniform drying there must be good contact over the entire surface of the material. This entails cutting fillets or slices of fish of uniform thickness. After several years' work, a process known as *Accelerated Freeze Drying* (AFD) has been evolved, the products of which are greatly superior to those obtained earlier. Although these new freeze-dried products reconstitute far better than ordinary air-dried fish, and can with care be made up into attractive dishes, the water absorbed in reconstitution is still held loosely, rather as in a sponge. Consequently the product can still be distinguished from fresh or good quality frozen fish, and the water taken up readily boils off, when, for example, reconstituted fillets are fried in deep fat. Furthermore, the new process requires considerable skill and is costly.

For these reasons it seems unlikely that any dehydrated fish such as cod or haddock will be able to compete in Britain with good quality frozen fish. Shrimps and prawns, which are in any event rather tough, may make satisfactory AFD products for sale in Britain.

AIR DRYING

The reader will find it easier to understand this section if he is conversant with the meaning of the terms *relative humidity* and *vapour pressure.* These are discussed in the chapter on Fish and Physics.

During the drying of fish two processes take place. Water evaporates from the surface and also moves towards the surface from within the fish. The removal of the water can be considered to take place in two stages. During the first stage, whilst the surface is wet, drying depends only on the state of the air surrounding the fish, in particular its speed, temperature and how much moisture it already carries. If conditions remain constant then the rate of drying remains constant. This stage of the drying is called the *constant rate period.*

The second stage of drying occurs when the surface moisture of the fish has evaporated because water can then only evaporate as quickly as it reaches the surface from within. As the amount of water in the fish becomes less, the water molecules take progressively longer to reach the surface from the deeper-lying parts and drying hence becomes progressively slower. This second stage is known as the *falling rate period* of drying. These two periods will now be considered in more detail.

CONSTANT RATE DRYING

The rate of drying during this period depends only on the speed at which water molecules can be carried away from the surface of the fish. When a fish dries in a current of air, the water is carried away as vapour by the air; clearly, if the air is already saturated then it cannot hold any more water and no drying will occur.

The air can be imagined to consist of three separate layers. In contact with the surface of the fish is a very thin stationary layer;

above this is a slowly moving layer, moving more rapidly in its outer zones, where it joins the main stream of turbulent air.

The stationary layer is saturated with water vapour, which is continuously passing into the slowly moving layer. The rate of drying depends upon the thickness of this slowly moving layer and the dryness of the main stream of air (see Figure 6·2).

In the discussion that follows it is convenient to consider separately the factors affecting the drying of the fish, and those influencing the drying properties of the air passing over the fish.

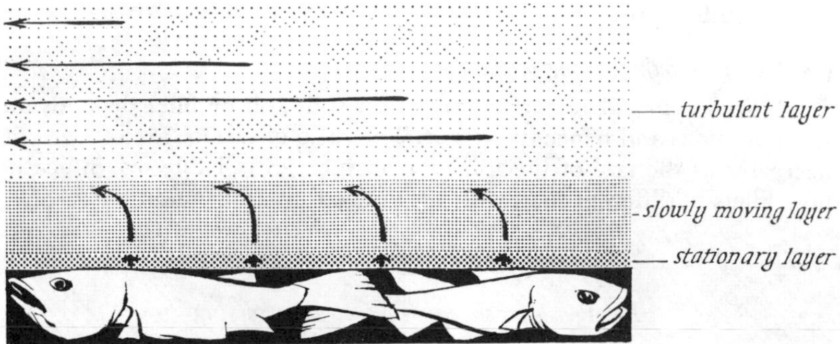

FIGURE 6·2 Air movement over the surface of fish being dried

When the speed of the air is increased the sluggish layer becomes thinner and water vapour can move out into the main stream much more quickly. Hence the higher the speed of the air flowing over the fish, the more quickly water evaporates.

The evaporation of water is accompanied by cooling; this is known as *evaporative cooling*. After a short time the temperature of a fish that is being cooled by evaporation reaches a steady value. At this point the heat supplied to the fish from the air is balanced by the heat removed by evaporative cooling. This steady temperature, which is below the air temperature, is the same as that indicated by a wet bulb thermometer exposed to the same conditions as the fish.

The difference in the temperatures indicated by wet bulb and dry bulb thermometers placed in a stream of air is called the *wet bulb depression*.

The magnitude of the wet bulb depression is directly related to the difference between the actual water vapour pressure of the air, and the water vapour pressure of saturated air at the same temperature. The rate of evaporation of water depends directly on this difference. It follows, therefore, that since the rate of evaporation from the surface of wet fish is directly dependent upon the difference in water vapour pressure of the air and of saturated air at the same temperature, the rate of evaporation is directly related to the wet bulb depression.

Furthermore, it will take longer to dry a thick fish to a given percentage weight loss than it will a thin one. This clearly follows from the fact that a thin fish has proportionately a greater surface area than a thick one, and water, during constant rate drying under given conditions, evaporates from equal areas at the same rate.

To summarize, the rate of drying during the constant rate period is dependent on the surface area of the fish, the velocity of the air and its wet bulb depression.

(1) *Factors affecting drying of the fish*

(*a*) *Surface area.* The surface area of a split fish or fillet depends on its dimensions and hence its weight. The ratio of surface area to weight decreases as the size of the fish or fillet increases. Table 6·1 indicates how fillets of different sizes will dry under the same conditions.

TABLE 6·1

Relationship between relative rate of drying and weight for cod fillets

Weight of fillet lb	*Rate of drying relative to rate for 1 lb fillet*
1	1
2	4/5
4	2/3

Thus if a 1 lb fillet dries at a rate of 1% loss an hour, a 4 lb fillet dries at 2/3% loss an hour under the same conditions. Or, to give another example, if a 1 lb fillet lost sufficient weight in three hours, a 4 lb fillet would take 4½ hours to lose the same percentage weight under the same conditions. Hence there is a need to sort fish into groups of appropriate weight range if reasonably uniform loss of weight within a batch is to be obtained.

(*b*) *Air velocity.* If the air velocity is raised, the rate of drying during the constant rate period also increases, but not quite in proportion, as will be seen from Figure 6·3. If the air velocity is doubled the rate of drying is only increased by about three-quarters. For example, if fish were drying at a rate of 4% loss an hour in air travelling at a velocity of 400 ft a minute, the rate of drying if the velocity was increased to 800 ft a minute would be approximately 7% loss an hour.

(*c*) *Wet bulb depression.* As already stated, the rate of drying of fish is directly proportional to the wet bulb depression of the air. This means that during constant rate drying, if the wet bulb depression is doubled the rate of drying will also be doubled.

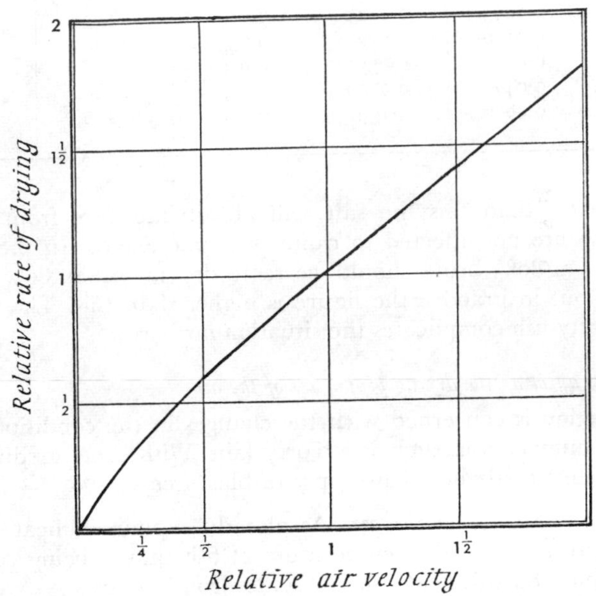

FIGURE 6·3 Effect of air velocity on rate of drying

Much experimental work has been carried out on the measurement of the rate of drying of fish over a wide range of temperatures, humidities and air velocities. From many thousands of results, a mathematical formula has been worked out which enables the rate of drying occurring under any given conditions to be calculated and Table 6·2 has been constructed covering the range of conditions likely to be found in practice.

The presence of salt affects the vapour pressure of water, and a solution containing salt has a lower vapour pressure than pure water. This means that whilst unbrined fish will continue to dry even in an atmosphere which is almost saturated, salt fish will not dry in air above about 76% relative humidity. If the relative humidity of the

TABLE 6·2

Rate of drying of fish during the constant rate period

Air speed in ft/min	Rate of drying in lb/ft² in an hour at wet bulb depression of:				
	4 deg F	*8 deg F*	*12 deg F*	*16 deg F*	*20 deg F*
50	0·0036	0·0073	0·0109	0·0146	0·0182
100	0·0063	0·0127	0·0190	0·0253	0·0317
200	0·0110	0·0221	0·0331	0·0441	0·0552
600	0·0266	0·0531	0·0797	0·106	0·133
1000	0·0400	0·0800	0·120	0·160	0·200
2000	0·0696	0·139	0·209	0·278	0·348

air is greater than this the salt will absorb moisture from the air. Brined fish are not affected to quite the same extent. In theory, fish brined in an 80% brine should cease to dry in air at 80% relative humidity but in practice the figure is higher than this. The presence of oil in fatty fish complicates the situation further.

(2) *Factors affecting the drying properties of the air*

This section is concerned with the changes in the conditions of the air in any tunnel dryer, such as a Torry kiln. With some modifications, the comments made here can apply to blast freezers.

(*a*) *The temperature of the air.* As already explained, heat must be provided to maintain the temperature of fish that is being cooled by evaporation, and this heat usually comes from the air.

For every pound of water evaporated from the fish, a definite amount of heat, about 1100 Btu, termed the *latent heat of evaporation,* must be provided. The extent of cooling of the air will depend upon the quantity passing over the fish and the rate of drying as well as the amount of fish in the dryer.

(*b*) *The relative humidity of the drying air.* The air must become moister as well as cooler since water is evaporated from the fish. The uptake of moisture by the air, as well as the cooling effect, means that the relative humidity increases and therefore the wet bulb depression decreases, resulting in a progressively lower rate of drying downstream.

FALLING RATE DRYING

Many of the drying operations carried out in the fish industry are confined to the constant rate drying period. Where, however, the product is very dry, as, for instance, fish meal or salt fish, the process of diffusion of moisture from the centre of the commodity to the outside becomes an important part of the drying process. Although only constant rate drying normally occurs in the smoking of white fish, falling rate drying is important in the cold smoking of fatty fish such as herring.

(1) *Factors affecting the commencement of the falling rate drying period*

The duration of the constant rate period of drying depends upon the rate of water loss. More water can be evaporated from fish during the constant rate period if drying is slow than if drying is rapid. This situation can be complicated, however, by the fat content of the fish. For samples of the same size under the same drying conditions, the constant rate period is shorter for fatty fish, such as herring or salmon, than for lean fish such as haddock or cod.

To summarize, falling rate drying commences after a period of constant rate drying, the duration of which depends upon:

(*a*) the rate of drying;

(*b*) the temperature of the fish, which affects the rate at which water can move through the flesh (diffusion);

(*c*) the thickness of the flesh;

(*d*) the initial water content;

(*e*) the fat content.

Table 6·3 gives the approximate duration of the constant rate drying for cod fillets of a uniform thickness of about ¼ inch. Approximate figures are given in Table 6·4 for the duration of constant rate drying for herring; it will be seen that these are shorter than those for cod given in Table 6·3.

TABLE 6·3

Length of constant rate period for white fish ¼ inch thick drying at 80°F

Rate of drying % loss/hour	*Duration of constant rate period hours*
2·5	23
5·0	9
6·0	7
9·0	3

TABLE 6·4

Length of constant rate period for herring of different fat content drying at 86°F

Rate of drying % *loss/hour*	*Duration of constant rate period in hours* *Fat content* (%)			
	5	10	15	20
2·5	19	12	9	7
5·0	5	3	2	2
6·0	3	2	1	1
9·0	1	¾	½	½

(2) *Factors affecting the change of drying rate*

When the water on the surface has been evaporated, the rate of evaporation is thenceforth controlled by the speed at which water diffuses through the surface from the layers beneath, that is, it is independent of the state of the air above the fish. In other words, the rate of drying does not depend either on the velocity of the air passing over the fish, or on its humidity, provided it is not saturated.

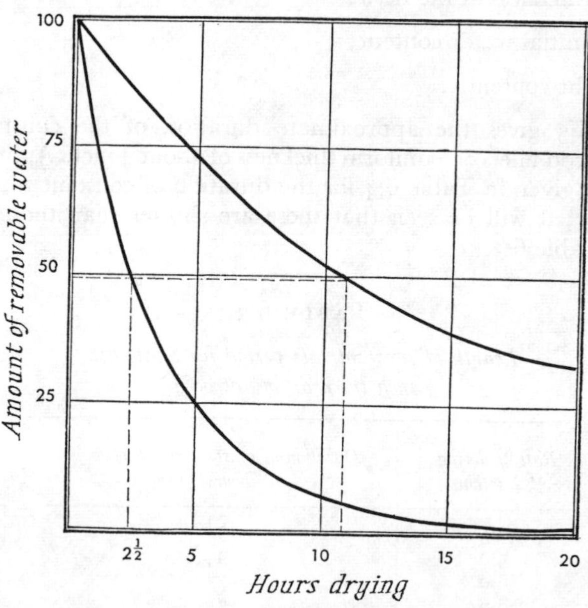

FIGURE 6·4 Drying curves during the falling rate period

The speed of the diffusion of water, and hence the rate of drying, is governed by the nature of the fish, its thickness, and its temperature. It is possible to represent in mathematical terms the weight changes occurring in fish as it dries. Simply, this mathematical expression states that the actual speed of drying depends on the amount of water that can still be removed.

As the water content of fish becomes less, drying becomes progressively slower and slower, until eventually it ceases, when about 95% of the total quantity of water originally present has been removed. This is shown in Figure 6·4, which represents the drying curve at 80°F for white fish ¼ inch thick. From the graph it can be seen that it takes 2½ hours to evaporate half the water originally present; this remaining water is reduced by half again after a further period of 2½ hours' drying. In general, this means that the time taken to evaporate half the water present at any stage is constant and is a measure of the rate of drying. This period of time is called the *half loss period.* The half loss period becomes longer as the thickness of the fish increases, because in thicker fish the water has further to diffuse to the surface. The drying curve for fish ½ inch thick, for which the half loss period is 10½ hours, is given by the upper line in Figure 6·4.

Fat in the fish retards diffusion of water and therefore the half loss period at any given temperature is longer for fatty fish than for white fish of the same thickness.

The final water content of fish drying in the falling rate period depends only on the relative humidity of the air. It finally becomes impossible to evaporate any more water by means of normal air drying, and at this stage about 5% water still remains. In fact, dry fish will absorb moisture from moist air, and Table 6·5 gives the water content of dried cod or similar fish when stored in air at different relative humidities.

TABLE 6·5

Effect of humidity on the final water content of non-fatty fish

Relative humidity %	*Water content* %
20	7
30	8
40	10
50	12
60	15
70	18
80	24

Two things are apparent from Table 6·5. It is pointless to dry fish to a low water content if it is then to be stored in air of high relative humidity, since the fish will absorb moisture from the atmosphere until it reaches an equilibrium moisture content. Also, it is unnecessary to operate a dryer at a relative humidity lower than that in equilibrium with the final water content desired. The fish will not dry any quicker and power will be wasted in heating unnecessary volumes of air.

PROBLEMS OF AIR DRYING

Here only the factors controlling the throughput of a kiln for the cold smoking of fish will be discussed. In general, this also applies to other types of dryer, but in British practice these are of lesser importance.

The temperature of the air after it has passed over all the fish in a kiln is obviously lowered and it may be necessary to incorporate additional heaters at intermediate positions down the length of the kiln. This is done in the larger sizes of Torry kiln. In any case, however, the air at the outlet end of the kiln, although cooler than the air that passes over the first rows of fish, is usually much warmer than the inlet air taken into the kiln. In the warmer months of the year this temperature difference may be only 10 Fahrenheit degrees, but in the cooler months the temperature difference can amount to 30 Fahrenheit degrees. In a large kiln this would represent a considerable quantity of waste heat escaping up the chimney if there were no recirculation.

In consequence, in the Torry kiln up to 80% of the air is recirculated, the balance being drawn in from the factory. Less heat, therefore, needs to be supplied to the kiln. Recirculation of the air has the added advantage of permitting control of the humidity in the kiln by admitting more or less air from outside and venting correspondingly less or more moist air.

Unfortunately, if the rate of drying is too great there is a tendency for the product to dry mainly from the surface layers, and after removal from the kiln the surface of the fish becomes moist again. As an example, the average rate of drying for small cod fillets should, it is thought, not exceed about 4% weight loss an hour, if this condition is to be avoided. Under practical conditions this could mean that the fish at the inlet end of the kiln, meeting the driest air, lose about 6% an hour, and those at the outlet end about 2% an hour, the average being 4% an hour. Since it is impossible, during constant rate drying, for all the fish along a Torry kiln to dry at the same rate, it is necessary to reverse their positions halfway through the process, so that finally all the fish have been dried to the same extent.

The changes in the drying conditions in a large 120-stone Torry kiln are shown diagrammatically in Figure 6·5. The temperatures and humidities given are those that might occur during normal commercial smoking of fish, although clearly conditions vary from one

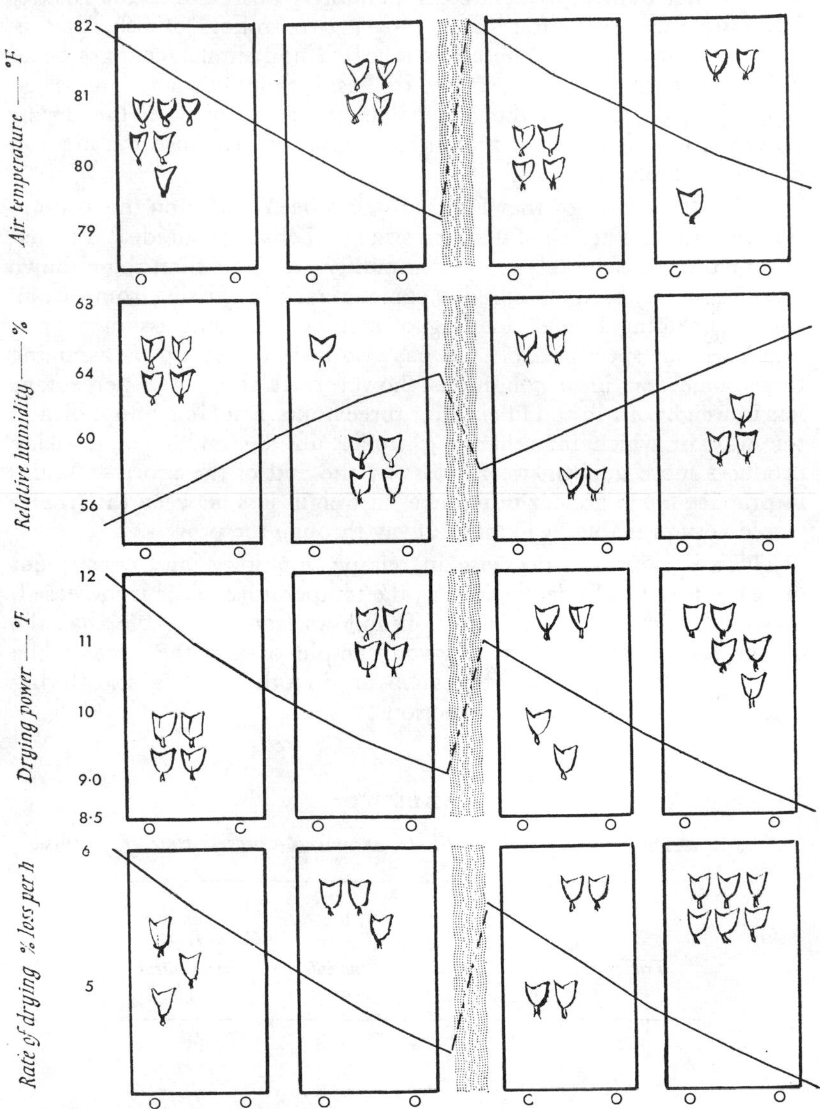

FIGURE 6·5 Changes in drying conditions in a 120-stone Torry kiln

kiln to another from day to day according to numerous variables including outside air temperature and humidity, and the amount of surface area of fish in the kiln.

The air temperature falls as the air passes through each trolley, and the wet bulb depression correspondingly decreases as the relative humidity increases. After passing over two trolleys of fish, that is, halfway down the kiln, the air is reheated and similar changes occur as it is blown through the next two trollies. It should be noted, however, that after reheating to the original temperature of 82°F the drying power of the air does not regain its original value, since the air now contains more water.

It should be stressed that Figure 6·5 has been drawn on the assumption that the fish are all of uniform size and behave in an ideal manner. In practice, considerably steeper humidity gradients than those shown have frequently been measured in commercial kilns, arising from various causes including heavier loading of trollies than that assumed here, and lower air speeds. Table 6·6 has also been drawn up by assuming these somewhat ideal conditions, however. It shows the percentage loss in weight of 1 lb cod fillets after three hours' smoking, and indicates the way in which interchanging trollies in the middle of smoking produces more uniform weight loss by the end of the process. Again, in practice much greater differences in weight loss between the trollies would very probably be found halfway through the process.

There is a marked decrease in relative humidity, and consequent increase in wet bulb depression, as the temperature of air is increased, as will be seen from Table 6·7. It follows from this table that the higher the temperature of a given sample of air, the greater the increase of the wet bulb depression or, in other words, the drying potential in the constant rate period.

TABLE 6·6

Loss in weight of trolleys of cod fillets, assuming a process time of 3 hours

Trolley No.	*% loss/hour*	*% loss after 3 hours (no interchanging)*	*% loss after 3 hours (with interchanging)*
1	5·6	16·8	15·2
2	4·8	14·4	15·0
3	5·2	15·6	15·0
4	4·5	13·5	15·2

There is a maximum temperature to which fish can be raised without cooking it. In commercial practice it is therefore usual to restrict fish smoking kiln temperatures to about 85°F. As already mentioned, during constant rate drying the fish temperature is lower than the dry bulb temperature and, for fish containing no salt, approximates to the wet bulb temperature. As an example, air at 82°F and 61% relative humidity has a wet bulb depression of 10 Fahrenheit degrees and in these air conditions the temperature of the fish would be about 72°F. Brined fish would be a little warmer than this, and salt fish warmer still. During falling rate drying there is a much smaller evaporative cooling effect and the fish temperature is nearer to the dry bulb temperature. The maximum recommended air temperature for drying salt fish is 80°F.

TABLE 6·7

The effect on the relative humidity and wet bulb depression of heating air initially at 60°F and 65% r.h.

Air temp. (Dry Bulb) ° F	*Relative humidity* %	*Wet bulb depression deg* F
60	65	6·5
65	54	9·5
70	45	13·0
75	38	16·0
80	32	19·0
85	27	22·0

TABLE 6·8

Variation in rate of drying with load

Load lb	*% loss/hour*	*Average hourly output in lb fish cured to 12% weight loss*
2750	2·7	607
2500	3·1	645
2000	4·0	666
1680	4·6	637
1000	5·9	488

The relationship between the weight of fish and its rate of drying in a kiln operating under a given set of conditions is complex. Halving the size of the load does not necessarily double the rate of drying. Table 6·8 gives the rates of drying obtained in loads of various weight of similar fish, smoked in a Torry kiln of 120 stones nominal capacity. These figures were obtained from a kiln during normal commercial operation.

Achieving the maximum rate of drying is not the whole story. The producer is generally concerned with obtaining the greatest economic output from his kiln. This is not necessarily obtained when the drying rate is near the maximum possible. For example, the maximum throughout in this instance is obtained with a load of about 2000 lb.

In a 120-stone kiln approximately 80% of the air is recirculated. In other words, 20% of the air is continuously replaced by air from outside the kiln. The humidity of this air is important. The changes in the humidity of air with season of the year can have a significant effect on the rate of drying obtained in the kiln and steps must be taken to increase the quantity of inlet air, that is, reduce the amount of recirculation, when the air is damp.

Care must be taken to ensure that the inlet air is as dry as possible. It has been found, for example, that if air is drawn into the kiln when the factory floor is wet, drying may be significantly slower.

7

Freezing and cold storage

. . . 'tis fix'd as in a frost;

ALEXANDER POPE

THE purpose of freezing fish, either fresh or processed, is to obtain a commodity that can be stored for some months and will then give a thawed product that has been changed hardly at all by the process.

Preservation of fish by freezing is not a new idea; the first British patent for freezing fish was taken out in 1842. Commercial freezing of fish began in America in the 1860s, and a number of attempts to establish the process in Britain were made in late Victorian times. Unfortunately not enough was then known about how to make a good quality product and how to store it for any length of time, with the result that very poor quality frozen fish was sold to the public. Frozen fish thus obtained a bad reputation, which the mediocre products made before 1939 did little to improve. Only since the war have frozen fish become acceptable although some of what are sold are still of very inferior quality.

The potential advantages of freezing over traditional methods of preservation such as salting, smoking and drying are very great. The product is almost unchanged by the process, so that fresh fish, properly frozen, stored and thawed, is virtually indistinguishable from iced fresh fish. Gluts can be preserved against times of scarcity, and good quality fish can be supplied to places where fresh fish is a novelty, and at times when fresh fish is not readily available. There is little doubt that the proportion of the British catch that is frozen will continue to increase.

This chapter sets out the principles of freezing and cold storage, describes good practice, and explains what can go wrong if proper care is not taken.

WHAT HAPPENS WHEN FISH IS FROZEN

Every pound of ice at its melting point needs 144 British thermal units just to melt it, without raising its temperature above 32°F. In other words, 144 Btu are locked up in every pound of water, and this quantity of heat has to be taken away in order to freeze the water. Fish is largely water, from 60% to 80% depending on species, and the process of freezing converts some of this water into ice. Freezing the water in fish can cause undesirable changes in the proteins and fats that make up most of the remaining 20 to 40%, and good practice is essential to keep these changes to a minimum.

Although pure water freezes at 32°F, a fish does not begin to freeze until its temperature reaches about 30°F. This is due to salts and other chemicals that are present naturally in the muscle. As the temperature falls below 30°F more and more water is frozen and the remaining liquid becomes an increasingly stronger solution of these chemicals. Even at 23°F, however, when freezing is often thought to be completed, over 20% of the water in fish muscle is still unfrozen. While most of the water has been converted to ice by the time that the temperature reaches 0°F, a little probably remains unfrozen even at very much lower temperatures. Nevertheless most of the water in the fish is frozen by the time the temperature has fallen to 23°F. The interval between 30°F, when freezing begins, and 23°F is termed the *critical range.*

A block of fish does not freeze uniformly, because heat is removed only from the outside of the block, either by the cold air in a blast freezer or by the cold liquid in a plate freezer or an immersion freezer. The heat at the centre of the block has to travel outwards to the surface before it can be carried away. Thus the block freezes from the outside inwards. The time that the heat takes to travel from the middle to the outside of the block depends upon the thickness of the block; a slab four inches thick takes longer to freeze than one two inches thick, under the same conditions.

When the centre of the block is frozen, the block may be removed from the freezer, but the temperature at the centre is still much higher than the outside, and it will be an appreciable time before the temperature is the same throughout the block. The surface temperature of the block on removal from the freezer may be well below that of the cold store in which it is to be kept, but by the time the temperature has become uniform throughout the block its average temperature may be above that of the store. Mere inspection of the surface of the block will not show whether or not the centre is frozen.

As the temperature of fish muscle falls below its freezing point, ice crystals begin to form throughout the tissue. The size of these crystals depends upon the rate of freezing. In fish muscle that is cooled rapidly from 32°F to 23°F, in less than half an hour, for example, very small ice crystals are formed within the structure of the microscopic thread-like cells of the flesh. When the frozen muscle is examined, very little change can be seen, even with a microscope, and very little fluid or drip oozes out of the flesh when the product is thawed. But with prolonged time in the critical range, much larger ice crystals form in the spaces between the cells, disrupting the muscle structure. These changes are readily seen in the frozen products by microscopic examination and in the case of very slowly frozen fish by normal visual means. The texture, after thawing, of slowly frozen muscle is much poorer because of such damage.

Flavour and texture of frozen fish can be further impaired because slow freezing results in the concentration of salts and enzymes in solution within the cells; these enzymes become more active in stronger solution and produce spoilage changes that cannot be reversed when the flesh is thawed. It is best, therefore, to reduce the temperature of the fish as quickly as possible to a point where these reactions are reduced to a minimum.

Tests have shown that only when the time taken to cool the centre of a block of fish from 32°F to 23°F exceeds four hours do people notice any change in the texture of the thawed cooked product. In some circumstances even longer cooling times have had no discernible effect on texture and flavour. But times in excess of two hours have an adverse effect both on appearance and on suitability of the fish for subsequent filleting and smoking.

With these facts in mind, it was decided by the Ministry of Food in the early 1950s that *quick freezing* should be defined as a rate of freezing at which no part of a fish or packet of fish takes more than two hours to cool from 32°F to 23°F, thus ensuring an adequate margin of safety. Since then this definition has been incorporated in various British codes of practice and regulations covering the freezing of fish and other foods. *Sharp freezing* has no precise definition; indeed, some so-called sharp frozen fish is frozen very slowly.

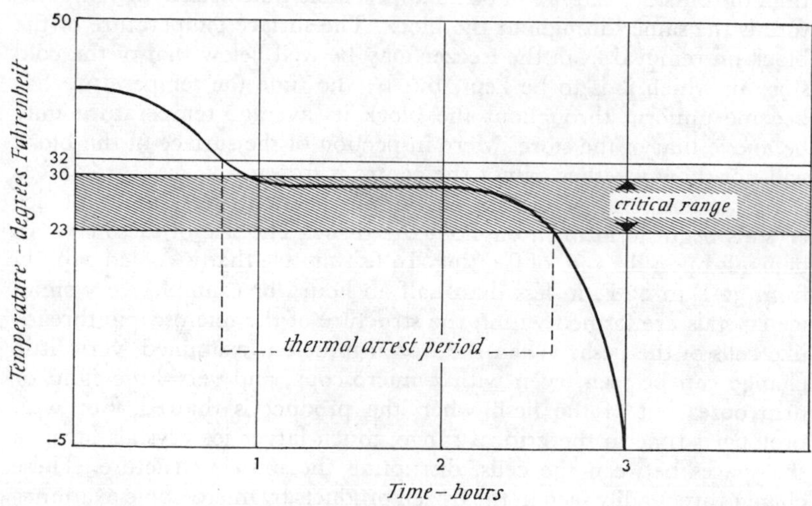

FIGURE 7·1 Ideal freezing curve for fish muscle

If freezing rates are to be compared it is usual to measure the time required to cool fish from 32°F to 23°F. Total cooling times cannot be compared unless the fish in each process begin and end at the same temperature.

Apparently contradictory behaviour of fish during freezing has been demonstrated at the Torry Research Station; the higher the initial temperature of the fish the shorter is the time taken for the centre of a block to cool from 32°F to 23°F, the *thermal arrest* period. One would expect that cool fish at 32°F when loaded into the freezer would reach 23°F much more quickly than fish loaded at say 50°F. This is, in fact, so; the higher the initial temperature of the fish, the longer the total freezing time. But since the warmer fish have to be in the freezer much longer than the initially cool fish before the centre temperature reaches 32°F, a greater proportion of the block has by that time reached 23°F than in the initially cooler fish. The time through the thermal arrest period at the centre is consequently less. This fact is of no practical advantage, however, as it is obviously undesirable to keep fish at such high temperatures before freezing. Spoilage would be rapid, and unnecessary refrigeration power would be needed to reduce the temperature of the fish to freezing point.

Spoilage of fresh fish is attributable to two main causes, bacterial action and autolysis. When fish are frozen bacterial action is reduced, and below about 15°F bacteria are prevented completely from growing and causing spoilage. Some of the bacteria are killed, but others may become dormant, survive freezing and become active again as soon as the temperature of the fish rises.

Autolysis, or breakdown by self digestion of the flesh after death, is due to the action of enzymes in the fish tissues. These chemical changes, that can bring about irreversible changes in odour, flavour and appearance, continue to some degree at temperatures below that at which bacterial action stops, and still go on at a very slow rate at minus 20°F. Therefore it is necessary that there should be systematic turnover of stock in cold store to prevent deterioration of product over long periods. At storage temperatures above minus 20°F the chemical changes occur more rapidly and storage life of the product is shorter. Recommended storage times and temperatures are given in Table 7·1 on page 150.

HOW TO MAKE A GOOD PRODUCT

Freezing and cold storage of fish can never improve quality. If the process is properly carried out, initial quality can be maintained within limits so that the difference between the final thawed product and the fresh article is small. To produce frozen fish that conform to this high standard, attention must be paid to a number of important points; occasional neglect of these principles is often the explanation for the very poor quality frozen fish that are sometimes sold.

FRESHNESS

Stale fish have a strictly limited shelf-life and deteriorate very rapidly when frozen and cold stored. Even though they may be considered edible to begin with, they may be quite unacceptable by the time they reach the consumer. The unpleasant flavour and texture that such frozen products may have are not solely due to the initial staleness; changes due to freezing and cold storage will be superimposed to make the product uninviting and disagreeable. The storage properties of frozen fish of various kinds given in Table 7·1, apply only to perfectly fresh fish. When fish are kept in ice for longer than a few days before freezing, storage life in the frozen condition becomes progressively shorter.

(1) *White fish.* The best quality frozen cod is obtained from fish that have been kept in ice for no more than three days before freezing. Whole fish treated in this way and then thawed after suitable storage can be handled in the same way as fresh fish. They can be filleted and the fillets can then be distributed in the normal way for retail, or they can be brined and smoked to give a product with a normal gloss. As time of stowage in ice before freezing is lengthened, the fillets cut from the thawed whole fish become progressively softer and more ragged and take on a dull, opaque appearance. Although still retaining an acceptable eating quality, within limits, they become no longer suitable for distribution in the normal way and for smoke curing. In general, frozen fillets are less ragged than the fillets cut from frozen whole fish, but they are usually duller and do not produce a smoke cure of the same high quality.

Other types of white fish give results similar to cod, but with some differences. Hake, for instance, already a soft fish, do not stand up to the effects of freezing nearly so well as cod, and if kept for more than a day in ice, the fillets cut from the thawed whole fish would not be acceptable. With haddocks, the limit is two days, but flat fish such as plaice and lemon sole can be kept for about five days in ice before freezing without serious effect on the quality of the fillets cut from the thawed fish. In addition, the quality of frozen fish of all kinds is affected by seasonal variations in intrinsic condition. Spent fish, for instance, do not give high quality products when frozen. Moreover, since fish spoil much more rapidly at higher temperatures it is essential in all cases to cool the fish quickly to ice temperature and to maintain them at that temperature if the best results are to be obtained.

It is often impossible to obtain acceptable smoked products from commercial grades of frozen fish simply because the fish were not fresh enough to begin with. This may be because the fish were kept too long in ice at sea, as will be the case with distant water fish, or kept at too high a temperature as might be the case with fish taken from an inshore vessel that uses no ice at all.

The freezing of fish within a few days of catching can only be effectively carried out on a large scale by freezing at sea and this process is now well established commercially. On some ships the fish are filleted by machine and then frozen in conventional freezers; on others, the fish are frozen after gutting, and perhaps heading, in a special freezer designed for the purpose. The process of freezing at sea is being used more and more to replace icing as a means of bringing back fish from distant grounds in an acceptable condition and the products obtained, frozen fillets and frozen whole fish, find a ready market. Storage of the products at minus 20°F right from the start is essential.

In most cases, shore-based freezing plants are too far from the more prolific fishing grounds to obtain supplies of fish fresh enough for the production of the highest quality frozen products suitable for reprocessing when thawed. These establishments thus rely upon supplies of market fish for the production of a variety of frozen fish packs, some for sale to caterers, hospitals and schools in industrial packs, weighing 5 lb upwards, and others for retail sale through frozen food cabinets as consumer packs, weighing 8–16 oz. In this connection there is a need for adequate control of the quality of the raw material used for freezing. It is true that the poorest fish available, carefully frozen and cold stored under the best conditions for short periods and then distributed quickly, may reach the consumer ultimately in a somewhat better condition than if they had been transported in ice. However, with this class of fish, there is always a risk of supplying the consumer with fish of a poorer eating quality than that of the iced fish available at the fishmonger. This kind of situation can arise in areas that are normally supplied with inshore or near water fish and can obviously harm the freezing industry. It is clear that the aim should be to freeze only fish that when thawed will be at least the equal in eating quality of the average quality iced fish available throughout the country. The freezing of fillets cut from stale fish should be avoided at all costs.

(2) *Fatty fish.* Herring, mackerel, sprats and salmon are included in this group. In Britain herring are frozen in very much larger quantities than any of the other species.

Almost all British-caught fatty fish are taken by ring nets or drift nets. They must be frozen very soon after landing; if they are stored, even in ice, for more than 24 hours before freezing, they can deteriorate seriously in quality both before and during cold storage. They are not gutted when they are caught, and the digestive juices begin to attack the flesh immediately the fish die. Certain types of herring, for example feedy herring caught in early summer off north east Scotland, cannot be kept, even in ice, for as long as 24 hours without belly walls bursting because of self-digestion. Herring invariably spoil more quickly than white fish. In addition, however, it has been observed that certain off flavours, not necessarily rancid ones, will develop during the storage of herrings frozen after more than a day in ice, even at minus 20°F.

Herrings for freezing should be iced, preferably as soon as they are caught, and certainly as soon as they are landed. They should be frozen not more than 12 to 18 hours after catching, depending both upon whether or not they have been iced at sea, and also upon the type of herring.

(3) *Smoked fish.* Stale smoked fish should never be frozen. Smoked fish are often kept for several days awaiting sale; if still unsold they are then frozen. This practice cannot be too strongly condemned. Smoked fish should be frozen as soon as they have cooled after removal from the kiln.

TREATMENT BEFORE FREEZING

Where it is thought desirable to brine fillets or split-fish before freezing, in order to reduce drip after thawing, they may be immersed in chilled brine containing 10 to 15% pure salt, that is, a 40° to 60° brine. They should be properly drained before freezing, or there may be a considerable loss of weight after thawing that could be wrongly attributed to drip. If impure fishery salt is used for the brine, salt fish odours may develop during cold storage. There is some evidence that treatment with polyphosphates before freezing may reduce drip losses after thawing, and this treatment is being used commercially both in North America and also in Great Britain.

Fatty fish should never be brined before freezing.

THE FREEZING PROCESS

Fish should be frozen as soon as they are taken out of ice. For instance, fish laid out on freezing trays should not be allowed to warm up and dry out before loading into a freezer.

The fish must be frozen as rapidly as possible, and the temperature of the product reduced to that of the cold store in which it is to be kept.

The dangers of slow freezing have been outlined earlier in this chapter; spoilage does not stop when the fish are apparently frozen. Bacteria continue to grow at temperatures down to about 15°F and chemical and physical changes can still go on rapidly; some, indeed, more rapidly in fish that are only partially frozen than in wet fish. Slowly frozen fish looks much whiter and more opaque when thawed out than the original fresh fish; it may also be tough, stringy and fibrous.

Modern types of quick freezing plant are capable of cooling all classes of unwrapped fish and closely wrapped packs from 32°F to 23°F in two hours or less, provided that they are used in the manner recommended by the manufacturer, that they are working properly and that the thickness of the product does not exceed about 4 inches. Design and operation of freezing plant are described later in this chapter. That section also discusses the possibilities of freezing thicknesses greater than 4 inches.

Having cooled the whole of the product from 32°F to 23°F within the recommended time of two hours, it is then necessary to reduce the temperature still further until the average temperature is at least as low as that of the cold store. Since the recommended storage temperature is minus 20°F, this means that the temperature at the centre of the block should be minus 5°F. This ensures that when the temperature has become uniform throughout the block, its average temperature will be no higher than minus 20°F, since the outside of the block will be at or near the temperature of the refrigerant, in the region of minus 30°F to minus 40°F. The product then imposes no extra load on the cold store refrigeration plant, and is stored safely from the start.

Under these conditions, cooling from 32°F to 23°F in two hours and then reducing the warmest part of the product to minus 5°F may mean a *total* freezing time of three to four hours for a block 4 inches thick.

TREATMENT AFTER FREEZING

As soon as fish are removed from a freezer, they should be glazed or wrapped, unless they have been packaged before freezing, and transferred immediately to low temperature cold storage.

(1) *Glazing*. Evaporation of moisture from the surface of fish in cold store causes damage to the product by dehydration, and also promotes oxidation of fats. It is necessary to provide a protective coating to reduce this effect as much as possible, and except in the case of consumer packs this is normally done by glazing the fish. Blocks can be glazed by dipping them in cold water or by brushing all surfaces with cold water as soon as they are taken out of the freezer. A thin protective skin of ice forms immediately. A thicker skin can be obtained by dipping several times. The weight of the glaze on a block of fillets may be from 2% to 7% of the weight of the block, depending on the surface area and size of the block.

The water in the glaze evaporates during storage without reducing the water content of the product, and so long as the ice skin is renewed at intervals during long term storage, the product will remain protected.

Patented systems for coating blocks with other substances such as alginate jellies are claimed to have advantage over ice glazes in that the blocks do not need reglazing during storage. This has not been satisfactorily demonstrated in Britain, and the technique is not used commercially.

The temperature of the frozen block may rise by three to four Fahrenheit degrees for every 1% by weight of water added as glaze, although it may be some time after glazing before the block temperature becomes uniform. If the centre of the block has been reduced by freezing to a temperature of minus 5°F and it is to be stored at minus 20°F, there is a sufficient reserve of cold in the block to compensate for the slight warming effect of glazing.

The temperature of the glazing water is not very important, provided the block is dipped for no more than a second or two and provided that at least 20 seconds elapse between dips. Water near to freezing point is reported to give at times a rough and unattractive glaze.

Block temperatures measured immediately before or after glazing should be regarded with extreme caution, since the effects of the initially cold surfaces of the block and the added warmth of the glaze counteract each other and can produce fluctuations in temperature at the centre of the block before the temperature becomes uniform.

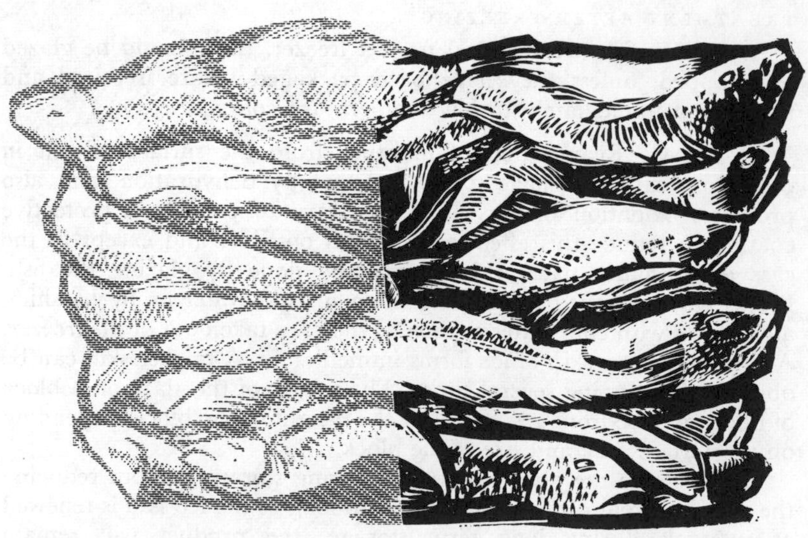

FIGURE 7·2 Ice glaze on block of whole frozen fish

(2) *Packaging.* For the protection of retail consumer packs, impermeable wrapping should be provided. As far as possible the packaging should be airtight to prevent oxidation of the product, and should prevent water vapour evaporating from the fish during storage. The degree of impermeability obtained depends upon the expense that can be allowed for packaging and on the storage life expected of the product. Among the more usual materials for packaging fish are waxed cartons, with or without a plastic lining material, and aluminium foil. The type of retail pack also depends a great deal upon whether the product is to be wrapped before freezing, when the effect of the wrapping upon freezing time must be considered.

Many industrial and catering packs consist of fibreboard outer cases for fish that have already been protected by glazing. Large single fish are more conveniently glazed than wrapped. Interleaving of good quality waxed paper between large blocks of frozen fish will allow them to be divided while frozen.

Fish that are to be further processed after cold storage are better glazed rather than wrapped.

(3) *Transfer to store.* The time between unloading the product from the freezer and putting it into cold store should be as short as possible, and intermediate steps such as glazing should be carried out in premises kept as cool as possible. Since there is a tremendous temperature difference between frozen blocks and the air in fish working premises, even on a very cold day, any delay will result in serious warming of the product. Frozen fish warms more than twice as fast as unfrozen fish, by as much as a degree a minute even in a chilly room.

Care must be taken not to damage frozen fish during transit from freezer to store. Although the product seems robust, it is easily harmed by rough handling; damage may not show until it is thawed. Mechanisation of the glazing and packaging processes can help in getting the product to the store in good condition.

STORING THE PRODUCT

THE STORAGE LIFE OF FISH

Even if fish are properly frozen within a few hours of catching and then stored properly at minus 20°F, they will not keep indefinitely. Bacteria will remain dormant, but autolytic changes will still go on slowly. Oxygen can attack the fat in fish and give unpleasant rancid flavours, especially if the surface of the fish dries out. Drying makes the fish unattractive to look at and unpleasant to eat.

If storage temperature is above minus 20°F then these changes take place more quickly and the life of the product is reduced. Temperature of storage is the most important factor affecting the storage life of frozen fish. There is increasing evidence to show that the poor quality of much of the frozen fish on the market is due to storage at too high a temperature.

Initial freshness influences the storage life; stale fish spoils much more rapidly during cold storage than fresh fish. Table 7·1 shows the potential storage life at various temperatures for fish in good condition. It is based upon results of experiments carried out at Torry Research Station over a number of years. All the samples were prepared from very fresh cod, haddock, plaice, lemon soles and herring. The fish were stowed in ice for not more than 24 hours between catching and freezing. All but the smoke-cured fish were well glazed, packed in wooden boxes lined with parchment paper, and then stored. Store temperatures were maintained within 1 Fahrenheit degree of those stated in the table. Samples were tasted and compared with corresponding fresh fish at regular intervals.

The figures in the columns headed 'good' give the period within which the cold stored article is for all purposes as good as fresh. The columns marked 'inedible' indicate the time when the product becomes so distasteful to a consumer accustomed to fresh fish as to be uneatable. The figures cannot be more than approximations of the limiting periods.

TABLE 7·1

The cold storage life of fish

Type of fish	*15°F*		*minus 5°F*		*minus 20°F*	
	good	*inedible*	*good*	*inedible*	*good*	*inedible*
white fish (gutted)	1 month	4 months	4 months	15 months	8 months	more than 4 years
herrings (ungutted)	1 month	3 months	3 months	6 months	6 months	more than 1½ years
smoke-cured white fish	1 month	3 months	3½ months	10 months	7 months	more than 1 year
kippers	3 weeks	2 months	2 months	5 months	4½ months	more than 9 months

Fatty fish, for example herring and salmon, do not keep so well as white fish, such as cod, which have a very low fat content. Smoke-cured fish store less well than unsmoked fish.

FACTORS LIMITING STORAGE LIFE

(1) *Protein changes.* Fish proteins become *denatured* and permanently changed during freezing and cold storage. The speed at which denaturation occurs depends very largely upon temperature. At temperatures not far below freezing point, 28°F for example, serious changes occur rapidly; even at 15°F the changes are so rapid that an initially good quality product can be spoilt within a few weeks. Badly stored fish are easily recognizable; the thawed product, instead of being glossy and translucent, is opaque, white and dull. The firmness and elasticity of the properly stored product are replaced by sponginess and in very bad examples the flesh may break up. Juice can easily be squeezed out. Cooked samples, instead of having the succulent curdiness of cooked fresh fish, at first feel wet and sloppy in the mouth, and on further chewing become dry, fibrous and tasteless.

When smoke cured, such inferior fish sag and gape and have an unattractive, matt surface. Smoke-cured products made from properly stored frozen fish acquire an attractive glossy pellicle because the brine dissolves some of the protein, and this solution subsequently dries on the cut surface of the fish; denatured protein is insoluble in brine so that the surface of poorly stored fish remains dull after smoking.

Protein denaturation is more obvious in white fish than in fatty fish. Nevertheless, herring stored at 15°F for more than a month or so do not split well, and kippers made from them are dull. The dullness may be masked to some extent by dye and by oil in the fish.

Mention has been made of the possibility of brining fish before freezing. White fish fillets treated in this way sometimes show less ill-effect from high temperature storage. Less *drip* runs out of the thawed fillets, and the texture is better than of fillets not so treated. If impure commercial grades of salt are used for the brining process, pronounced salt fish odours develop in fillets stored at 15°F. It is claimed that brining before freezing is also of value for fillets that are to be stored at low temperatures of about minus 20°F. Protein denaturation takes place very slowly at this temperature but a small amount of drip from the thawed product is unavoidable. It is suggested that preliminary brining reduces this drip still further and prevents further loss of juice when the thawed fillets are kept for any length of time before cooking. The value of brining fillets destined for low temperature storage is somewhat doubtful, however, especially as drip after thawing can be considerably reduced by brining the fillets just before or just after they are thawed. There is then no fear of development of salt-fish odours provided they are cooked within a few hours.

Fatty fish should never be brined before freezing, for the reasons given in the following notes on fat changes.

(2) *Fat changes.* The fat of fish may become unpleasantly altered during cold storage. Fish oils readily combine with oxygen. Some of the enzymes normally present in fish muscle, particularly those in the red strip of muscle just under the skin of fatty fish, assist this reaction. Herring, salmon, mackerel, sprats and pilchards are particularly prone to this type of spoilage. Badly affected fish have a most disagreeable odour and flavour. The fat becomes gummy, and the flesh develops a yellow, rusty appearance. Flavour may vary from that of a mild cod liver oil, which some people find not unpleasant, to an acrid, burning or painty taste which is definitely objectionable.

These changes take place more rapidly at higher temperatures of storage, and are sometimes accelerated by the presence of small amounts of certain chemicals, such as salt, that probably increase enzyme activity. For this reason fatty fish should never be brined before freezing. Brine-frozen Norwegian herring have been known to be rancid by the time that they enter this country.

Chemicals that can slow down or delay these reactions are called *antioxidants*. Some occur naturally, in plants, for example, and in wood smoke. Attempts have been made in recent years to produce artificial ones that could be used commercially. Such substances must be harmless to man and animals before they can be introduced for the treatment of food; in addition they should have no pronounced odour or flavour of their own, nor develop any during use. Various tests with antioxidants on British-caught herrings have not shown results of any promise. Antioxidants added to the glaze of whole fish have proved ineffective. In addition, soaking fillets in a solution of antioxidants did not improve storage life in any way.

The development of rancidity in the fat of frozen fatty fish limits storage life but it can be difficult to control in practice. As already indicated rancidity occurs primarily as a result of attack by atmospheric oxygen. Consequently it is found that simple factors that will affect the degree or ease of attack will affect the development of rancidity. The beneficial effect of glazing has already been mentioned; in addition herrings frozen in a compact block keep better than herrings frozen singly and the larger the block, the better the keeping quality. Whole herrings, too, keep better than fillets, possibly because the process of filleting opens up the fish and allows free access of oxygen. Drying and the addition of salt accelerate the development of rancidity and over-ride any antioxygenic effects attributable to antioxidants in the smoke picked up during the preparation of kippers or smoked salmon. It has been observed that because of this, frozen kippers and probably frozen smoked salmon, too, have a much shorter storage life than the raw fish. Owing to the popularity of these products and particularly of kippers in relation to herrings, the development of rancidity in smoked products can be of serious consequence, particularly since it is not possible to protect these fish by means of a glaze. However, it has been shown that by packing these products in bags of a suitable plastic material, sealed under a good vacuum before freezing, it is possible to obtain very considerable extensions of the storage life of frozen herrings and also kippers. The material used needs to be impervious to moisture and atmospheric oxygen; a foil laminate and certain plastic laminates have proved the most effective in this respect.

It is necessary to bear in mind that there is an enhanced risk of food-poisoning organisms developing in vacuum packed products stored at high temperatures, that is above 40°F, for anything more than a few hours. Properly packed herrings and kippers, and probably other fatty fish and their products, too, can be kept at minus 20°F for periods in excess of a year without any signs of rancidity or other change in flavour. The success of the process depends very much on a proper vacuum being produced in the bag and on the efficiency of the seal. Equipment is available for carrying out these operations successfully.

The success of vacuum packing with small consumer packs of fatty fish suggests that ways of extending the process to the larger packs of whole fish should be explored. The adoption of the vertical plate freezer for the bulk freezing of herrings could, with distinct advantage, be combined with vacuum packing.

(3) *Dehydration changes.* Frozen fish may dry slowly in cold store even under good operating conditions. This is undesirable for reasons other than the obvious one that the product will lose weight. Most important, drying accelerates denaturation of the protein and oxidation of the fat; both effects are described above. Texture and appearance are adversely altered.

Frozen fish that have suffered severe drying in cold store have a white, toughened, dry and wrinkled appearance on the surface that is characteristic of the condition known as *freezer burn.* The skin of the thawed fish may have a similarly dry, wrinkled look, and if drying has been exceptionally severe, the flesh beneath can become spongy and as light as balsa wood.

DESIGN AND OPERATION OF FREEZING PLANT

GENERAL CONSIDERATIONS

144 Btu of heat must be extracted from one pound of water at 32°F in order to turn it into one pound of ice at 32°F. One Btu must be removed from a pound of water to cool it by one Fahrenheit degree from, say, 50°F to 49°F, but only $\frac{1}{2}$ Btu has to be removed to cool one pound of ice by one Fahrenheit degree.

Since fish are at the most four-fifths water, rather less heat has to be removed to freeze them than to freeze the same weight of pure water. In order to freeze one pound of cod fillets from 40°F to minus 20°F, only about 135 Btu must be removed as against about 190 Btu for one pound of water.

Fish flesh, even when frozen, is a relatively poor conductor of heat. Very low refrigerant temperatures are therefore necessary to remove heat rapidly enough. In practice such temperatures range from minus 20°F to minus 50°F. In spite of this, only relatively thin layers of fish can be frozen sufficiently quickly to comply with the specification that the centre be reduced from 32°F to 23°F within two hours. Theoretically it is not possible to freeze in this manner slabs of fillets more than $5\frac{1}{2}$ inches thick, or single whole fish greater in thickness than 7 inches, when the temperature at the surface of the fish is reduced to minus 40°F.

In the early days of refrigeration, fish were frozen in closely stacked boxes standing in a cold store. The rate of freezing was slow, spoilage during storage was severe and the products were of deplorable quality. Later, boxes of fish were spread out singly on slatted shelves in the cold store, or in some cases, on refrigerated grids, while cold air was blown around the boxes. Known as sharp freezing, this process is still sometimes used for boxes of smoked fish. This method may result in quick freezing if the fish and the packaging are thin enough and the air is blown about in sufficient amount at a low enough temperature.

It is easy to overload a sharp freezer by putting in too much warm, unfrozen fish at one time, or by stacking the boxes too closely together. Air movement and temperature may vary in different parts of the store, so that freezing times vary correspondingly. Since considerable handling is needed in and out of store and a great deal of space is needed to lay out the product in this way, special quick freezing plants have been developed.

It must be emphasized that the temperature of the air in a refrigerated space is no guide to the suitability of that space for freezing fish. As an example, a well insulated cold store built to hold 20 tons of fish at minus 20°F can be run satisfactorily by a compressor of from 3 to 4 horsepower. But if one ton of wet fish is to be quick frozen in an hour to a temperature of minus 20°F, the compressor would have to absorb about 100 horsepower. The first compressor may be dealing with a heat leak into the store of about 10 000 Btu an hour, while the second is extracting perhaps as much as 4 000 000 Btu an hour from the fish, when losses are included. There is very little reserve of cold in the walls and pipes of a cold store, and wet fish introduced into a store already containing frozen fish can easily cause the store and its contents to rise rapidly in temperature, to the detriment of the frozen fish. Wet fish introduced in this way, in the ratio of 1 pound of wet fish to every 2 stones of frozen, could result in the temperature of the frozen product rising about 10 degrees in order to absorb the heat released by the wet fish during freezing.

TYPES OF QUICK FREEZER

The three most important methods of freezing fish are by direct contact between the fish and a refrigerated metal plate, by blowing a continuous stream of refrigerated air over the product, and by immersing the fish in a low temperature liquid. The first two methods will be described in some detail, while the third, at the present time little used in Britain, will be given only brief mention.

(1) *Air blast freezers.* In a properly designed air blast freezer, the temperature, speed and degree of agitation of the air that passes over the fish should be everywhere reasonably constant. This will give uniform freezing of the product. Still air is a poor conductor of heat, and has a low heat capacity, so that fairly high air speeds are necessary to avoid uneconomically low temperatures. High air speeds, however, mean powerful fans which heat the air. This excess heat has to be removed by the refrigeration machinery.

Air blast freezers can be divided into two main groups; those in which the product moves through the freezer during the process and those in which the product remains stationary.

The first group is more suited to mass production with continuous runs of standard packs having similar freezing times, whilst the second one is more flexible but requires very close supervision to maintain maximum output without overloading.

Where the product moves through the freezer on trucks, these may move across the air blast or upstream or downstream. Where they move across, loading and unloading do not interfere with air flow.

Good air distribution can be achieved with a number of small fans. The length of the air path is short, so that air speed need not be very high. A small disadvantage is that much of the heat load has to be carried away by the cooling coils near the end at which the trucks enter. Maximum use of all the cooler surface is possible only if two lines of trucks move in opposite directions through the tunnel. Such an arrangement may be difficult to fit in with factory layout.

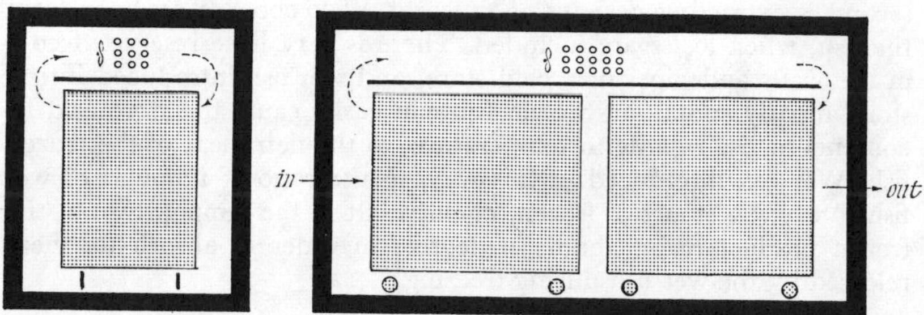

FIGURE 7·3 Types of air blast freezer: (*left*) Cross flow type; (*right*) Series flow type

Air blast freezers, because of the fans employed, consume more power than equivalent plate freezers and they generally occupy more space. Blast freezers can accommodate all shapes and sizes of product, and are particularly suitable for freezing large single fish and blocks or packs of irregular shape. Good performance from air blast freezers is obtained by freezing the produce in open trays without wrapping, and when the trays are filled sufficiently to avoid pockets of dead air above the surface of the fish. On the whole, the air blast freezer is less sensitive to irregularities in shape and size than the plate freezer, but is more expensive to operate.

Construction of air blast trucks makes them more difficult to overload than a sharp freezer, in terms of weight or thickness of warm produce. Nevertheless, it is possible to misuse the freezer by removing the produce before it is properly frozen, or by poor control of packing and loading. The only type of freezer that cannot be unloaded too soon is the continuous belt freezing tunnel, provided that the belt speed has been set by competent engineers and cannot be altered by unauthorized people.

Frost accumulates in air blast freezers, especially on the upstream side and near the warmest fish. Tunnels should be constructed so that they can be easily inspected and the frost removed, as these accumulations reduce the heat transfer from the evaporator to the air, causing the air to warm up. The frost can also restrict the flow of air through the tunnel, thus slowing down the freezing rate and reducing the output of the freezer. Frost is produced from three sources; 1% to 2% by weight of the unwrapped product may evaporate as moisture into the air stream, warm air can leak in from outside, and moisture may diffuse from outside through cracks and openings. Periodic defrosting of air blast freezers is essential if performance is to be maintained. Defrosting by heating with the freezer doors closed is preferable to the more common practice of turning off the flow of refrigerant and throwing open the doors. This natural defrosting can be a slow and messy process, and can damage structure and insulation through heavy condensation.

The choice between batch and continuous blast freezers depends very much on the types and quantities of product, but it can affect the design and cost of the freezing plant because the size of both condenser and motor depends on the type of operation. The load on a continuously operated plant is reasonably constant throughout the run, whereas with batch operation the whole freezer is often loaded simultaneously with warm produce; the initial rate of heat extraction with which the machinery has to cope can then be very high.

Trays for use in air blast freezers should transfer heat efficiently, be emptied easily, and be robust. Normally they are required to produce a pack that is rectangular or nearly so. Since fish swell when they are frozen, it is difficult to reconcile robustness and ease of unloading with a requirement that the pack be perfectly square. Strongly made trays are frequently damaged during unloading through insistence on the use of trays with no taper on the sides. Trays with a taper of about one in eight can be emptied quite easily by applying a cold water spray on the underside for a few seconds and then giving a gentle tap on the edge. It is also possible to use inserts with a tapered tray if a truly rectangular product is essential.

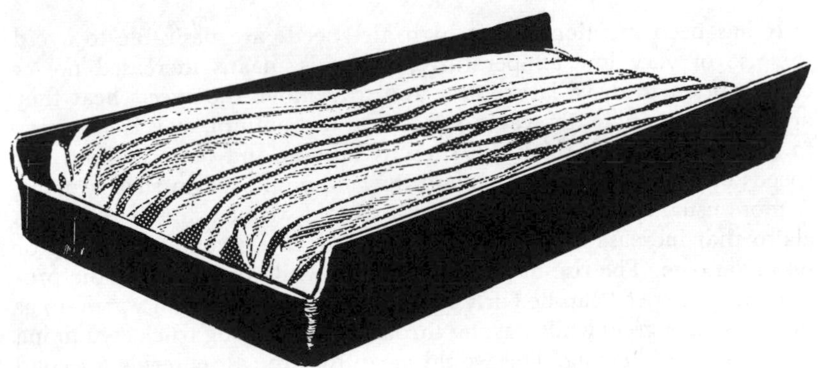

FIGURE 7·4 Tray for use in air blast freezers

Trays should have those edges that lie across the airstream reduced in height so that air flows over the product in close contact with its surface. The sides of the trays that lie along the airstream may be made higher than the product to facilitate tapping out the block when unloading. Fourteen-gauge to 16-gauge aluminium is suitable for the manufacture of trays (see Figure 7·4).

Trays should be spaced on the trucks so that the distance between the upper surface of the product in one tray and the bottom of the tray on the shelf above is between 2/3 and ½ of the thickness of the product. The trays should be distributed evenly across the tunnel so that nowhere over the cross-section is the resistance to air-flow greater than anywhere else. If trays are packed together in one part the air will simply bypass the product altogether (see Figure 7·5).

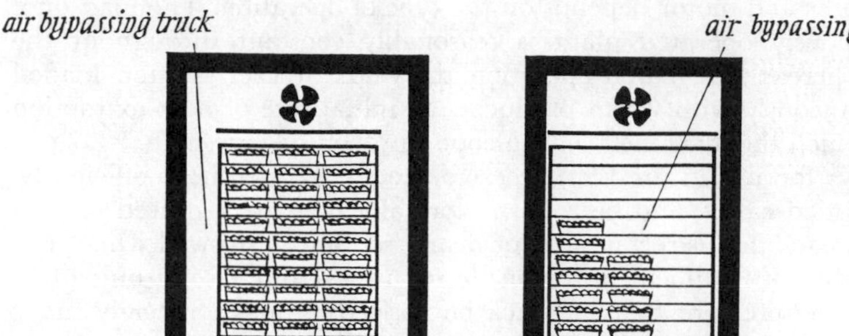

FIGURE 7·5 Bad practice in loading air blast freezers: (*left*) Well filled trolley that does not fit the tunnel; (*right*) Trolley that fits the tunnel but is unevenly loaded

It has been mentioned that high air speeds are desirable to avoid the use of very low temperatures, but this means increased power consumption to drive larger fans and to remove the excess heat they generate. Sometimes, however, space is at a premium, or it is necessary to freeze very large, thick fish such as salmon and large halibut. Air speeds of 40 feet or more a second may be employed in such a plant. A more usual speed is about 20 feet a second, while some investigators claim that increase of air speed above 10 feet a second gives little or no advantage. The rise in air temperature while passing over the product at slow speed can be fairly large; for instance, air can warm up as much as 10 degrees while passing through a 3-feet-long truck containing single layers of herring. This would mean that the air entering a second truck downstream would be at a temperature of minus 10°F, and performance would be reduced; air speeds in air blast freezers should be more than 20 feet a second in order to maintain a reasonably constant temperature throughout the working section.

(2) *Plate freezers.* The fish are frozen between pairs of metal plates through which the refrigerant passes. Both horizontal and vertical plate freezers are made, the former being the type commonly used in land installations for freezing prewrapped packs and unwrapped fish in lidded trays or moulds. The vertical type has been developed in recent years for use on board ship.

With the horizontal type of plate freezer the flat rigid plates can usually be moved by hydraulic or pneumatic rams to improve contact with the product. Packaging both insulates the product and hinders close contact between it and the plate. Moving the plates together by gentle force partially counteracts these effects.

In the past, the flow of heat from the product surface to the refrigerant was hindered by the design of plate, and hence the refrigerant temperature had to be lower than would be required with an ideal plate. Plates are now available that approach much more closely the ideal, with internal labyrinth passages so designed that the working surface is everywhere in contact with well agitated refrigerant. Because

FIGURE 7·6 Horizontal plate freezer

of these improvements in heat transfer, the plate freezer can compete with air blast freezers operating at the same evaporating temperature, and in some cases can freeze the same thickness of product faster than by air blast in spite of the difficulties of good surface contact. The choice between the two types is now usually made for reasons other than speed of freezing.

The horizontal plate freezer consumes less power than the equivalent size of blast freezer, and generally occupies less space. It works best with wrapped packs that have flat, parallel, upper and lower surfaces, or with unwrapped products that can be gently compressed into shape.

Poor contact between product and plates results in slow freezing, and can occur in several ways. Waxed cardboard cartons may be loosely packed so that an air space is left between the top of the product and the box; this may happen accidentally or it may be done deliberately to avoid the use of too many sizes of carton. Freezing then takes place mainly from one side, since the trapped air acts as insulation on the upper face of the product. If the layer of fish between the plates is not of uniform thickness, or packages of slightly different height are loaded, then freezing will proceed fastest at the high spots where the plate makes best contact with the fish. The condition will be further aggravated since the high spots will expand on freezing and force the plate even further away from the rest of the product. The most serious cause of poor contact, however, is the formation of ridges and nodules of ice on the plates. These accumulate wherever the plate is exposed to the air, between packs or trays, and when the freezer is reloaded with a fresh batch of produce the ridges prevent close contact with the plate. Defrosting arrangements are essential to eliminate this source of trouble.

While it is difficult to overload a plate freezer by putting in packs that are grossly in excess of the practical maximum thickness, it is

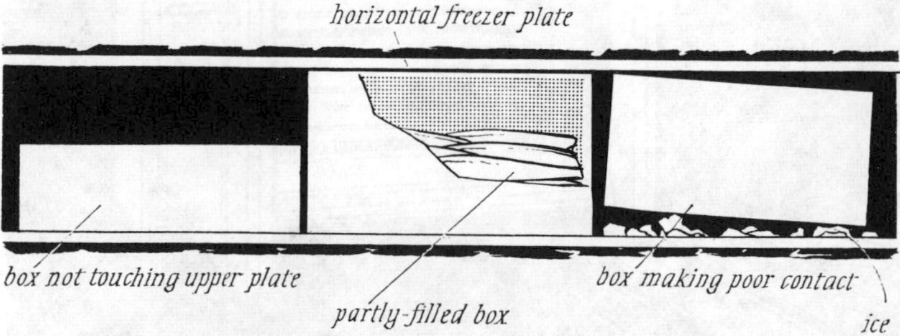

FIGURE 7·7 Causes of poor performance of horizontal plate freezers

possible to misuse the plant by removing the product before it is properly frozen, or through insufficient control of the packing and loading operations.

Horizontal plate freezers in single units are used for batch production, but a group of such freezers can be operated on a continuous basis by loading in rotation so that the refrigeration load on the machinery is always the same. For long runs of products of similar shape and size this can offer economies in cost of plant by keeping down the size of the motor and condenser required.

FIGURE 7·8 Vertical plate freezer being loaded at sea

The thinner the fish or pack to be frozen, the higher the operating temperature can be, with consequent savings in running costs, but these savings are offset to some extent by increased expenditure on packaging and labour.

Horizontal plate freezers operate at an apparent heat transfer coefficient of from 8 Btu a square foot every hour for each Fahrenheit degree of temperature difference in older models up to about 13 in the latest designs. The effect of wrappers and irregular contact may reduce these coefficients by 2 or 3, resulting in freezing times for a 2-inch thick slab of fish of from $1\frac{1}{2}$ to 2 hours.

Vertical plate freezers were developed in this country to provide an efficient means of freezing fish at sea in the confined space available aboard fishing vessels and with the minimum of labour. They are also of value for freezing large industrial packs and for other special applications.

Wrappers are not normally used, and the whole fish in its natural wet state makes reasonably good contact with the plates as it packs down under its own weight. The area of contact may also be increased by squeezing the plates together to compact the block. Hence performance is usually better than with the horizontal type, provided that the plates are regularly cleaned. Frozen blood and slime cannot readily be removed without damaging the plate surface unless there is some means of frequent positive defrosting. Natural defrosting by shutting down the plant and allowing it to warm up is generally too slow; plants should have adequate defrosting arrangements as part of the standard equipment.

Vertical plate freezers are more versatile than horizontal ones in their method of removing the frozen block, and can be designed to unload from the side or the bottom, or even the top. The higher performance makes it economically possible to freeze thicker packs, such as whole gutted white fish. The ease of loading from the top without the use of trays or containers permits economy in manpower. It is not difficult to prevent overloading of the freezers, and they can be operated in groups so that the load on the machinery is uniform during continuous operation; but careful supervision is needed to ensure that the product is never removed from the freezer before freezing is complete.

An efficient vertical plate freezer with refrigerant at minus 40°F can freeze a block of whole white fish and reduce the temperature at the centre to minus 5°F in less than 3 hours under ideal conditions. Freezing times of less than $3\frac{1}{2}$ hours are thus possible when operating under practical conditions, with a total time, from the start of freezing one batch to the start of freezing the next, of less than 4 hours including allowances for loading and unloading.

(3) *Immersion freezing.* The disadvantages of freezing fish in air can be lessened or avoided by freezing in liquids of which the heat capacity and conductivity are much greater. Some of the liquids that have a freezing point lower than that of water are suitable for this purpose, and these include refrigerated brine made from a solution of common salt in water.

Such a brine cannot cool fish much below 0°F but fish immersed directly in the brine can be frozen very quickly since excellent contact is made between fish and refrigerant. The fish absorb some salt during freezing, and not only may this accelerate spoilage during cold storage, but may make the thawed product too salty for the taste of the consumer. There are also difficulties in handling and storing large numbers of fish that have been frozen individually. For these reasons, brine freezing has fallen into disfavour in this country, although most early attempts at freezing fish were made by this method. At the present time it is used mainly for freezing large individual fish such as Pacific salmon and tuna, where the product is destined for canning, particularly in North America and Japan. Other liquids with freezing points lower than that of salt solutions are occasionally used for immersion freezing of fish products, a mixture of water, sugar and alcohol being an example. With all such liquids, there is always the difficulty of avoiding contamination of the product that would make it less acceptable to the consumer.

(4) *Hybrid freezers.* There are various hybrids of air-blast and plate-freezing plant that have been tried from time to time. For instance, fish may be frozen in trays laid upon refrigerated shelves, over which cold air is blown; an air blast freezer tunnel may be fitted with trucks that are virtually horizontal plate freezers, the trays being clamped between metal plates through which refrigerated air can pass at high speed.

FREEZING TIMES AND OUTPUT

Prediction of freezing times by calculation is at best an inaccurate process. The mathematics are complex, and even approximate solutions are tedious to determine. The use of such methods should be restricted to the planning stages of new enterprises and the answers treated with caution.

Freezing time is influenced by the rate of heat transfer at the surface of the product, its shape and size, and the temperature of the air or refrigerant. For any particular product and freezer, the freezing time depends on the temperature of the air or refrigerant. If the output of the freezer is fixed, the operating temperature depends then upon the capacity of the machinery. There is only one operating temperature at which a given compressor can match a given weight of produce formed of one particular kind of pack. The lower the operating temperature, the lower will be the capacity of the compressor, but the rate of flow of heat from the product will increase.

Assuming some particular operating temperature, it is fairly easy to estimate the freezing time for a particular pack in a given type of freezer. It is then possible to work out the size of machinery that is needed to give the output of product that is required. This may mean that only part of the freezer space available can then be used for that particular pack, in order to avoid the danger of overloading.

If the size of the freezer is fixed, for example in plant that already exists, then output can be estimated for a given operating temperature and the compressor size matched to cater for the load.

But if both freezer and compressor are already fixed in size, then it is more difficult to decide at what temperature a balance will be achieved, and what the throughput will then be. These can only be determined step by step on a trial and error basis. The most reliable guide to what the compressor is capable of is from knowledge of its performance on other products at known operating temperatures.

Actual measurement of freezing time for a particular product is the only safe method. Two times are of interest, the time taken to reduce the temperature at the centre to minus 5°F, and the time taken at the centre of the pack for the temperature to pass from 32°F to 23°F. The first is known as the total freezing time, and governs the output of the plant. The second is the thermal arrest time, and determines whether or not the product has been frozen in a proper manner. The time taken through the thermal arrest should not exceed two hours. The relationship between the two is not always constant, and depends upon the initial temperature of the product, the thickness of the block and the rate of heat transfer during freezing. The temperature of fish entering the freezer should be 40°F or below, and any product that is warmer than this should preferably be precooled before freezing begins. Higher starting temperatures will reduce output and encourage spoilage before freezing.

If freezing takes place equally from both sides of a block, the centre will be the slowest to freeze. Temperature measurements should, therefore, be made at the centre. Sometimes freezing is unequal, for example when a carton in a plate freezer is not completely filled, or when a whole fish is being frozen singly and bone structure and belly cavity upset the uniformity of the material. The part of the product that will freeze last is then more difficult to determine, and the temperature-measuring device has to be placed with care to give an answer that is meaningful.

Suitable equipment for temperature measurement is described later, but examples are shown here of the kind of freezing curve that can be drawn from measurements taken during freezing (Figure 7·9). Curve 1 is typical of that part of the fish which takes longest to freeze. The temperature remains almost steady whilst the bulk of the water is changing to ice, and then falls very quickly indeed. Provided that the refrigerant temperature has remained fairly steady throughout the freezing process, readings can be taken directly from such a curve to give both total freezing time and thermal arrest time.

Temperatures represented by curve 2 have been measured at a point close to the surface of the block, and freezing has taken place so quickly that there has been no pronounced pause at about 30°F. The surface temperature of the product has quickly approached the temperature of the refrigerant. Such a curve, and any curve that represents some point in the block that is well away from the slowest point, should not be used for determining freezing times.

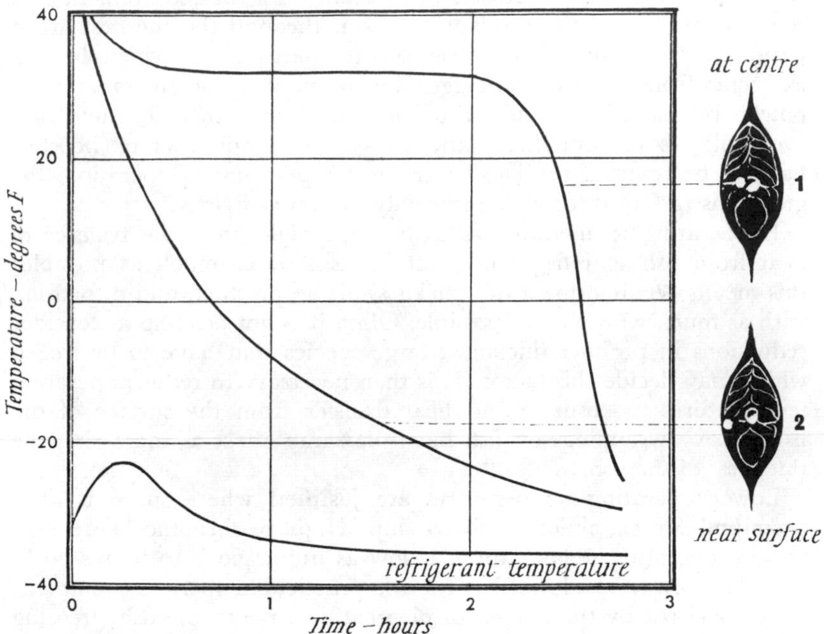

FIGURE 7·9 Freezing curves at the centre and near the surface of a fish

It should be mentioned here that since the water content of smoked fish is lower than that of wet fish, less heat has to be removed during freezing, and because of added salt freezing begins at a lower temperature. The preservation and denaturing effects of smoking allow a longer period of thermal arrest without adverse effects during freezing. Kippers, for example, may be allowed to take up to 12 hours to cool from 32°F to 23°F. If smoked fish are to be frozen quickly, much denser packing than is usually employed is required. More usually smoked fish are frozen in wooden boxes of various shapes and sizes, with paper linings. Kippers freeze more rapidly in open boxes than in closed boxes. There is no reason why smoked fillets cannot be packed in the same way as wet fillets for freezing and there is a method of packing kippers for rapid freezing in which the fish are packed singly with slight overlapping. Finnans could be packed in a similar way.

It is now necessary to discuss in more detail some of the factors that affect freezing times and output, as a guide to the choice of the most suitable operating conditions. In theory, it is best to operate at the highest temperature possible, consistent with the requirement that the temperature of the product be taken through the thermal arrest period in two hours. The lower the temperature at which heat is extracted from the fish, the bigger the compressor and the greater the power consumption required to do the work. Indeed, mere size eventually is of little help, and two-stage compressors or boosters have to be resorted to. The lower the temperature of operation, the greater is the need for a more heavily insulated freezer.

If operating temperature is to be kept high, then the transfer of heat from fish to refrigerant must be assisted as much as possible; this means keeping down the thickness of the product, and dispensing with as much wrapping as possible. Often it is not possible to consider reductions in product thickness; large species that have to be frozen whole may decide this factor. It is then necessary to reduce operating temperatures to ensure rapid heat transfer from the surface of the fish. Heavy wrappings on fish have the same effect as increasing the thickness of the fish.

Lower operating temperatures are justified where space is at a premium, for example on board ship. High performance here may require operating temperatures as low as minus 60°F to minus 70°F. There is, however, a lower limit to the practicable operating temperature, governed by the danger of physical damage to the fish; freezing in very low temperature liquids causes structural damage when the already frozen outer layers of fish are stretched by the inner layers as they expand during freezing. Experiments are being made to determine the safe lower limit of temperature.

The output of a freezer cannot be described as so many tons a day, or hundredweights an hour, unless the type and size of pack is specified; a definition without such information is meaningless. Output will not be achieved unless the freezer is loaded with the correct amount of product in the correct size of pack. If a particular pack takes x hours to freeze at a given operating temperature, and the compressor is capable of freezing y tons of that pack in an hour at that temperature, then the freezer must be loaded at all times with x times y tons of fish to give the required output of y tons an hour. For example, an air blast freezer takes 2 hours to freeze a 2-inch slab of fillets and the compressor is capable of freezing 2 tons of fillets an hour at that particular operating temperature. During continuous freezing, that freezer will be holding 2 times 2, that is, 4 tons of fillets in 2-inch packs at any one time. If the product is changed to kippers in boxes that take 12 hours to freeze, then to maintain the output of the freezer at 2 tons an hour, the freezer would have to have room for no less than 24 tons of boxed kippers at one time.

If the size of the freezer is made a compromise between the two extremes, then it cannot cope with a full load of the faster freezing packs of fillets, and freezing will be slower. If, on the other hand, the freezer is made just big enough to hold 4 tons of fillets, its capacity for boxed kippers may be little more than a ton. Then with the reduced product load of kippers the operating temperature would fall, but the capacity of the machinery would also be reduced at the new low temperature, with the net result that the ton of kippers would be frozen in only a little less than 12 hours. Output of kippers compared with fillets would therefore be much reduced and freezing more costly. It is, therefore, desirable that the product that freezes fastest be the one that is produced in greatest quantity. Freezers can then be made just big enough to hold the right amount. There is still a risk that freezers can be overloaded with packs that freeze more quickly than the one for which it was designed. Packs that freeze more slowly can be produced only in reduced quantity. An exact knowledge is needed of the freezing times of all packs at various temperatures.

A balance also has to be struck at a point where savings made by reduction in thickness of a pack are matched by extra expense and difficulty involved in wrapping, handling and freezing the same weight of product. Under ideal conditions freezing time should vary as the square of the thickness of the product; that is, if a three-inch slab takes four hours to freeze, a 1½-inch slab should take only one hour. In other words, if the thickness is doubled, the freezing time is increased fourfold. Thus, theoretically it is possible, with fish in 1½-inch slabs instead of 3-inch slabs, to freeze twice the amount in the same time. But time will be lost in loading the freezer twice, so that the output will not be quite doubled. In addition, the number of surfaces of fish has been doubled and as there is a slight resistance to the transfer of heat wherever there is an exposed surface, this is equivalent to a small increase in thickness of the product. The doubling of this resistance increases slightly the proportion of the time taken to freeze the thinner slabs. Wrappers can add further to the freezing time in a similar way; two 1½-inch blocks with wrappers have a greater effective thickness than one 3-inch block with wrappers.

All of these factors may result in, for example, a cycle time for a load of 3-inch blocks of about 4¼ hours, as opposed to a cycle time of 3½ hours for two batches of blocks 1½-inch thick. There is, therefore, still some saving to be made by freezing the thinner block, but this advantage might disappear entirely if, for example, it were decided to reduce still further to three 1-inch packs; increase in handling time, the effect of extra wrapping and additional exposed surfaces, together with labour and packaging costs, would probably completely outweigh savings made by reducing the theoretical freezing time. Therefore, there is an optimum thickness for every type of pack which depends upon the performance of the freezer. Absence of wrappers, mechanical loading, and high air speed or good plate contact, depending on the type of freezer, all favour the thinner pack.

Once the freezing time for a particular pack under specified conditions has been established by measurement of the temperature at the centre, close supervision of packing, loading and unloading time and machinery performance should ensure that the process continues satisfactorily. It should not be necessary to use temperature measurement as the primary means of control; trouble might not be apparent for some hours and a temperature record of a small proportion of packs would mean very little.

TEMPERATURE MEASUREMENT IN FREEZING PLANTS

The most convenient, cheap and accurate method of measuring freezing times is by means of thermocouples attached to a suitable indicator or recorder. Resistance thermometers, though accurate, are bulky, easily damaged, and relatively expensive. Glass spear thermometers are quite useless for measuring temperatures in frozen fish, since the protective metal case conducts heat far too readily and results in false readings. The principle of the thermocouple is described in the chapter on instruments. Here the technique is explained of using them in freezing plants.

Thermocouple wires should be as fine as possible, particularly at the junction of the two wires; wire thickness greater than 26 swg is not recommended. Junctions are better welded than soldered. Cellulose varnish may be used to insulate the bared wires from each other right down to the actual junction; the varnish also helps to stiffen the thermocouple sufficiently for insertion into the product. The thermocouple ends are expendable, and the wires may be cut close to the surface of the frozen block once measurements are complete, leaving enough wire showing to enable the sample blocks to be detected and segregated from normal blocks leaving the freezer. The junction can then if required be made between the shortened pairs of wires.

An indicator or recorder that reads temperature directly is to be preferred to a potentiometer that gives the answer in microvolts. The instrument should read to within half a Fahrenheit degree. Care must be taken that if a separate selector switch is used all parts are at the same temperature to avoid errors in readings. Two-pole switches should be used, and if the instrument is not temperature-compensated and reference junctions are used, these junctions should be separate for each thermocouple and immersed in oil. Temperature-compensated instruments must not be used inside refrigerated spaces. Thermocouple wires should be kept away from other electrical wires.

Thermocouple wires conduct heat; they should always be put into a block of fish so that for several inches along their length away from the junction they are at about the same temperature. If there is a temperature gradient, heat will be conducted away from the fish around the junction point. The thermocouple might show, for example, that the fish in the middle of the pack is apparently freezing much faster than it really is. This means in practice that the thermocouple should be laid parallel with the freezer plates or the cold air stream.

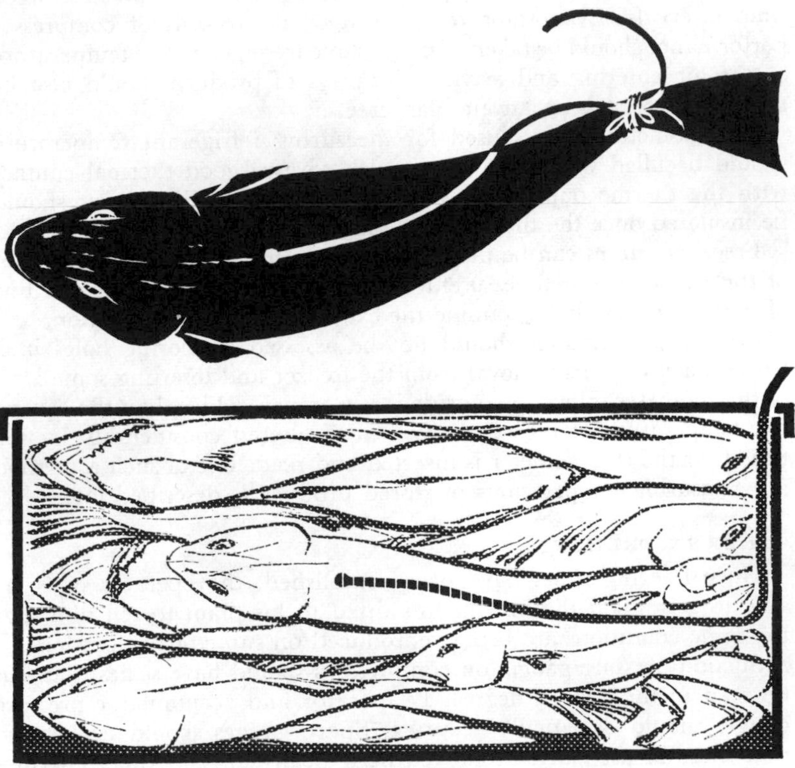

FIGURE 7·10 Positioning of thermocouples in fish to be frozen: (*top*) Location of thermocouple within a fish; (*bottom*) Location of the fish near the middle of a block

Thermocouples may be built into a pack before it is frozen, by laying them between fillets, for example, or by piercing a hole in the assembled pack or in a single fish. The hole should be made with a sharp probe of about the same diameter as the thermocouple, so that the end of the hole is judged to be at the centre of the thickest part. The probe is withdrawn and the thermocouple inserted, taking care that no air space remains between it and the end of the hole. The thermocouple should then be tied carefully in place, and the wires fastened to the freezer structure.

Several thermocouples should be used in any one run to ensure that at least one or two are in the centre to produce reliable freezing curves. Since performance will vary slightly from place to place in the freezer, thermocouples should not be confined to one section of the plant under test. Each thermocouple should be read from ten to twenty times during a freezing run. Readings should be taken at fixed time intervals. Evaporator temperatures and records of compressor performance should be taken with the same frequency. The temperature of the air entering and leaving the trays of produce should also be taken if the test is on an air blast freezer.

Thermometer pockets used for measuring refrigerant temperature should be filled with methylated spirit to give good thermal contact with the thermocouple. The pipeline adjacent to the pocket should be insulated once the thermocouple has been inserted.

Freezing curves can be plotted as the run proceeds, and at the end of the run it is sometimes useful to cut the pack open along the line of the thermocouple to examine the exact position of the junction.

Freezer performance should not be assessed by boring holes in a frozen pack after its removal from the freezer and inserting some kind of thermometer. Since the outside is very much colder than the centre, the temperature at the centre will have altered considerably by the time that the thermometer is inserted and read. Use of such a method for measuring temperatures of stored produce is described later.

OTHER INSTRUMENTS

Once freezing times have been established, the operator must be able to rely upon the instruments fitted to his plant to tell him that the same conditions are being reproduced on subsequent occasions.

Suction pressure gauges on compressors should have scales that can be read to the nearest degree. Evaporator and accumulator pressure gauges should be equally accurate. Spare gauges should be kept so that those in use can be removed periodically for checking. Thermometer pockets should be provided, for test purposes, adjacent to all pressure gauges. Sight glasses and liquid level indicators should be fitted where appropriate, and all electrical machinery should have ammeters and revolution counters. Air blast freezers should have air temperature indicators or recorders that have quick response and are accurate to 1 degree, and all freezer working spaces and machinery rooms must have accurate clocks. Dummy clocks are often useful to remind operators of loading and unloading times.

A few inches of fine wool fastened to the end of a slender rod will help the operator of an air blast freezer to examine the strength and direction of air flow at any point in the tunnel. A number of streamers can be used to study air flow over shelves and trays of fish without recourse to a more elaborate instrument for measuring air speeds.

DESIGN AND OPERATION OF COLD STORES FOR FROZEN FISH

Having discussed the design and operation of plant that will freeze fish quickly, attention may now be turned to the design of proper storage for the product. Only those aspects of design that specifically affect the quality of frozen fish are discussed. This is not a guide to specification of cold stores in general.

The fish must enter at the temperature of the store, be maintained at that temperature without fluctuation and must not dehydrate during storage. Storage temperature must be low enough to prevent spoilage of the product within the maximum time that any batch of product is likely to stay in the store. Since this temperature is different for different products the present trend is towards the adoption of minus 20°F as the maximum safe and economic temperature at which fatty fish and white fish may be stored together for long periods.

REDUCING DEHYDRATION

The temperature of the product must be kept as steady as possible, since fluctuations increase damage to the flesh. Rises in temperature also encourage dehydration; how this happens must be explained.

Whenever there is a difference in temperature between the product and some other part of the store there will also be a difference in water vapour pressure, and water vapour will migrate from the product to any colder surface. The greater the temperature difference, the faster the water will be transferred. Even though the water at minus 20°F is in the form of ice, this transfer still takes place. Thus if cooling pipes or plates in a cold store are operating at a temperature only a fraction of a degree below that of the fish, the pipes will gradually become covered with frost and the fish will slowly dry out, resulting in the condition described as freezer burn. Movement of air within the store can accelerate the transfer of water vapour; good cold store design and practice reduce the temperature difference and restrict movement of air.

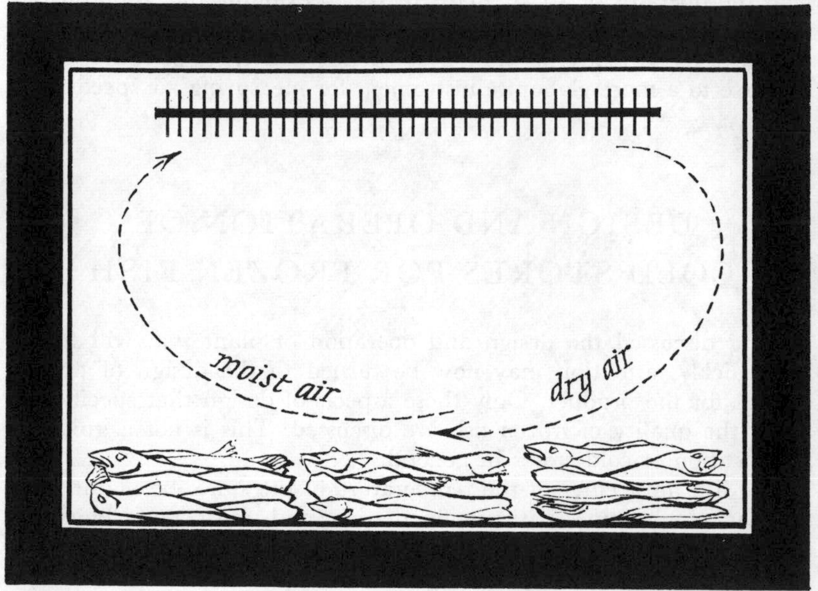

FIGURE 7·11 Dehydration in cold stores

The protection of the product against dehydration by glazing or wrapping has been described; this section is concerned with how the design of the cold store can help. In a well designed store the temperature of the product and the air surrounding it will be very nearly the same, although cooling pipes or coils will be at a lower temperature to absorb heat entering the store through insulated walls and open doors, heat from lights and fans or from men working in the store. Desiccation can be greatly reduced by restricting the quantity of heat entering the store.

In very small stores, and in transit stores, the biggest single source of heat may be due to the opening of doors, but in general most of the heat comes through the insulation. The heat leak depends mainly upon the type and thickness of the insulation, the method of construction and the surface area of the store. Size and shape are important here; a small chamber will have a bigger heat leak in proportion to the amount of product it can hold than will a big chamber of the same shape, if method of construction and insulation thickness are the same. For example, a room 45 feet by 30 feet by 15 feet will hold more than three times as much produce as a room 30 feet by 20 feet by 10 feet, but has not much more than twice the surface area. Smaller chambers should be more heavily insulated to avoid excessive dehydration. For rooms of the same capacity, shape can affect surface area; for example, a room 100 feet by 100 feet by 10 feet has $1\frac{1}{2}$ times the surface area of a room 100 feet by 50 feet by 20 feet, for the same internal volume.

One possible solution to the heat leakage problem is the use of jacketed stores. If an air space is provided between the inner lining of a cold room and the insulation, cooling surfaces within this air jacket can absorb all the heat that passes through the insulation before it reaches the room itself. The lining between the room and the air jacket must be completely vapour tight, since there is still a vapour pressure difference between the product and the jacket cooler; the air in the room can then remain saturated with water vapour and prevent evaporation from the fish. The air-jacketed store can be regarded as an impervious wrapper for a room full of packs in place of individual wrappers for each separate pack. It is unlikely that jacketed construction is necessary for stores of 1 000 tons and upwards having insulation equivalent to 10 inches of slab cork, since weight losses over normal storage periods in such stores are reported to be small. Jacketed construction to prevent desiccation may be most useful only for smaller stores, where insulation thickness is more often determined by the need to prevent excessive weight loss, rather than by consideration of the balance of capital and running costs as between insulation

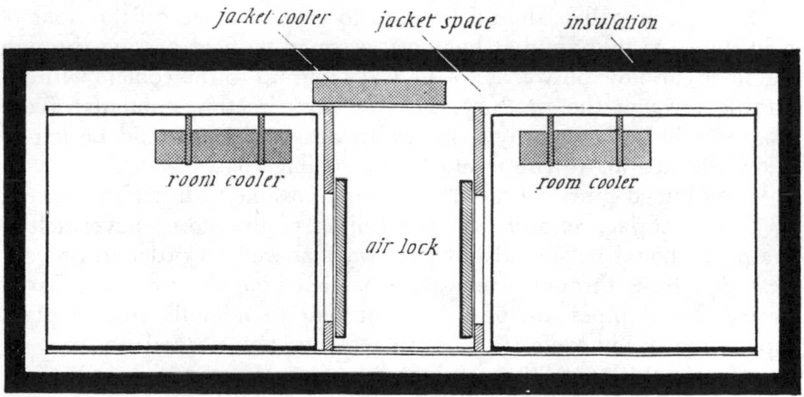

FIGURE 7·12 Jacketed cold store construction

and refrigerating power; jacketed construction might well, therefore, result in cheaper storage by reducing insulation thickness to the economic optimum. Jacketed stores are in use in Canada and USSR, although there are many difficulties in putting theory into practice.

If the temperature of an existing store is lowered, without adding to the insulation, more heat will leak in and drying will increase. Early objections to the use of minus 20°F storage arose through difficulties of this kind. If the insulation is increased so that no more heat gets in at the lower temperature than did at the higher temperature, then less drying takes place than before owing to the way that water vapour pressure varies with temperature.

If the air is blown over cooling coils in a cold store to obtain better heat transfer between air and pipe, the rate of water vapour transfer is increased to the same extent. If air is moved over the coils, it will move over the produce as well and may cause severe local drying; air that has been warmed by passing over walls or floor may be agitated locally and warm nearby produce instead of giving up its heat to the cooling coils. The fans used to circulate the air will also add to the heat load in the store. It is therefore preferable to reduce the temperature difference between produce and pipes by using a cooler with as large a surface area as possible in still air, rather than a small cooler with fans. It can be argued that fans may help to reduce the temperature of newly loaded produce rapidly to that of the cold store; all frozen products loaded into a cold storage chamber should be at the temperature of the chamber, and the question of cooling new produce should not arise. If frozen produce has become warmed during transit, it should be placed in a special small cooling room, or quarantine chamber, for 24 hours before putting it in the main store; this prevents any ill-effects on the rest of the contents of the store.

Dunnage or pallets should be used to keep produce off the floor of cold stores, so that air that has been warmed by heat passing through the floor can flow outwards to the walls and up to the coolers without passing through the produce. For the same reason, contents of the store should be kept away from walls; an air space should be left to allow warmed air to rise freely to the cooling coils.

When finned pipes are used, it is often possible to fit sufficient area of cooling surface entirely on the ceiling of the store; nevertheless, the pipes should be spread out over walls as well, in order to prevent heat that leaks through the walls from reaching the produce. Badly frosted finned pipes are often less efficient than badly frosted plain pipes, and rapid defrosting arrangements are needed to prevent severe dehydration of the product.

Any structural member that passes through the insulation must be well insulated; large amounts of heat can enter cold stores through pillars and supports. Cold store lights produce heat; it is therefore necessary to use lights of high efficiency and to switch them off when not required.

When a cold store is opened, cold air pours out, rather as if the store has been full of water, and warm air rushes in to take its place. Some well designed large stores have only one door, but if a store has more than one, then only one door should be open at any one time. Fans often increase the amount of heat that enters an open door. There are two main devices for minimizing the exchange of warm and cold air when the door is opened; the air lock and the air curtain.

Large refrigerated air locks are not common; the modern tendency is to provide a non-refrigerated air lock that is very small in size compared with the main store. The design is such that both inner and outer doors cannot be open at the same time. The exchange of air whenever a door is opened is then very small. Ideally, an alarm should ring if a door is open for too long.

In theory a curtain of warm air blown downwards from above the the doorway should prevent exchange of warm and cold air while maintaining a completely free passage. In practice the non-recirculating type in common use is difficult to adjust so that it works properly. It should be regarded as a useful aid when the door is opened at intervals but never as a device that will allow the door to be left open throughout a working day.

Products stacked near a door that is often used should be screened in some way. Not only will the product be warmed by the incoming air, but also by moisture that may condense on the product. Although water can condense on the product near the door, the warm air that has come into the store can increase the difference in temperature between product stacked elsewhere and the cooling pipes; dehydration can thus still go on in parts of the store remote from the door, even though water is condensing close to the door. Some of the condensation will form as frost on pipes near the door and reduce the effective cooling area; frequent defrosting may be necessary.

Hatches may be used to feed the product into or out of the store; this means that the larger doors need be opened much less frequently.

INSULATION

For the temperatures at which fish are kept in cold stores, only the 'solid' microcellular insulating materials can be recommended. These include granulated cork slab, expanded ebonite and foamed plastic slabs. Fibrous or loose granulated materials, or those with visible air spaces such as aluminium foil, are not quite so suitable because strong convection currents of air can be set up within the insulation with temperature differences of 80 degrees or more across the cold store wall.

Fibrous materials are convenient for filling in spaces where rigid slabs can be fitted only with difficulty, such as between frames and deck-beams of ships; it is then best to complete the wall thickness with rigid insulation over the toes of frames and beams.

Even with the 'solid' materials, air movement within the insulation can be considerable. Effective fitting, sealing and overlapping of slabs can do much to eliminate this. Since most insulating materials contract markedly as they get colder, precautions should be taken to ensure that the layers of insulation on the cold side of the wall do not crack or open at the joints.

Moisture will always tend to pass from the warm air outside a cold store wall to the colder air inside. There it will condense on the cold pipes and frozen produce. This is because the water vapour pressure outside is very much greater than the pressure of the moisture in the air within the store. But some of the water that passes through an unprotected cold store wall will freeze at the point in the insulation where the temperature is at 32°F. The insulation gradually becomes saturated with water and ice and its insulating value falls rapidly, until eventually the insulation may be completely destroyed. There is therefore a need for a vapour barrier. Any such barrier on the inside face of the wall will make matters worse, since water vapour that has managed to enter the insulation through the warm face cannot pass through and is trapped. What is required is a well constructed vapour barrier on the warm face of the insulation and as little obstruction as

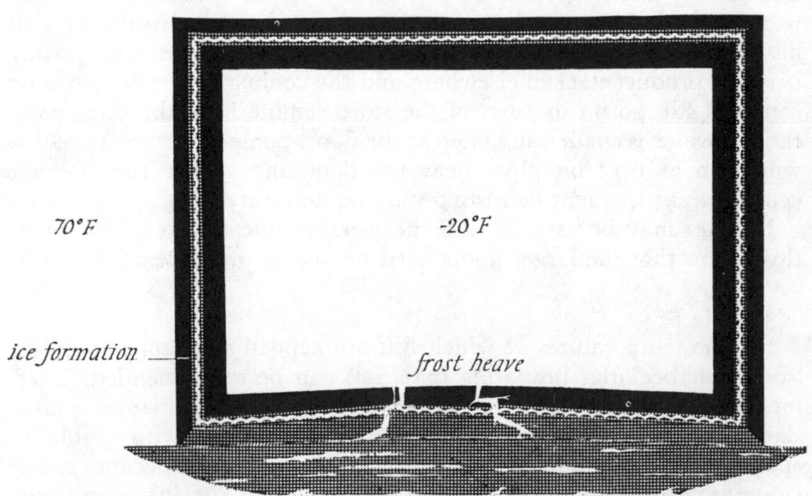

FIGURE 7·13 Ice formation in cold store insulation

possible on the cold side to the emergence of any water that has managed to get in. The insulation should then work efficiently for many years.

Insulation that is already waterlogged or full of ice can sometimes be dried out by improving the outside vapour barrier and circulating very dry, cold air through grooves cut in the inside face of the insulation. Unless such a system is hermetically sealed from the contents of the store, and this is difficult to do, the produce also will tend to dry rapidly.

When a cold store is built directly on the ground, the soil beneath is cooled. If the store is large enough and cold enough, the ground under the centre of the store floor may be permanently below 32°F. This may lead to gradual accumulation of ice that can distort the floor and impair the effectiveness of the insulation; it may even wreck the building. This phenomenon is known as frost heave. Thorough ventilation of the space under the floor, or underfloor heating, will prevent this. Thermostatically controlled electric heating is most usual for this purpose, either embedded in, or laid on top of, the structural concrete base.

Supports for pipes or linings should not be embedded in insulated walls if this is avoidable. Walls should be pierced as little as possible by materials that conduct heat well.

TEMPERATURE CONTROL

Temperature in small cold stores is usually controlled by a thermostat that starts and stops the compressor. The differential of such a thermostat is often rather wide as this is usually governed by the number of times an hour that the motor may be started. If electrical equipment above the minimum standard is specified, the thermostat differential can be narrowed to give closer control of store temperature.

In larger installations, where several stores or freezers may be served by the same machinery, the thermostat usually controls the flow of refrigerant to the cooling coils. In this case, there is no objection to making the thermostat differential as small as possible, but the pipe coils should be arranged to be self-draining to avoid large quantities of liquid refrigerant being trapped in the coils after the flow has been shut off.

In very large installations, an experienced refrigeration engineer can maintain very steady store temperatures over long periods by continuous modulation, and this can nowadays be achieved automatically.

The operator must keep records of air temperature in cold stores. A remote-reading thermometer or thermograph should be provided, accurate to within 1 deg F at minus 20°F, and sensitive to changes of a quarter of a degree; the scale should be large enough for these changes to be read. The instrument should respond rapidly and have the minimum of protection around the sensing element in the store. A daily chart is preferable to a weekly one. The sensing element should be placed 1 foot or more below roof grids, away from wall grids, and shielded to some extent from both. In large stores, two thermometers should be fitted, one with its element near and above the door, the other in a corner well away from the door. If the instrument near the door fails to record the effect of opening the door, then the instrument is too sluggish or insensitive. Ordinary glass thermometers, or better still thermocouples, can be used to make detailed checks of temperatures at other points in the store, so that the operator can learn to interpret the readings of the remote thermometers.

When good practice is observed the temperature records for the store should be sufficient evidence that avoidable deterioration has not taken place. Sometimes, however, the operator wants to know the temperature of the frozen fish on receipt or on delivery. This is done by boring a hole in a sample block and inserting a thermometer. Great care is needed if the answer is not to be seriously wrong.

The hole may be bored while the block is in the store. The thermometer should not be inserted and read inside the store because heat may flow in either direction between the centre of the fish and the air of the store, depending on which is the warmer. The thermometer will then give an incorrect reading above or below the true temperature of the block. The thermometer should be inserted in the block outside the store, where any error in a reading is known to be always above the true temperature. A large warm thermometer of the mercury-in-glass type is not suitable, as its heat capacity is too large. A very small, light thermometer is most suitable, such as a thermocouple of 26 swg wire, or a Torry resistance probe thermometer of less than 16 swg outside diameter. The block is now warming up rapidly, while the thermometer is cooling down towards block temperature. Thermometer readings will at first fall and then begin to rise again. The technique is to take readings every few seconds after insertion until the temperature begins to rise. Provided that the thermometer is a fairly good fit in the hole, the lowest reading taken will not be more than two or three degrees above the true temperature of the block.

Temperature-sensitive capsules have been made which change colour when the temperature rises above a predetermined point. These can be inserted in frozen packs to show whether storage has remained within the temperature range specified.

TRANSPORT OF FROZEN FISH

Frozen fish that are delivered to a fish frier, a school or a hospital may be eaten within a few hours of arrival, and no harm is done if they are partially thawed on receipt. The customer may prefer that fish thaw out on the journey, as when a fishmonger buys frozen smoked fish. But fish that is being transferred from one cold store to another, or from cold store to retail display cabinet, should warm up as little as possible *en route*. Some rise in temperature is inevitable, and this section is concerned with the problem of keeping the rise to a minimum.

Frozen fish should be carried in insulated containers. The most highly insulated container in common use is the A F container of British Railways. It has a capacity of 193 cubic feet and is roughly a cube in shape. It is insulated all round with 9 inches of expanded ebonite and the heat leak is stated to be only about 10 Btu an hour for every Fahrenheit degree of difference in temperature between outside and inside. If such a container were filled with a full load of $2\frac{1}{2}$ tons of frozen fish at 0°F and travelled for 24 hours with an outside temperature of about 60°F, the fish would arrive at a temperature of about plus 5°F. Not all of the fish would have been at 5°F on arrival; some would have been warmer than this, some cooler.

Many frozen fish products are carried by road, often in containers larger than the one described. Containers with capacities of 600 cubic feet are common, holding about 7 tons of frozen fish. Insulation

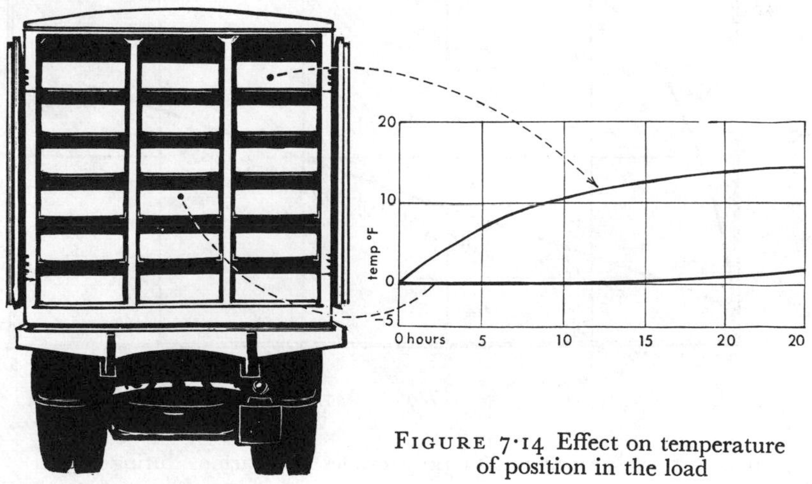

FIGURE 7·14 Effect on temperature of position in the load

is often only 3 inches thick, and rarely more than 5 inches. A 7-ton load of frozen fish at 0°F would suffer an average temperature rise of 10 degrees during a 24-hour journey in a container with 3 inches of insulation, with the air outside at 60°F. The same load carried in a container having 5 inches of insulation would rise only 6°F under the same conditions.

Packages at the edges and corners of the load will warm much more than those at the centre of the load, but the extent of this temperature difference is not commonly realized. Figure 7·14 shows the results of temperature measurements made across the middle of a load in the same container, though not on the same journey; here the temperature rise was almost entirely in the outer one-foot layer of the load, which in this case was packed firmly against the container wall without an airspace. Packages in the middle of the load warmed very little during the 24 hours. It must be remembered, however, that the outer one-foot layer represents a very considerable part of the total; for example, in a full container that measures 16 feet by 6 feet by 6 feet, it is a little over 60% of the total load.

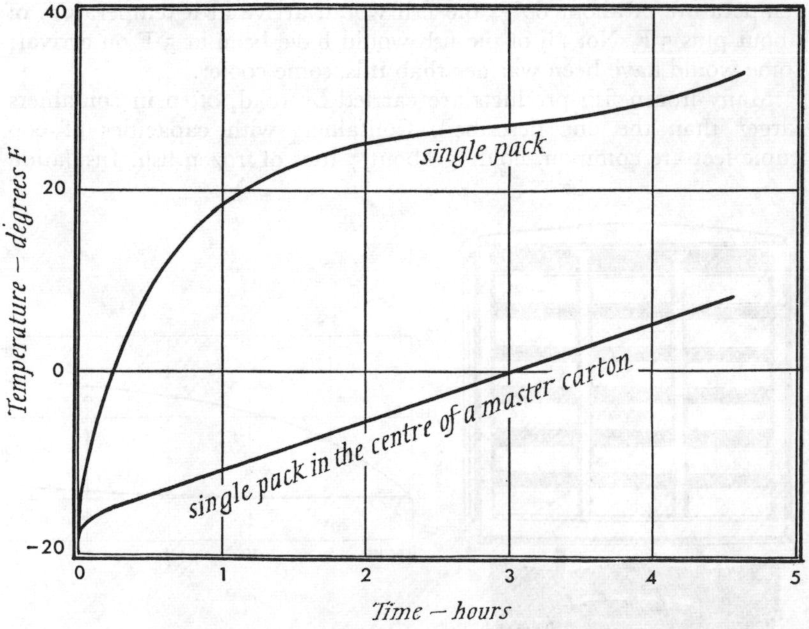

FIGURE 7·15 Warming of single packages and cartons during transit

Mechanical refrigeration or solid carbon dioxide cannot reduce this warming of the outer layer unless the cold air is allowed to circulate properly. Tests were made on a road container that had ample refrigeration equipment to deal with the likely heat leakage into it, but the cooling effect was confined almost entirely to the space above the load. Packages near the edges and corners became warmer during the journey, although those at the top were actually cooled a little. A clear space must be left between load and walls so that cold air or gas can circulate freely all round the load. If possible the cargo should be on pallets or dunnage so that a clear space is left beneath the load as well. Several European railways have an exchange system for standard pallets so that the pallet may stay with the load throughout the journey. A fan can be used with advantage to assist natural convection.

The use of liquid carbon dioxide, liquid nitrogen or liquid air for cooling containers is a recent development that shows some promise as a means of combating the heat leak. Again it is important to leave space for the cooling medium to circulate.

The container must be loaded and unloaded, and at these times the load can rise in temperature more than it does during the journey. Frozen fish standing on a loading bay often rise from 0°F to 15°F, and the damage is not obvious at the time, since fish at plus 15°F look and feel very much like fish at minus 15°F. The time taken to load a 6-ton container may vary between 20 minutes and 2½ hours; fish loaded in the shorter time rose only a degree or so above 0°F, the temperature of the cold store from which they had come, but fish that were delayed for the longer time had risen 10 degrees before they were loaded.

The size of a package affects the speed at which it warms; the smaller the pack the faster it warms. Figure 7·15 shows laboratory measurements made on a single consumer pack and on a carton of the same packs.

Faster handling in and out of cold stores and containers can do more to maintain the quality of frozen fish in transit than any improvement in container design.

8

Thawing

> O, that this too, too solid flesh would melt,
> Thaw and resolve itself into a dew!
>
> WILLIAM SHAKESPEARE

A CONSIDERABLE, and increasing, quantity of frozen fish is being used in Britain for further processing, and this frequently involves a thawing operation. This chapter discusses some of the factors that affect the rate of thawing, and also describes some of the methods at present in use in the British fish industry. It is impossible to say which is the best method, since this will depend upon many factors, including the size of the thawing operation in question, and whether rapid thawing is required. A further point of some importance is the time and labour required for cleaning the plant.

After thawing, fish should be kept chilled with ice until required, since it spoils in the same way as wet fish. It should not be allowed to remain in the thawer after thawing is completed. Partially thawed fish should not be stored for long periods, although it is sometimes advantageous, for example when there is a brief delay between thawing and filleting, to remove fish whilst it still contains some ice. This may be very necessary in conduction thawing, when there is such a wide variation in size that the small fish may be overheated while the large ones are still thawing.

Provided there is not too much ice left in the fish, and especially that the backbone is not encased in ice, an experienced filleter finds no difficulty in cutting satisfactory fillets from a partially frozen fish.

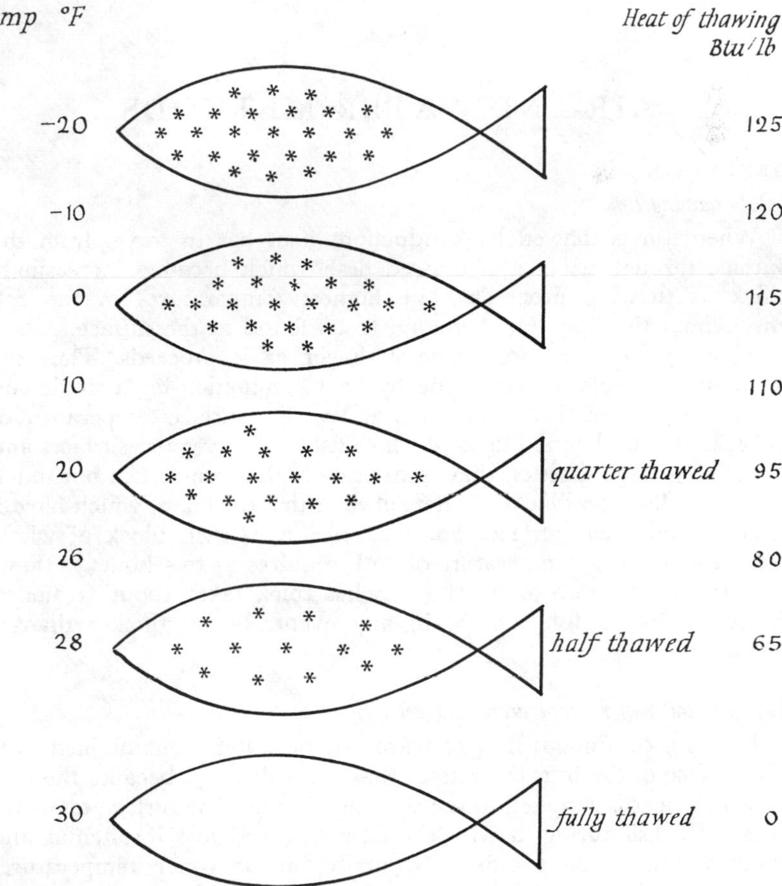

FIGURE 8·1 The amount of heat required for thawing

Filleting is, however, somewhat slower and this can be an important factor in processing.

Broadly, the recommended methods of thawing can be divided into two groups; those in which heat is conducted into the flesh from the surface, and those in which heat is generated throughout the flesh, as in the electrical methods of thawing.

The amount of heat required for thawing is considerable and equals the amount extracted in freezing. White fish flesh at minus 20°F requires about 125 Btu to thaw every pound, and fatty fish almost as much. The initial temperature of the frozen fish, if below plus 10°F, makes little difference to the heat required for thawing. This is shown in Figure 8·1.

AIR AND WATER METHODS

GENERAL

(1) *Thawing time*

When fish is thawed by conduction, heat has to travel from the surface through a layer of thawed flesh which becomes increasingly thick as thawing proceeds. The highest temperatures in the fish throughout thawing are, therefore, to be found at the surface.

Thawing becomes progressively slower as it proceeds. Thus the minimum possible thawing time by heat conduction for a single cod 4½ inches thick at the widest part, and with a surface temperature of 70°F, is about 4 hours. Blocks of whole fish, with irregular surfaces and containing large cavities, thaw more slowly than single fish but much more rapidly than blocks of fillets of the same thickness, which have a smooth, unbroken surface. For example, a 4½-inch block of whole fish with a surface temperature of 70°F requires 4½ to 5 hours to thaw, whereas a solid slab of fillets 4½ inches thick takes about 10 hours. Blocks of frozen fish should, in any event, be no thicker than is necessary.

(2) *Air and water temperature and velocity*

Unless a continuous flow of warm air or water is maintained over the surface of the fish, the rate of thawing will drop, because the rate at which heat is supplied to the surface will fall. The surface temperature of the fish during thawing should not exceed 70°F if softening and spoilage are to be avoided. Normally air or water temperatures between 60°F and 70°F are used to give rapid thawing rates and good quality products.

If the rate of flow of air or water is sufficient to maintain this surface temperature, then the rate of thawing is dependent upon the rate at which the flesh will conduct heat to the frozen deeper-lying parts. There is little advantage in raising the flow rate much above this point since no significant reduction in thawing time is obtained. These limiting flow rates are, for air, roughly 1200 feet a minute and, for water, roughly 1 foot a minute.

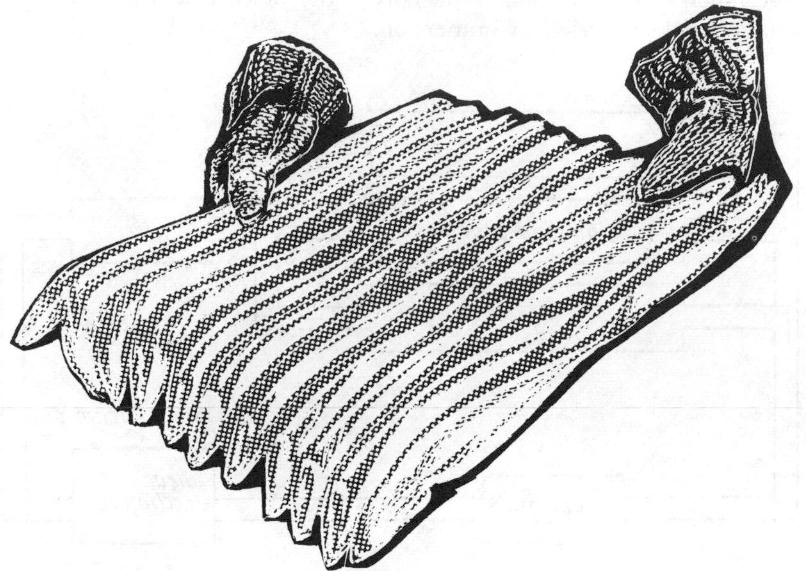

FIGURE 8·2 Block of frozen whole fish

SLOW THAWING

Slow thawing by, for example, laying out blocks of fish in the factory overnight, can produce acceptable results, although some evidence suggests that slightly better quality is obtained when thawing time is short. Blocks of fillets may lose some moisture and there may be some deterioration in the outer layers, especially since it is difficult to separate the fillets before thawing is complete.

Nevertheless where, for instance, the scale of operation is small, slow thawing may be the preferred method. The rate of thawing can be increased by using tap water in a water thawer, without heating, or a cheap batch air thawer can be made with low fan and heater capacity.

In thawers of large size, however, conditions for maximum thawing rate, consistent with high quality, should always be maintained to conserve space and save time. Slow thawing is not always very satisfactory for frozen herring intended for kippering since the fish becomes soft and more liable to damage in the splitting machine.

THAWING IN WATER

Whole fish, both white and fatty, thawed in water under controlled conditions are satisfactory. Fillets, however, may be waterlogged and lose much of their flavour, especially if the block is a thick one and requires a long period of immersion.

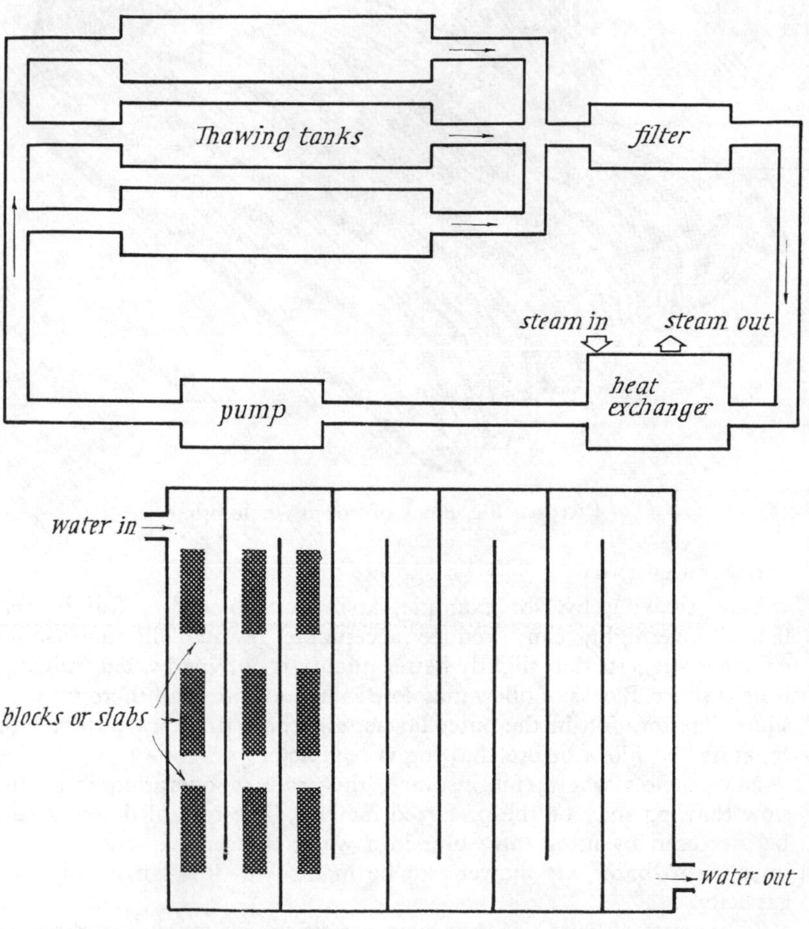

FIGURE 8·3 Batch thawing in water

Although recirculation of the water may be the best method to adopt in industrial continuous thawers for whole fish, there are problems of filtration of water contaminated by fish slime and blood which have not yet been solved. It would in any event be necessary to change the water periodically or add water continuously and fit some suitable overflow device.

A batch thawer presents fewer difficulties and two suitable arrangements are given in Figure 8·3. The fish would have to be placed in the tanks in wide wire-mesh baskets or other suitable hygienic containers.

THAWING IN AIR

(1) *Still air*

As already indicated, this is really practicable only on a small scale and under close supervision. Even here, however, overheating and drying are sometimes difficult to prevent. On a large scale the method is generally uneconomic because of the amount of handling involved, and the considerable area of floor or racking required. Stacking one layer above another is useless, because it further slows thawing by reducing the exposed surface.

Drainage must be provided to prevent juice from one block draining over blocks below. This is necessary both because of discoloration and possible spoilage of the lower blocks and also because the cold liquid may further slow the rate of thawing.

(2) *Keeping the air moist*

In air blast thawers it is essential to maintain humid conditions or serious drying occurs. Drying not only spoils the look of the fish, but also causes loss of weight and, hence, money. Furthermore, it takes longer to thaw fish in dry than in moist air.

The air passing over the fish should therefore be completely saturated with water vapour. It is better to use some means of humidification of the air than an elaborate system of water sprays for wetting the fish directly. The system of humidification adopted will depend upon the design of the thawer. One of the main reasons for humidifying the air is that moisture from the air condenses on the cold fish thereby giving up a very considerable quantity of latent heat. A water spray, on the other hand, gives up very much less heat to the fish as the water is cooled. Moreover, it is difficult to produce a spray in such a way that it will reach all exposed surfaces of the fish.

(3) *Industrial air blast thawers*

An air blast thawer using cheap fuel is a simple means of thawing fish on any required scale. Various types of air blast thawer could be built, both batch and continuous, all employing recirculated air. The Torry kiln can be adapted for batch thawing.

In a system with a heat exchanger in the airstream, the air can be humidified by a water spray after the heater (see Figure 8·4) with recirculation of the water. Another method would be to pass the air through a spray of heated water. It might also be feasible to introduce steam directly into the airstream, but this would depend on the practicability of producing large quantities of steam.

The thawing time of blocks of whole fish can be significantly reduced by separating the fish after an initial thawing period. This can be done by hand or mechanically.

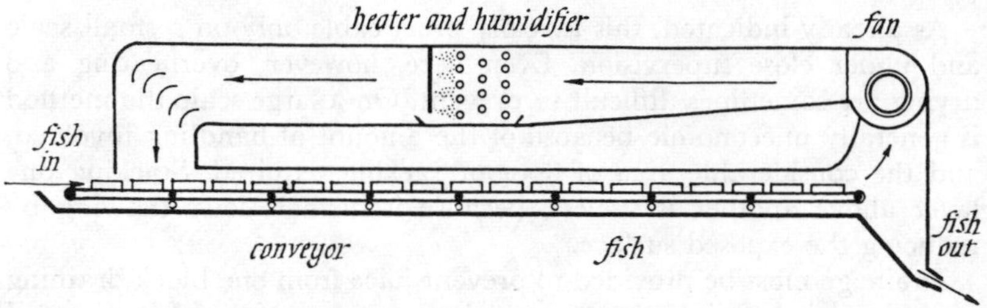

FIGURE 8·4 Continuous air blast thawer

(*a*) *Thawers using cross flow of air.* Continuous thawers using cross flow of air, that is at right angles to the direction of the fish along the conveyor, can be made compact by placing conveyors one above the other.

Powerful fans are necessary to maintain high air speeds and hence a high rate of thawing. In a thawer designed to thaw 1 ton of fish an hour, a recirculation rate of roughly 100 000 ft^3/min is required. The fan power required depends on the resistance offered by fish and heaters, but in any event will account for an appreciable fraction of the total energy costs of operation. Some heat is produced by the fans heating the air, so that not all the energy used by the fans is wasted. Indeed, under some conditions, overheating of the air might occur.

(*b*) *Thawers using parallel flow of air.* Continuous thawers using parallel flow of air, or, for that matter, water, have not yet been designed and operated, at least in Britain. The main advantage of such a system, in which the blocks would move in the same direction as the airstream, is the lower rate of recirculation that can be employed thus giving economies in power consumption and fan size. Better use could be made of the total available latent heat obtained when moisture condenses. A thawer with parallel flow and efficient humidification and designed to thaw 1 ton of fish an hour would require an air flow of only about 20 000 ft^3/min compared with the 100 000 ft^3/min of a cross flow thawer of the same capacity.

Figure 8·4 shows a parallel flow thawer in its simplest form. The blocks could be separated by vibration of the conveyer at the appropriate point. The disadvantage of this system is that the thawer would be rather long, but the total length could be reduced by reversing the direction of air flow in sections one above the other.

ELECTRICAL METHODS

The maximum speed of thawing in air or water is limited by the rate of conduction of heat through the fish. Where it is necessary to thaw fish very quickly, then methods other than those depending on normal heat conduction must be employed. At present the only practicable methods are those depending upon electrical heating, that is, electrical resistance heating, dielectric heating and microwave heating.

ELECTRICAL RESISTANCE THAWING

This method depends on the same principle as an electric fire. When an electric current flows along a wire or, indeed, through any material, then the material will become warm. It is difficult to produce heat in materials that do not conduct electricity easily, that is, are good insulators, but fairly good conductors, whether in the form of wires, slabs or, indeed, any other shape, are easily heated by the passage of electricity.

Provided the electrical properties of the material are uniform throughout, the flow of electricity will be uniform and the heat will be produced uniformly. This principle is used in a method of thawing fish recently developed at Torry and called electrical resistance thawing.

In practice, it is not always easy to obtain uniform conditions throughout a block of fish. For one thing, frozen fish at a low temperature does not conduct electricity very readily but as it becomes warmer its electrical resistance becomes progressively less: the change in resistance as the fish becomes warmer is so great that quite small differences in temperature between two nearby regions in a block result in most of the electric current passing through the slightly warmer region, so that whilst the cooler region develops practically

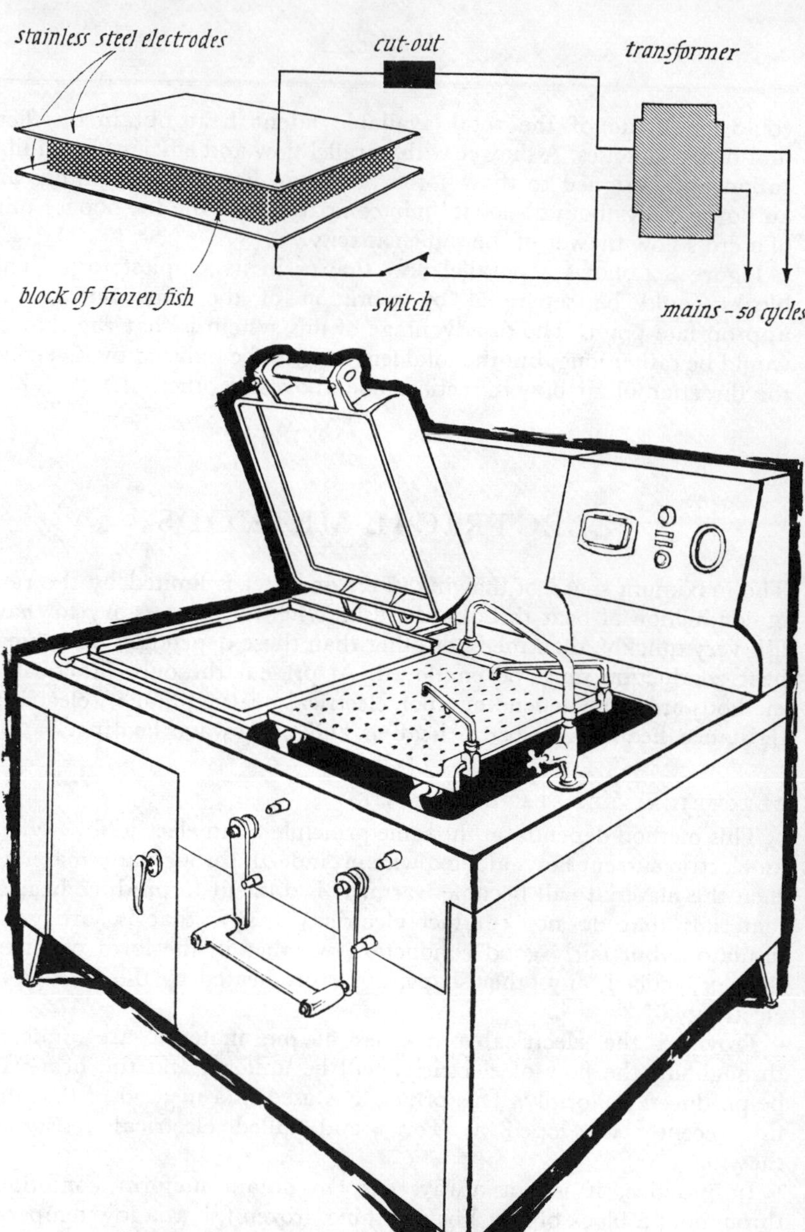

FIGURE 8 5 (*top*) Diagram of electrical resistance thawer; (*bottom*) Single station electrical resistance thawer

no heat at all the other is completely cooked in a short time. Another difficulty is that electricity does not always flow uniformly through all parts of a block of fish even though its temperature may be uniform.

However, the conditions for thawing fish uniformly have been found and a technique developed which is very easy to use in practice. It is necessary to have a block of regular shape and uniform thickness, for example a block of fillets; single fish cannot be treated by this method, and the procedure for blocks of whole fish is a little complicated, and is not at present recommended. The block must be at least five times as wide as it is thick and air spaces within the block should be kept to a minimum.

The temperature of the block is raised by immersing it in running water, usually tap water, or by placing it between two heated plates. Under these conditions the temperature of the block rises rapidly and about one-third of the total quantity of heat required to thaw the fish passes quickly into the block. At this stage the average temperature of the block is about 25°F and electricity will pass through it relatively easily. Both electrical and conduction heating can be employed together when heated plates are used.

The block is then sandwiched between two stainless steel plates, or electrodes (see Figure 8·5), and an alternating current, supplied by a transformer, is passed through the block. The voltage from the transformer depends on the type of block being thawed but it is always low enough to be completely safe. The part of the circuit containing the electrodes is fully isolated from the mains supply.

The current is allowed to flow for sufficient time to supply enough heat to complete the thawing. The current is switched off automatically when it reaches a pre-determined level, corresponding to that at which the fish is thawed. The actual level depends on the size of block, that is, its area and thickness, and whether the fish is whole or filleted.

Usually it is convenient to choose equal times of water immersion and electric heating. The *total* thawing time varies between about 20 minutes and 1 hour according to the type of block.

Equipment is now available commercially. A typical single station thawer is illustrated in Figure 8·5.

Electrical resistance thawing is two or three times faster than is possible by air or water thawing. Because of low initial cost and running cost it is a very economical method. It cannot, however, be used for irregularly shaped blocks because good electrical contact and a uniform flow of current cannot be obtained. The method is at present most suitable in catering, where suitable blocks are available. Semi-continuous methods for industrial use are currently being developed.

DIELECTRIC THAWING

Dielectric heating provides another means of thawing frozen fish; in the most usual arrangement, blocks of frozen fish are carried on an endless rubber conveyer belt which passes between one or more pairs of electrodes one above and the other below the conveyer. To each pair of electrodes is applied an alternating voltage of high frequency, typically 5000 volts and 80 million cycles a second. The blocks pass through without touching the electrodes (see Figure 8·6).

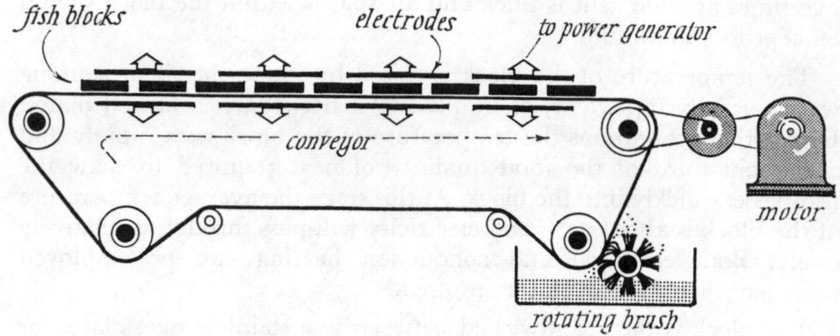

FIGURE 8·6 Diagram of dielectric thawer

The fish absorb electrical power which is turned into heat. The heat is produced almost uniformly throughout each block and in principle there is no limit to the thickness of the block that can be thawed.

In practice, however, there is a limit to the rate at which heat can be pumped into a given weight of fish. If a certain critical value is exceeded, then a condition known as *runaway heating* occurs. Regions that are slightly warmer than the rest of the block absorb more electrical power and so become warmer still. They then absorb still more power. Finally, cooking occurs in these relatively small regions, while the remainder of the block is hardly warmed at all. The power level at which runaway heating occurs determines the minimum thawing time and depends on the type of block being thawed. Nevertheless, this time is in all cases less than that required for electrical resistance heating and very much shorter than that required for air or water thawing. It ranges from about 10 minutes for blocks of whole herrings to about 40 or 50 minutes for blocks of sea-frozen whole white fish.

The actual thawing procedure varies slightly with the type of block. Blocks of herring and herring products can be thawed in one stage, that is by passage between only one pair of electrodes, but regular blocks of frozen white fish fillets and whole flat fish require a minimum of two stages. Whole fish and blocks of sea-frozen fish, which are somewhat irregular, must be passed between the electrodes immersed in water or ice in plastic trays since they are very liable to runaway heating. Even so they require a minimum of six stages of dielectric heating, and runaway heating of tails and belly flaps may sometimes occur. Since these portions are trimmed off in filleting, however, and the fillet is not affected, it is of little commercial importance.

One advantage of dielectric heating is that fish can be removed from the thawer at a uniform temperature, if necessary slightly below the the point at which fish is completely thawed. This may be desirable if there is delay between thawing and filleting. Sometimes, complete thawing is not required; thick blocks of whole kippers which take a long time to thaw in air, may be passed rapidly through a thawer so that they absorb in three minutes about one-fifth of the total heat needed to thaw them completely. The individual frozen kippers can then be easily separated for repacking before despatch. Blocks of whole flatfish can be similarly warmed so that the individual fish can be separated; these will then thaw relatively rapidly in air because they are thin. Frozen herring intended for kippering is found to give the highest yield of first quality split or boned herring, higher even than that obtained from similar unfrozen fish, if removed from the thawer at a temperature of about 32°F.

Boned herrings, herring fillets, and white fish fillets often tend to stick together when they are thawed, and it will be found helpful, therefore, to use blocks in which the cut surfaces of each layer are in contact with the skin surfaces of the layer below.

MICROWAVE THAWING

Mention should perhaps be made of a method of thawing fish with the type of radiation used in radar. This form of radiation is known as *microwave* and is produced at very high frequencies, typically at about three thousand million cycles a second. It is possible to force heat into fish by means of multiple beams of microwaves at a rate at least ten times that possible by dielectric heating, and thin pieces of fish have been thawed experimentally in a matter of seconds. The main drawback, apart from high cost, is that the strength of the beam is reduced as it goes deeper into the fish flesh so that the outer layers are warmed more than the inner ones. The method does not appear to offer any advantage over those already available.

HYBRID METHODS

Combination of two or more methods of thawing may be termed *hybrid* thawing. The effectiveness of hybrid methods should be judged on the basis of cost and convenience and, of course, quality.

For example, the combination of water thawing and electrical resistance thawing is very convenient, at least for small scale operation, and in practice only about one-third of the heat required for thawing is provided by electricity. The hybrid hot plate/electrical resistance thawing is even more convenient and is certainly quicker, but it uses at least twice as much electricity.

In dielectric thawing, where the process of producing high frequency electricity is only about 70% efficient, a very definite saving can be obtained by hybrid thawing. The best method is to heat the fish in water sufficiently to supply about one-third of the heat of thawing and then to complete the thawing by dielectric heating. A suitable arrangement is shown in Figure 8·7. Here cooling water at 90°F from the high-frequency generators flows into a tank containing the frozen blocks and is partly recirculated. The tank size and the temperature and rate at which water is recirculated is designed to maintain a sufficient supply of blocks to the dielectric thawer. A supplementary heater is necessary to warm the water during the start-up period before all the generators are in operation and to add to the heat from the cooling water if necessary during normal running. Thus the total high frequency power required to thaw a given weight of fish is only two-thirds of what is normally needed, most of the remaining third being supplied by heat from the generators which would otherwise be wasted.

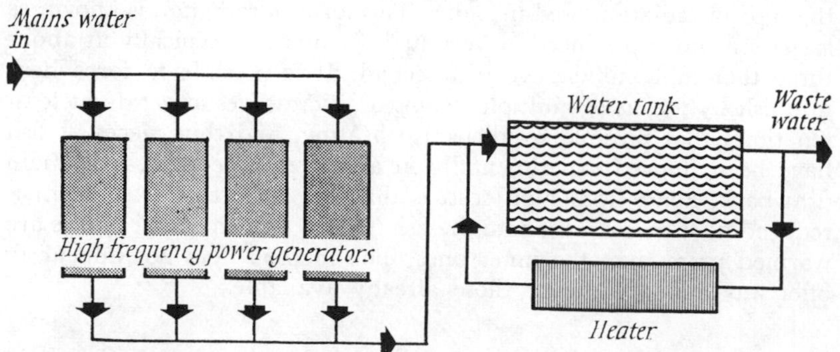

FIGURE 8·7 Hybrid thawer employing warm water and dielectric heating

9

Canning

> Tinned fish cannot be said to be as good . . . as fresh, but it is very useful for sauces, mayonnaises, etc., and answers well for little dishes where a good sauce can be introduced.
>
> MRS. BEETON

If the term preservation is limited to methods by which the fish is maintained, or can be restored to, its original state at the end of the storage period, then canning cannot be regarded as true preservation. During canning, heat processing alters the nature of the material, and new products are formed, whose nature may be changed further by various treatments of the fish before canning, or by the addition of substances such as sauces to the can. The main object of fish canning is to yield a product that may be stored for a considerable time, at the end of which it will still be interesting and safe to eat.

Fish deteriorate after death due to the action of enzymes and bacteria. Both enzymes and bacteria can be permanently inactivated by heat, and provided that re-infection does not occur, heat-processed fish should keep indefinitely. This is the principle on which the preservation of fish, and indeed all foods, by heat processing and canning is based.

In order to achieve a fish pack of satisfactory keeping quality the following conditions must be realized:

(1) the contents of the can must be sterile, that is, free from active bacteria and enzymes;

(2) the inside surface of the can must be resistant to attack by any part of the contents, and the outside surface must be resistant to corrosion under reasonable storage conditions;

(3) the lid of the can must be sealed to the body to prevent entry of air, water, and contaminants carried by these.

All these conditions are not always fully realized in practice, but nevertheless packs are prepared commercially where storage life can be confidently predicted to be at least two years.

In the following sections, some of the practices adopted by fish canners are described, and critically examined.

THE SUPPLY OF RAW MATERIAL

Herring, pilchards and sprats or brisling account for most of the fish canned in Britain; herring is the major species canned. In recent years small quantities of imported frozen salmon have been canned.

Much of the fish used for canning is caught in near waters. Herring are caught mainly by drift net, pilchards by drift net and ring net, and sprats and other small fish by ring net. The saying, 'The freshest fish make the best pack', has been proved sound by years of experience, and every effort should be made to ensure that the fish being processed are fresh. In recent years boxing of herring at sea in quarter-cran or six-to-the-cran aluminium boxes has been shown to have a markedly beneficial effect on the appearance and keeping quality of the fish. The herring tears readily in the belly region, especially if the fish were feeding heavily before capture. Boxing prevents excessive pressure on lower layers of fish by those above, which would otherwise lead to bursting of the belly wall. This, apart from causing the fish to look unsightly, accelerates deterioration by releasing digestive juices on to the surface of surrounding fish. These remarks apply also to pilchards.

If the cannery to which the fish are consigned is close to the port of landing, it is not necessary to treat them in any way on the quayside; where there is a delay between the landing of the fish and its arrival at the cannery, it is advisable to ice thoroughly to reduce the risk of spoilage. This is especially true in warm weather. It is common practice, at least in the herring industry, to sprinkle the fish with salt when they are transferred from the boat's boxes to those of the cannery, especially when this is situated some distance from the port. The value of this procedure has not been investigated scientifically, but since often only a small quantity of salt is used, it is doubtful whether any benefit results, and icing at this stage would provide a more efficient method of preservation. If the fish cannot be handled immediately on arrival at the canning factory, they should be kept well iced. Some canneries have chill stores, where the fish can be held at temperatures just above freezing point. It must be borne in mind, however, that if the fish are warm when placed in the store it may take hours for them to attain the desired temperature. Although it is bad practice to store herring for any length of time, it is sometimes necessary. A combination of ice and chill storage is probably the most efficient method of storing the fish for short periods of two or three days. Freezing and cold storage offer a means of storing fish over considerably longer periods without significant loss of quality, so regulating supplies. On thawing, this fish is quite suitable for canning.

TREATMENT OF THE FISH BEFORE CANNING

On arrival at the factory, the fish may be treated in one of several ways, and brief descriptions of the methods for preparing various products are given later. Certain procedures are common to most packs, however.

NOBBING

In the case of larger fish, such as herring and pilchards, the head and gut are removed, but not the roe or milt. This applies more especially to herring, which are caught mainly when the roes are maturing and of considerable size. Pilchards are captured mainly when the roes are small. This process of removing head and gut in one operation is called *nobbing*, and was previously carried out by hand, but nowadays the work is done entirely by machines capable of handling 180 fish a minute when fed by two women. The fish are placed in pockets on an endless belt which feeds them to a circular cutting wheel. This partially severs the head which, together with the gut, is then drawn away from the body by two corrugated rollers. If desired, the tails can be severed in the same operation. These nobbing machines are efficient but it is advisable at a subsequent stage to examine the fish to ensure complete removal of the gut. This is very important and is conveniently done by the packers.

FIGURE 9·1 Nobbing machine

WASHING AND DE-SCALING

The next operation is washing. Nobbing releases blood which must be removed because it causes brown staining in the processed fish. Washing also removes surface slime. Many types of washers are available. One, used for herring and sprats, is a revolving inclined latticed cylinder, the lower half of which is immersed in running water. The fish are fed in at one end and tumbled about by paddles as they pass along the cylinder. In another design, the fish are carried along a cylinder by means of an endless screw, at the same time being subjected to many jets of water. A third type of washer, which also removes scales, is used mainly in the pilchard industry. De-scaling is essential for pilchards, because the scales are hard and inedible.

Unfortunately they are difficult to remove and require harsh treatment. This machine consists of a circular tank with a rotating bottom fitted with four ribs, to ensure that the fish are moved around rapidly. Stationary baffle plates fitted to the walls cause the fish to tumble and rub both against one another and the walls of the washer. At the same time, sprays of water give the fish a thorough cleaning. This rough treatment tends to split the belly wall and should, therefore, be used only for pilchards. Herring sheds its scales easily; since these are

FIGURE 9·2 Herring washer

softer than those of the pilchard, they cause no trouble if they are eaten. In fact, some canners endeavour to retain them on the fish, as they act as a buffer between fish and the wall of the can, and so reduce skinning. Washing water, especially when not taken from the public supply, should be chlorinated.

BRINING

The fish are immersed in a concentrated solution of common salt for a predetermined length of time. Salt is absorbed by the flesh and imparts a desired flavour to the finished product in which a salt content of about 2% is acceptable. Brining has other beneficial effects, mainly that it toughens the skin of the fish; when unbrined fish are canned, much of the skin adheres to the can. Brining also brightens the appearance of the fish, removing any remaining slime.

Brining is at present a batch operation, and a saturated solution of salt is generally used. The salt should be pure and should not contain appreciable quantities of magnesium chloride, a common contaminant of unrefined salt. A salt containing much magnesium chloride increases the risk of struvite formation. *Struvite* is a chemical called magnesium ammonium phosphate which sometimes forms crystals resembling bits of glass in fish cans. To ensure uniform brining, the fish should be stirred frequently. The common practice of sprinkling each fresh batch of fish with a few scoops of salt is less satisfactory than frequent stirring. Brineometers should always be used to check the salt content of the brine at frequent intervals. These instruments are cheap and very simple to use, and ensure that the fish receive uniform treatment.

Brining times vary considerably, depending on various factors. Large fish require longer times than small fish. Fatty fish also require longer brining times than leaner fish. Fish that have been filleted or split for kippering require shorter times than nobbed fish. In practice, brining times varying from 15 minutes to 45 minutes are used, the actual time being decided by the experience of the operator.

Brines are used many times over, but it is important to remember that there are many bacteria that are able to multiply in high concentrations of salt, especially when the solution carries increasing quantities of oil, blood and pieces of gut. Thus brines that are kept in use for too long a period become a source of contamination and therefore they should be changed frequently. After brining, the fish are removed in wire mesh baskets and allowed to drain for several minutes before going to the packing tables.

PACKING

The choice of can used is governed mainly by the size of the fish being packed. In this country, the shallow, oval or oblong cans are generally preferred by canners to the tall, round cans, although the latter are used by some pilchard canners. Herring and pilchards are packed in oval cans of 14 oz or 7 oz capacity, depending on the size of the fish. Kippers are also packed in these cans, but sometimes 8 oz and 16 oz oval ones are used. Small herring, and sometimes large sprats, are packed in the 4 oz club can, while a series of oblong dingley cans, eighth, quarter, and half, with capacities of 2 oz, 4 oz, and 8 oz respectively, are extensively used for sprats and sild. Lacquers for fish cans are discussed later.

Cans should be washed immediately before filling as they rapidly accumulate dirt and dust during storage.

The design of can-filling lines varies considerably, but the aim should be to ensure a smooth supply of fish and cans to the packer while removing the filled cans as they are packed. Herring and pilchards have to be weighed out before packing, but this is not necessary for sprats or sild as their small, even size ensures even packing, and therefore uniform weight, in each can. In one design, women work in pairs at individual tables, to which supplies of fish and cans are carried by other workers. One woman weighs the required quantity of fish, while the other packs them belly upwards, head to tail, into the can. The

FIGURE 9·3 Types of can

can is then placed on a conveyer belt which carries it under the sauce dispenser. Weighing out the requisite weight of fish to within narrow limits is a skilled operation, especially when the fish are large. With very large herring, portions of fish are sometimes added to make up the weight, but this is bad practice and spoils the appearance of the pack. Only small herring should be canned. In another type of line empty cans are fed along a central belt, while two other belts, one on either side, move in the same direction. At one end of the line women weigh out fish on side tables and place them on the side belts. Lower down, the fish and empty cans are removed, the fish packed, and the filled cans replaced on the centre belt. Sauce or oil may be dispensed either before or after the cans are packed. This method is more flexible than the first one. If necessary, more women can weigh than pack or vice versa, and differences in the speeds of weighing and packing can be avoided. Occasional cans are removed from the packing line and their weight checked. Cans persistently outside the allowed tolerance suggest that one or more of the weighers is unreliable and their work should be checked. With sprats, where weighing is not necessary, trays of fish are taken to the packers sitting at long tables, who are also supplied with cans. Oil or sauce is dispensed either before or after packing, and sometimes half of the oil may be added before and half after packing the cans.

FIGURE 9·4 Packing line

EXHAUSTING

Small cans, such as the quarter dingley, once filled can be closed and processed directly. Larger cans, however, have to go through one further important operation before being closed, namely, exhausting. The object of exhausting is to produce a partial vacuum in the headspace of the can which will persist after the can is heat processed and cooled. The headspace of a can is that part which is not filled either with solid or liquid. When a can is open, the headspace is filled with a mixture of air and water vapour, and if the lid is then seamed on, the pressure in the headspace will be equal to that of the outside atmosphere.

If the closed can is now heated, the pressure in the headspace will increase as the air expands, the water evaporates to steam and solid and liquid contents of the can expand. Gases are also given off by the fish flesh which add slightly to the total pressure.

This internal pressure forces the ends of the can outwards and may lead to swelling and distortion. As a result, the seam of the can may be strained and may leak even after it is cooled. Because small cans are mechanically stronger than large ones they are unlikely to be damaged in this way. If, while the can is still hot, a small hole is bored in the lid, air, water vapour and other gases will blow out until the pressure inside is again equal to that outside. If this hole is sealed up immediately while the can is still hot, and the can is allowed to cool, the headspace pressure will drop *below* that of the outside atmosphere, because air that has blown out while the can is hot cannot be replaced. This difference in pressure between the can headspace and the atmosphere is called a partial vacuum.

REASONS FOR WANTING A VACUUM

There are two good reasons for wanting a partial vacuum in a can:

(1) As described above, the positive pressure in a can increases considerably during heat processing. Any partial vacuum present helps to reduce this pressure, and in doing so reduces the chance of distortion and damage to the seam;

(2) After processing and cooling, the vacuum causes the ends of the can to be collapsed, that is concave. Cans with swollen ends, known as blown cans, are immediately suspected of having become spoiled, as will be explained later. Cans not evacuated may appear to be blown, but in fact are still quite wholesome. This occurs when the headspace of the can expands due to considerable rise in temperature over that of packing, or due to the pack being exposed to lower atmospheric pressures, that is at higher altitudes. The vacuum prevents this occurring.

Another reason often given for requiring a vacuum is that this reduces the oxygen content of the headspace and therefore the extent of internal corrosion. Experiments on canned herring have shown that variations in the oxygen content of the headspace do not affect the rate of corrosion. As explained later, the flesh itself can supply oxygen for this purpose.

METHODS OF OBTAINING A VACUUM

The method mentioned in detail above by which the hot closed can is punctured and then re-sealed was the method initially used for achieving a vacuum in cans, and was known as *broguing*. Nowadays a different method is used, more suited to high-speed production lines.

(1) *Heat exhausting*. After packing, the lids of the cans are lightly clipped or clinched on to the body in such a way as to allow free passage of gases and vapours out of the can. The can and contents are then heated by passage through an exhaust box. Exhaust boxes vary somewhat in design but consist basically of a long, well-lagged box which is heated by steam and the temperature maintained at about 200°F to 208°F. The can passes through on a conveyor and is usually subjected to sprays of steam which play on to the sides. It remains in the box for a time determined by the speed of the conveyor. A vacuum of 8 inches should be aimed at, and for 7-oz and 14-oz oval cans, this usually requires 20–30 minutes. Where cans are stacked on top of each other, the time should be increased. The lid is seamed on the can immediately it emerges from the exhaust box, otherwise the contents will cool and no vacuum result. Here, as in broguing, the vacuum is produced by sealing the can while the contents are hot.

(2) *Vacuum seaming*. Another method of achieving a vacuum in the can is by the use of vacuum seamers. These machines close the can and while doing so draw air out of them and so create a vacuum. In British fish canneries, however, these seamers have been largely replaced by the vacuum box, together with a normal seamer. Vacuum seamers were found to be difficult to maintain and slow in operation, handling only 15 cans a minute compared to 60 cans on a normal seamer. High vacuum of the order of 12–15 inches mercury was achieved by this method.

Other methods, such as injecting steam into the can before seaming, or filling the can with hot fish or hot sauce, are used abroad.

CLOSING THE CAN

All fish cans prepared in this country are closed by the double-seaming method, and the operation is usually called *seaming*. A seal must be achieved that will prevent passage of contaminating material, carried either in air or water, into the can after it has been sterilized. Proper care and maintenance of seaming equipment is vital, and its performance should be checked at frequent intervals throughout the working day.

Only a brief description of the seaming operation can be given here. Figure 9·5 shows the main stages in the manufacture of the seam. The can, with lid placed or clinched on top, stands on a base plate which is raised so that a chuck fits into the countersunk portion of the lid (Figure 9·5*a*), holding both in position. The first operation roll is brought to bear on the unformed seam as in Figure 9·5*b* and forms a cover hook interlocking loosely with a body hook. This is followed by a second operation roll of a different shaped profile (Figure 9·5*c*) which flattens the seam and presses the layers of metal against each other. The sealing compound, usually a silicone rubber preparation, renders the seam tight. In the case of seamers closing round cans, the can and cover revolve, while the rolls remain stationary. Irregular cans, for example rectangular or oval ones, are kept stationary and the rolls travel around the seam.

Before the cans are heat processed they are washed to remove adhering solid and liquid. If this is left on it hardens and is more difficult to remove. After seaming, the cans pass through a washer where they are sprayed with hot water. Alternatively they pass through a bath of hot water to which detergent may be added to increase washing efficiency.

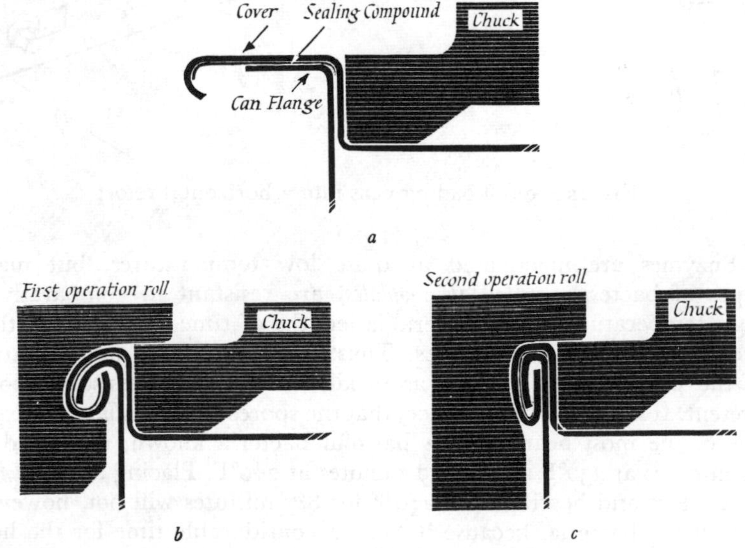

FIGURE 9·5 Closing the can

HEAT PROCESSING

The two objects of processing are to cook the fish in the can, and to inactivate all bacteria and enzymes present. In practice processing times and temperatures necessary to sterilize the pack are more than adequate to cook it also.

FIGURE 9·6 Loading cans into a horizontal retort

Enzymes are inactivated at quite low temperatures, but many types of bacteria, called *thermophiles,* are resistant to comparatively high temperatures for considerable lengths of time, especially if they are capable of forming spores. Thus the temperature and duration of the process must be such as to kill the most heat-resistant spores present. It is known, for instance, that the spores of *Clostridium botulinum,* one of the most heat-resistant harmful bacteria known, are killed in 32 minutes at 230°F and in 8·7 minutes at 240°F. Placing a can of fish in a retort and heating it at 240°F for 8·7 minutes will not, however, kill all the bacteria, because it takes a considerable time for the heat to penetrate the solid flesh of the fish to the centre of the can. Thus to ensure complete destruction of the spores, the centre of the pack must be held at 240°F for nearly 9 minutes. This statement is not absolutely correct as considerable numbers of spores will have been destroyed at temperatures from 200°F upwards, but is adequate for this discussion. Thus it is necessary to know the rate of heat penetration into the pack during retorting, and this is measured by sealing thermocouples into the filled cans, and following the change in temperature at the centre of the pack. Figure 9·7 shows the curve obtained with a 14-oz oval can

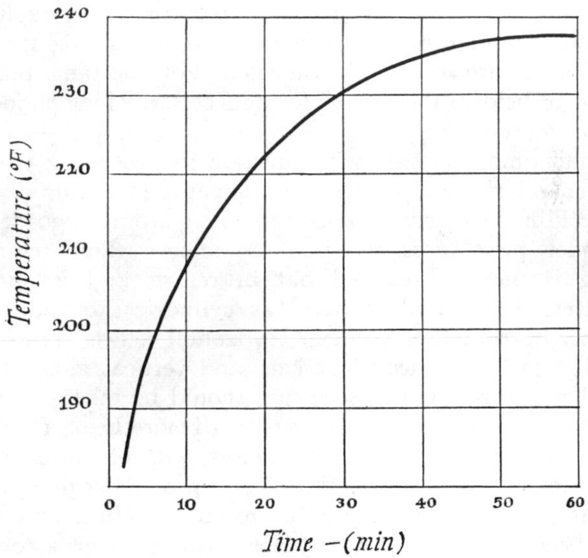

FIGURE 9·7 Temperature at the centre of a 14 oz oval can of herring during heat processing

of herring in tomato sauce, from which it will be seen that even after 60 minutes the temperature at the centre is only gradually approaching that of the retort. A temperature of 230°F at the centre is reached after 30 minutes and as stated previously a further 32 minutes would suffice to destroy *Cl. botulinum* spores, giving a total process of 62 minutes at 230°F. In practice a time of 60 minutes at 240°F is used, ensuring a safe margin. In the case of new packs, the adequacy or otherwise of certain processing conditions can be checked by inoculating cans with heat-resistant bacterial spores, and after processing, examining the can contents by microbiological methods to ensure that they are sterile.

It is important to remember that a process that is adequate under normal conditions may prove inadequate if, for some reason, the fish carry an unusually heavy bacterial load. It has just been stated that spores of *Clostridium botulinum* are killed in 8·7 minutes at 240°F. This is only an average figure, however, and in a given number of spores there will be some that are less heat resistant and some more heat resistant. The more spores present initially, the greater will be the likelihood of a few spores being present that will be sufficiently resistant to survive the process. Thus a given heat process must not be regarded as certain to give sterile packs under all conditions. This only holds provided that an adequate safeguard is taken regarding quality of raw material and standard of factory hygiene, thus ensuring that the fish being processed are not heavily infected. For the same reasons, fish should not be held in the factory for unnecessarily long periods at any stage prior to processing.

Processing time decreases with increase in processing temperature and in general shorter processes at higher temperatures are to be preferred. Fish packs are usually processed around 230–240°F, and steam under pressure is required to achieve these temperatures. Processing is therefore carried out in retorts and is often called *retorting*. Retorts may be box-shaped or cylindrical, cylindrical retorts being either horizontal or vertical. Horizontal retorts are loaded with cans stacked in large wheeled baskets, and vertical retorts loaded by hoists. When the retort is closed, care should be taken to replace all air inside with steam by adequate venting before being finally sealed, otherwise the temperature in the retort will be lower than that expected from the pressure reading. Total processing time should not include the time required for pressure to build up in the retort.

As has been explained earlier, a closed can develops a considerable internal pressure when heated which is only partly counteracted by the partial vacuum in the can. In the retort, however, it is largely balanced by the pressure of the steam, but if at the end of the retorting period the steam is shut off and the pressure suddenly released, the net pressure in the can will suddenly increase, causing further strain on the seam. It is common practice in this country to overcome this danger, at least partially, by releasing the pressure in the retort slowly over a period of about 15 minutes so that the extra strain is taken up slowly by the can. The cans are then either cooled under water sprinklers or allowed to cool in the air. The best method to avoid straining the cans is to use *pressure cooling*. Here, steam pressure is replaced by air pressure, and the cans, still in the retort, are cooled by water. When these are sufficiently cool, the air pressure is released.

When cans are cooled by water either inside or outside the retort, care should be taken to ensure that the water used is pure. As the cans are cooled, pressure inside changes rapidly and under such conditions even a correctly made seam may allow the passage of a trace of water which, if it is contaminated, may give rise to spoilage during subsequent storage. Chlorination of cooling water is therefore strongly recommended.

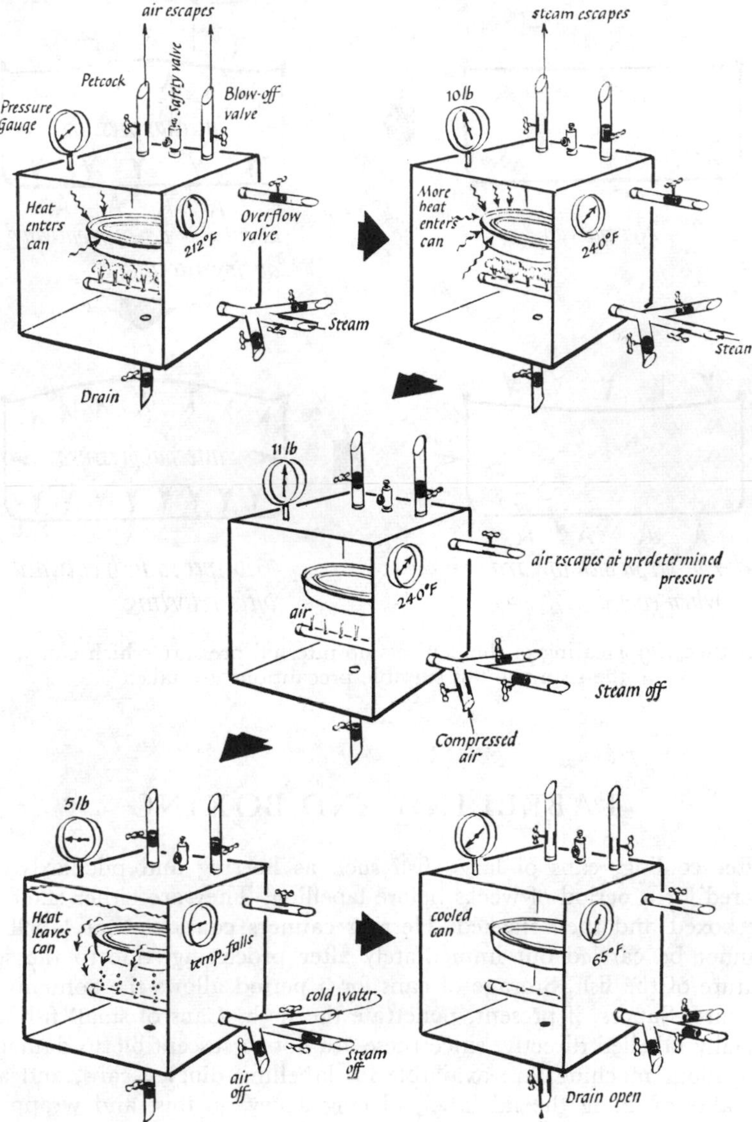

FIGURE 9·8 Operation of a retort with pressure cooling

Where cans are cooled by water, the outside surface of the can is not cooled below about 90°F. This ensures that residual water on the cans will evaporate rapidly and not lead to corrosion.

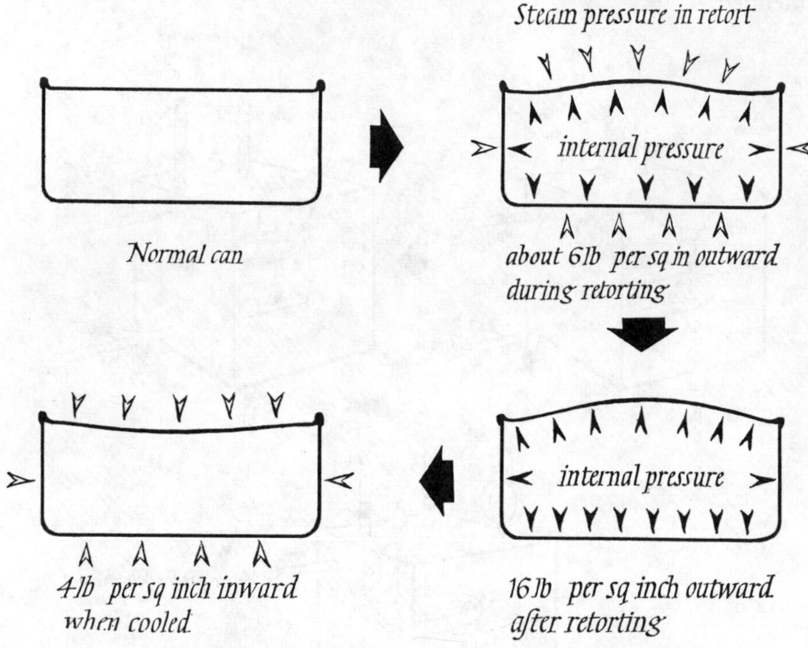

FIGURE 9·9 Heating of the can sets up internal pressure which can strain the seams unless suitable precautions are taken

LABELLING AND BOXING

After cooling, cans of large fish such as herring and pilchards are stored for a period of weeks before labelling. They are either stacked, or boxed and then stacked. Herring canners consider that labelling cannot be carried out immediately after processing, due to the soft nature of the fish. Storage of cans for a period allows the contents to mature. Sauces, if present, penetrate the flesh. Cans of small fish are usually labelled directly, since these are not so susceptible to damage. Ingenious machines are available for labelling dingley cans, and are capable of fixing the lid label, placing a key on this, and wrapping the whole in a greaseproof wrapper. Larger cans may have the top label pasted on by hand, and the side label by machine. Many canners label by hand, making use of female labour during off-season periods. In recent years, the introduction of decorated lids has cut down the use of paper labels.

CAUSES OF SPOILAGE IN CANNED FISH

There are two main types of spoilage in canned fish, bacterial and chemical. Bacterial spoilage is due either to underprocessing, or to subsequent infection of the can contents by infected cooling water or bad seaming. Chemical spoilage is caused by action on the can wall of various chemicals present in the fish or, quite frequently, by acid sauces present in the can.

DETECTION OF NON-STERILE PACKS

Bacterial spoilage of canned fish products may be divided into two types. In one, caused mainly by anaerobic spore-forming bacteria, swelling of the can occurs due to the production of gases such as hydrogen or carbon dioxide. Slight positive pressure inside the can is demonstrated by striking it against a solid object, when one end bulges out. This is called a *flipper*, and the end returns to its original position under slight external pressure. Greater pressure in the can causes the formation of a *springer*. Here, one end of the can is permanently bulging, and on pushing it back in, the other end bulges. In a *soft swell*, both ends are bulging, but can be moved slightly by thumb pressure. Finally, in the *hard swell* the internal pressure is so great that the bulging ends cannot be moved. Cans behaving in this manner are always assumed to be spoiling, although this behaviour, especially in its earlier stages, may be caused by too low a vacuum in the can, or by chemical spoilage. Where bacterial spoilage of this type occurs, the contents of the can disintegrate with the production of foul odours.

In the other main type of spoilage, bacteria produce *flat sours*. No gas is produced in the cans, but on opening, the contents are found to be inedible with an unpleasant flavour and odour.

Detection of non-sterile packs is carried out by microbiological methods. A sample of cans from each batch is taken and part of this incubated at 99·5°F and part at 131°F. If live bacteria are present, they will multiply under these favourable conditions. After one or two weeks, the cans are opened and examined for live bacteria.

PREVENTION OF BACTERIAL SPOILAGE

Microbiological examination serves to detect but not prevent bacterial spoilage. Indeed, it will not detect those cans which, due to reasons such as bad seaming, may at some future date become infected. Furthermore, bacteriological examination of cans is a lengthy and expensive procedure, and because of this only a limited number of cans out of a given batch can be examined. As a result, the chances are quite high of a significant proportion of infected cans in a batch passing inspection.

The best method of preventing spoilage is by careful attention to the whole canning procedure, that is *good process control*. Particular attention should be given to the quality of raw material, and purity of the washing and cooling water. The exhaust box and retort should be properly instrumented so as to ensure accurate knowledge of the temperatures and pressures involved. Seaming equipment should be well maintained and its performance checked by frequent examination of seams. Factory hygiene should be maintained at a high level. Work in South Africa has shown that rigid inspection of process control is a quicker, safer, and cheaper method of ensuring microbiologically sound packs than conventional microbiological testing.

CHEMICAL SPOILAGE

Cans used in this country are manufactured from tinplate, which consists of mild steel sheet, about 1/100 inch thick, coated with a very thin layer of tin. For many canned goods, the tin is sufficiently unreactive to be used without any further protection, but most fish products, especially those containing acid sauces, require cans with a second protective coat. This is a lacquer, which is applied to the tinplate before the can is formed. In this country lacquers prepared from polymerized fish oil are normally used. Most cans used in fish packing are seamless and are prepared by *deep drawing,* that is, the body of the can is stamped out of a sheet. This imposes considerable strain on the lacquer, as shown in Figure 9·10 and often leads to cracking and exposure of the underlying tin. Recently, herring canners have tried synthetic epoxy resin lacquers, with encouraging results.

Cans that contain acid sauces are particularly vulnerable to attack. If there is a flaw in the lacquer, the acid attacks the tin and may penetrate to the steel. Gaseous hydrogen is formed and this may cause the can to swell. In non-acid packs considerable solution of tin and steel may occur without the formation of hydrogen. The mechanism of this reaction is not fully understood, but a chemical compound in the fish, trimethylamine oxide, plays an important part. Some fish products, more especially crustaceans such as crab, lobster and shrimps, form volatile sulphur compounds during spoilage which cause dark sulphur-staining due to the formation of iron or tin sulphides, or both. When these products are packed, special lacquers

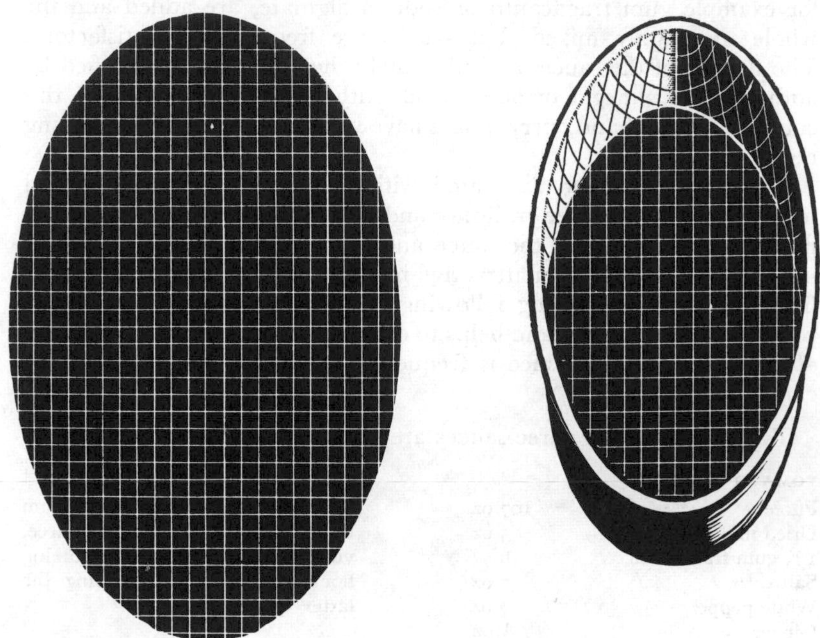

FIGURE 9·10 Deep drawn can showing strain on the lacquer

are required. One such lacquer, called a pigmented lacquer, contains zinc oxide, which reacts with sulphur compounds to give a white instead of black stain, and is therefore not offensive. In extreme cases, the cans may be lined with parchment.

SAUCES

The object of incorporating a sauce into a fish pack is to modify the flavour and improve the appearance of the finished product. The sauces most commonly used in this country are based on tomato purée. Many canners dilute the purée, which contains about 25%

solids, with an equal portion of water and use this as it stands. Purée must not be described on the label as sauce, but as purée. The addition of small quantities of oil such as cottonseed or groundnut into the sauce is considered to improve its appearance, but this introduces the problem of mixing water and oil. Materials called stabilizing agents, for example gum tragacanth or sodium alginate, are added and the whole thoroughly mixed, but results are frequently unsatisfactory. The flavour of the sauce, and ultimately the fish, can be modified by adding spices, vinegar or acetic acid. Although not often used in this country, mustard and curry sauces have been shown to give interesting results with herring.

One major problem associated with the use of sauces in canned fish is the dilution caused by liquor and oil exuding from the fish during processing. This dilutes the sauce and often discolours it, and results in the sauce becoming unattractive in appearance. Removal of liquor from the can by draining following exhausting reduces the amount present after processing and helps to conserve the attractive appearance of the sauce. This practice is frequently carried out in the pilchard canning industry.

Typical recipes for three sauces are given below:

TOMATO

Purée	107 oz	Mix dry ingredients with gum tragacanth solution, adding purée, vinegar and water. After mixing homogenize with oil, adding the latter gradually.
Dried milk	5 oz	
1% gum tragacanth	36 oz	
Salt	5 oz	
White pepper	½ oz	
Ginger	½ oz	
Spirit vinegar (6%)	2½ oz	
Water	1 pint	
Cottonseed oil	2 pints	

MUSTARD

Spirit vinegar	3½ pints	Mix the dry ingredients and vinegar, and mix to a thin paste. Then homogenize with oil.
Ground mustard seed	15 oz	
Turmeric powder	1 oz	
White pepper	1 oz	
Rusk	7½ oz	
Edible oil	¼ pint	
1% gum tragacanth solution	36 oz	

CURRY SAUCE

Spirit vinegar	5½ pints	Prepare as for mustard sauce.
Curry powder	15 oz	
White pepper	1 oz	
Rusk	15 oz	
Edible oil	¼ pint	

MAIN TYPES OF BRITISH CANNED FISH

The principles which apply to all fish packs having been discussed, brief mention will be made of specific canning methods used in this country. Roughly 8000 tons of herring, 1000 tons of pilchards, and 2000 tons of sprats were canned in this country in 1963.

HERRING

On receipt at the factory, the fish are nobbed by machine, washed with water to clean out the belly cavity and brined in a 100° brineometer brine for 10–30 minutes. The fish are then hand packed into oval cans. 14-oz cans contain 14 oz fish if packed alone, or 12-12½ oz

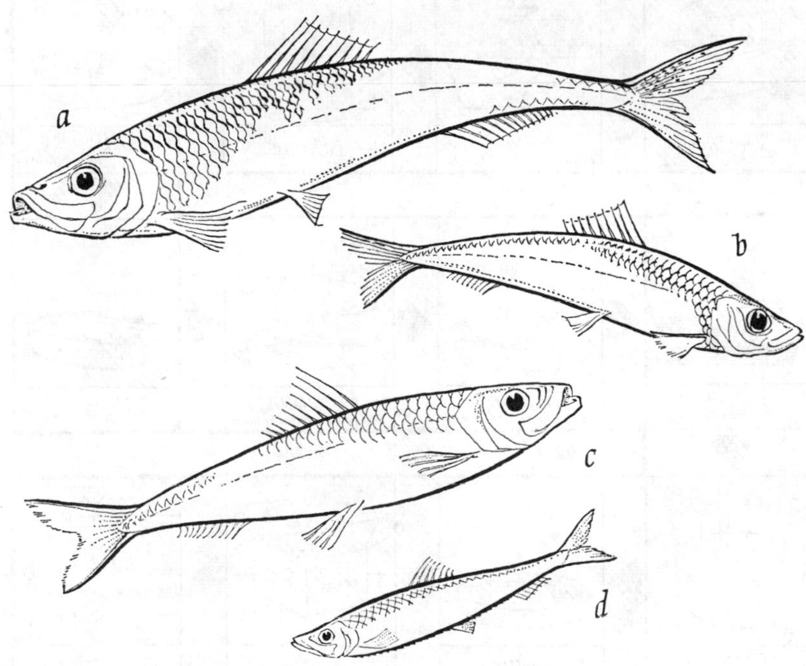

FIGURE 9·11 Main types of British canned fish: (*a*) herring; (*b*) young herring or sild; (*c*) pilchard; (*d*) sprat or brisling

fish and $1\frac{1}{2}$–2 oz sauce; 7-oz cans contain 7 oz fish or 6 oz fish and 1 oz sauce. After adding sauce the lids are clinched on, and the cans exhausted for 15–30 minutes at 200–208°F. A vacuum of at least 6 inches is required. After seaming, the cans are washed, and processed. 14-oz cans require 60 minutes at 240°F, 10 lb/in^2 steam pressure; 7-oz cans 55 minutes at 240°F. They are stored for a period of about one month and then labelled.

Most of the herring are packed in a tomato sauce. Small immature herring taken from the Moray Firth and sometimes from the west coast of Scotland are popular for preparing smaller packs. The preparation of these herring is similar to that of the larger fish.

The procedure outlined above is designed to produce a good quality, yet cheap, pack. Certain modifications may be introduced when the cost of the product is not of first importance. During cooking, considerable quantities of liquor are exuded from the fish. This liquor

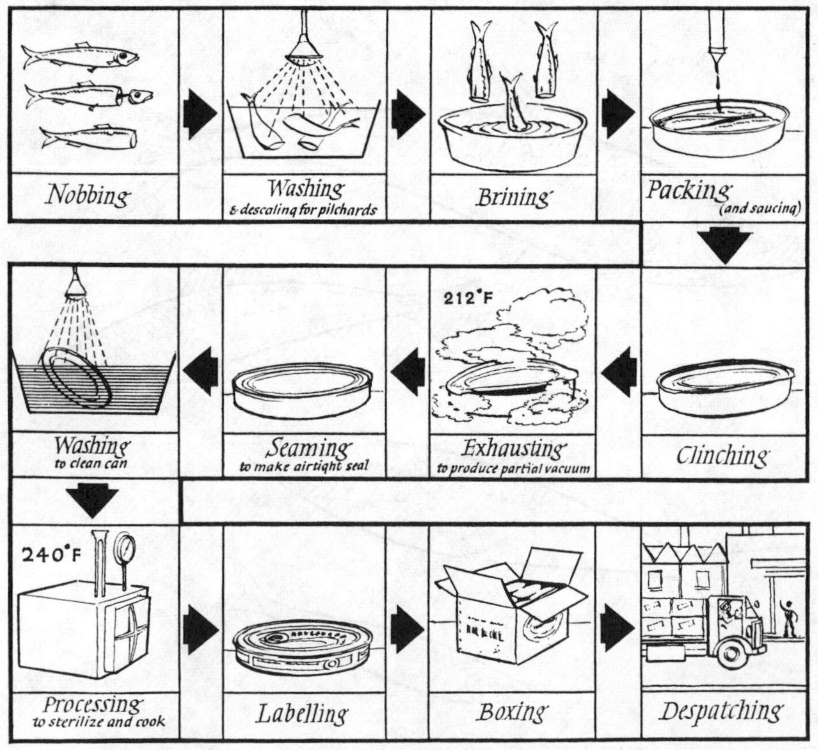

FIGURE 9·12 Sequence of operations in canning of herring

dilutes any added sauce and detracts from its appearance. The fish also tends to break up during rough handling of the finished product, especially in warmer climates. Much of this liquor can be removed before the can is closed if an extra operation is introduced into the line. After packing the fish into the can, it is passed through a heating chamber similar to the exhaust box and heated for about 20 minutes at 208°F. On emerging, the cans are mechanically inverted and drained; sauce is then added, the lid clinched on, and the can given a short exhaust of about 5 minutes before seaming.

Filleting as opposed to nobbing has certain advantages, the main one being the assurance of a perfectly clean belly cavity. Fillets require a shorter brining time compared to nobbed fish. It is preferable to fold the fillet to resemble the unfilleted fish before packing; this prevents excessive sticking of the fleshy surfaces to one another.

KIPPERS

The fish are split in the usual way and the head and tail also removed and are brined in a 70° brine for 10 minutes. Smoking is lighter than for normal kippers, because the colour darkens considerably during processing. In the old-fashioned kilns it was customary to smoke for about six hours. A smoking time of three hours at 80°F and 50–55% relative humidity is sufficient in the Torry kiln. This involves a weight loss of about 10–12%. Kippers are packed in 14-oz or 16-oz oval cans. An oval of parchment paper is placed between each kipper to prevent sticking. The cans are exhausted and processed as for normal herring. Due to the drying occurring in the kippering process, less liquor is liberated in the can during the processing, while the attractiveness of the product is enhanced by the softness of the bones.

PILCHARDS

All the pilchards canned in Britain are taken by drift net off the Cornish coast, the season lasting from July to March, with a peak in the autumn.

The procedure used is very similar to that for herring, most of the pack being prepared in tomato sauce. As has been mentioned elsewhere, the scales of pilchards have to be removed and this requires harsh treatment during washing. Most of the fish are packed in oval cans, a small proportion being in AI tall cans. Pre-cooking and draining of the fish in the cans is carried out by some canners who consider that a product results that is of a firmer texture and milder flavour. After packing, the cans are heated in a steam box for 10–15 minutes at 200°F and allowed to cool for 20 minutes before passage through a draining machine which inverts the cans over a screen. After draining, the sauce is added, and the cans exhausted and processed as for herring.

SPRATS OR BRISLING, AND SILD

Sprats belong to the same family as the herring, but even when fully grown are quite small. Sild are the young of the herring, and both are treated similarly for canning.

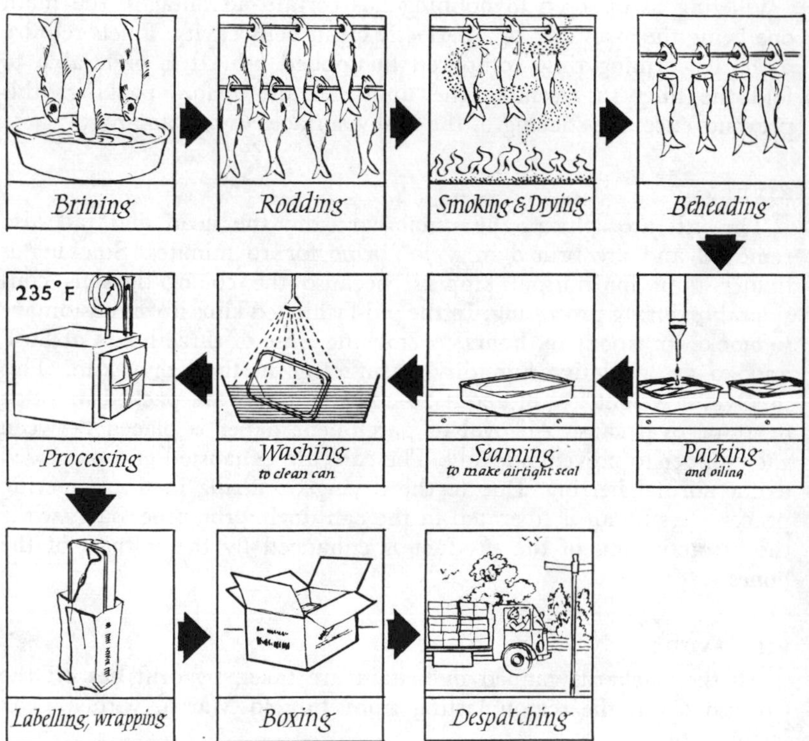

FIGURE 9·13 Sequence of operations in canning of sprats

Due to their size sprats are handled differently from larger fish. The main catching seasons are the autumn and spring. Sprats may be frozen and if properly stored, will keep for several months. It is customary to smoke the fish before canning.

On arrival at the factory the fish are de-scaled and washed in one operation, using a latticed cylinder type of washer. They are then brined for about 15 minutes in a saturated brine, 100° brineometer. After draining, the fish are *rodded.* This operation is carried out by women, and involves grading the fish into two or three sizes according to length. Fish of nearly equal length are placed head down in a series of slots and a thin metal rod or speat is run along through the eyes of the fish. The rod of impaled fish is then hung on a frame and the frames loaded on to trolleys. A Norwegian type of kiln is used for smoking. This is fitted with fans and flues and has three compartments, the first for drying, the second for smoking, and the third for cooking. The total operation takes 40–75 minutes depending on the size of the fish. The speed of the trolley passing through the kiln can be varied. Sprats can also be smoked in the Torry kiln. The fish are smoked for a period of 45 minutes, during which time the temperature increases from 80°F to 158°F. The fish lose about 10% in weight. It is important in the case of hot smoking to reverse the trucks and thus ensure even cooking.

After smoking and cooking, the fish are decapitated while still on the frame. The bodies drop on to a tray and are carried to the packer. Due to the small uniform size of the fish, they need not be weighed, as the weights of filled cans vary only slightly. The fish are packed head to tail, mainly in the ¼ dingley can, although larger cans are also packed. Edible oil or tomato sauce is dispensed into the cans, part before and part after packing. The cans are then closed and processed at 240°F for 55–60 minutes. They are washed and after drying may be labelled and boxed immediately.

MACKEREL

Small quantities of mackerel are canned in Britain. The fish mainly used are small young mackerel which swim with the herring shoals and are caught along with them in the drift nets. They are beheaded and gutted by hand, the belly wall being split to ensure the complete removal, by scraping, of the black membrane which lines the belly cavity. The fish are subsequently treated similarly to herring.

HERRING ROES

Some soft herring roes are canned. They have to be carefully selected to ensure that they are not torn, and are thoroughly washed in water to remove blood, which will otherwise discolour the product. They are packed in ¼ dingley or 7-oz oval cans and a small quantity of a good quality salt in the solid form added to the can. The larger cans are exhausted and then processed as for herring; they require 55 minutes at 240°F.

PET FOODS

Each year an increasing quantity of herring is used for the manufacture of canned foods suitable for feeding to cats and dogs. Some white fish is also used. The method of canning is quite straightforward; the whole fish after washing is minced, and may if desired be mixed with minced meat. Some packers cook the material before packing, others fill the raw flesh into the can, which is vacuum sealed and given a fairly severe process of about 90 minutes at 250°F. Suitable fillers and colouring matter are added as desired.

A BRIEF REVIEW OF OTHER METHODS OF CANNING

Apart from those mentioned in this chapter, many other fish and crustaceans such as crab, lobster, crayfish and shrimp are canned in various parts of the world, and the principles given above are generally applicable. Methods of handling will vary with the size of the fish. Thus salmon and tuna require different treatment. In the case of tuna the fish is pre-cooked and only the flesh packed, while salmon is packed as steaks. These fish, and especially salmon, lend themselves more readily to mechanical methods of can-filling than do small fish. In the case of these fish, also, it is the custom to add salt to the fish in the can, instead of brining before packing. In recent years a small amount of imported frozen salmon has been canned in Britain. A satisfactory product results provided that the raw material has been correctly frozen and stored. In the case of frozen salmon, trouble sometimes occurs during canning due to the formation of excessive curd, a creamy material which covers the surface of the flesh. *Curd* is protein that has been liberated from the cells of the flesh during freezing and thawing, and during processing coagulates on the surface of the steak. This can be reduced by treating the steak with a brine or weak tartaric acid dip before packing.

Crustaceans form an important part of the world pack of fishery products. All have one feature in common, namely, that they must be canned immediately after death, due to rapid deterioration accompanied by increasing sulphur-staining in the can. Even when canned with the maximum speed, this phenomenon is still a serious problem. The use of special lacquers and products has been discussed. Treatment of the flesh with a weak solution of citric acid is recommended. All the fish are cooked before canning, shrimps and prawns being peeled, and flesh removed from the shell and legs in the case of crab and lobster.

Several machines for the automatic processing of fish have been developed since the war, and have been applied mainly to the sardine industry. With most of these machines, practice is normal up to and including the can-filling stage, after which the machine takes over. The fish in the can are cooked in steam or brine or dried with hot air. The cans are drained before seaming. Many such machines are in commercial use on the Continent, but so far have not found approval in this country.

THE LEGAL ASPECT OF LABELLING CANNED FISH

Section 6 (1) of the *Food and Drugs Act,* 1955, and Article 4 of the *Labelling of Food Order*, 1953, are designed to ensure that the name given to a food or to describe a food gives the prospective purchaser a true indication of the nature of that food.

The Food Standards and Hygiene Division of the Ministry of Agriculture, Fisheries and Food has issued the following Code of Practice for the labelling of fish and fish products. This states, amongst other things, that fish sold under the names indicated below should belong to the species shown:

(1)	Herring	*Clupea harengus*
(2)	Sild	Young *Clupea harengus*
(3)	Pilchard	*Clupea pilchardus*
(4)	Sardine	Young *Clupea pilchardus*
(5)	Sprat	*Clupea sprattus*
(6)	Brisling	Young *Clupea sprattus*
(7)	Anchovy	*Engraulis encrasicholus*

Thus it would be misleading to label young herring as sprats or as sardines.

The above regulations apply only to the sale of canned fish in Britain. Other countries have their own regulations regarding the labelling of canned fish.

CONCLUSION

Recent years have seen the rise in popularity of quick frozen foods, including fish. Freezing is a form of preservation that has the advantage of presenting the foodstuff to the consumer in an apparently unchanged form. Canning, however, has sufficient unique advantages to withstand the encroachment of this new industry. Freezing requires elaborate plant and a chain of cold stores between the freezer and the consumer. A can of fish presents food ready-cooked in a form that can be easily transported and stored over a considerable period of time without regard to outside conditions.

Because of the uncertainty of the fishing during the past few years, the quantity of suitable fish available to the British canners has often not been sufficient to allow them to meet their normal commitments. This shortage also prevents them from extending their range of products, which would put them in a better position to compete in the considerable market enjoyed in this country by foreign canned fish products.

10

Fish meal and oil

. . . they accustom their cattle, cows, sheep, camels, and horses, to feed upon dried fish, which being regularly served to them, they eat without any sign of dislike.

The Travels of Marco Polo

A CONSIDERABLE proportion of the total fish catch is not saleable for human food. Some is discarded at sea in the form of guts, unwanted species and general trawl refuse. Of the landed fish, some will at times fail to find a market. This is particularly obvious in the herring fisheries, where glut and scarcity rather than steady supply is the common experience, but even with white fish the landings sometimes exceed demand. In addition, relatively small amounts of fish are unsaleable because they are condemned at landing by the Public Health Inspectors. Much of the fish bought for human food is processed in some way before reaching the retail stage, and this results in processing offal, for example heads, bones and fins from white fish filleting, or guts from kippering or herring canning.

Surplus fish and offal from processing must be disposed of. Far from being wasted, for example by simply being dumped, it is almost entirely utilized for the feeding of animals. Formerly much was used as fertilizer, but nowadays only negligible proportions are used in the manufacture of specialized fertilizers for gardens and nurseries. The manufacture of pet foods absorbs considerable amounts of surplus fish of good quality. But the really big outlet is in the manufacture of fish meal, which is a dry and easily-stored product forming a valuable ingredient of the rations of farm animals, particularly young pigs and poultry. Without it, or some equivalent source of high grade protein, the largely cereal-based diets of these animals would be inadequate for the rapid growth and productivity that they can achieve on a properly balanced diet.

The demand for fish meal in this country far exceeds the domestic supply, and the balance must be imported. The present home production of fish meal of all types is about 80 000–90 000 tons a year, most of which is white fish meal. Imports, which have been steadily climbing, are now almost 300 000 tons a year. Hence the production of fish meal from fish offal and surplus fish benefits both the national economy and the fisherman and should not be regarded as a salvage operation. If the raw material is of an oily nature, for example herrings, fish oil is recovered as an additional product and finds a variety of outlets, much of it going, after suitable refining treatment, into edible fat mixtures.

This chapter describes how fish meal and oil are made and draws attention to some of the difficulties and some of the more recent developments and trends in the industry.

THE RAW MATERIAL

The sources of raw material have been indicated briefly and it is now necessary to discuss its nature. White fish offal, chiefly heads and backbones, of which some 280 000 tons a year are collected, has, in recent years, made up about 80% of the total, while the poor herring catches have yielded only about 28 000 tons of surplus fish, that is about 10% of the total raw material for fish meal manufacture. The other 10% is derived mainly from edible quality but unsold white fish, only about 1% of the total raw material being condemned fish.

Only a little over half of the total 'inedible' portion of the white fish landings is recovered for fish meal manufacture. The rest is largely lost in retail and domestic waste. An increased supply of white fish offal would seem to depend on an increased proportion of fish being filleted at the ports. Material discarded at sea might be recovered in one of two ways, either by having a small fish meal plant on the trawler, or by having storage arrangements for bringing the raw material back. The first solution has been adopted by some foreign trawlers, but in this country only by one or two large vessels on which filleting and freezing operations are also carried out. The second scheme has been tried experimentally more than once, but so far without leading to a permanent development. Briefly, the difficulties of a trawler fish meal plant are associated with lack of space, while storage and unloading of the undried raw material pose problems that have not yet been sufficiently studied. Unwanted species would be ungutted and include the liver, often with a high oil content. Hence the plant would have to include a cooker and press as well as a dryer.

In general, not only in Britain but all over the world, the manufacture of fish meal from white fish waste on the one hand and from pelagic fish such as herrings, sardines and pilchards on the other, has developed along different lines and is carried out in different factories. This is partly because trawling for white fish and drift netting or other methods of fishing for pelagic fish are commonly based on different ports, but largely also because white fish offal is low in oil content whereas pelagic fish are rich in oil. In white fish such as cod, the oil is present almost entirely in the liver, which does not form part of the offal reaching the fish meal factory. It is only necessary to dry and grind white fish offal to obtain a satisfactory meal, but oily fish or fish waste must be treated to remove most of the oil before drying and grinding. Further, the physical properties of white fish offal on the one hand, and of oily fish after the stage of oil removal on the other, are so different that entirely different types of drying machines have been developed to handle them.

Thus in Britain we find some factories specializing in the manufacture of white fish meal and others in that of herring meal. However, there is a certain amount of overlap. Some white fish meal factories do receive a proportion of oily waste, and some herring meal factories receive a little white fish offal. Rather than divide this chapter in two, the industry will be treated as a whole, but frequent reference will have to be made to the nature of the raw material since it is crucial.

WHITE FISH OFFAL

White fish offal is deposited by the filleters in barrels or drums, which are subsequently delivered to a fish meal factory. It may come as a surprise to learn that in many cases it has been converted into meal long before the corresponding fillets reach the consumer. In some areas there are by-laws specifying the maximum time that fish offal may lie at the factory before being processed, and this period is typically 24 hours or less.

At the factory white fish offal is sometimes stored in the barrels until fed into the drying plant, sometimes stored in special bins, or in heaps on a concrete floor. Local regulations may again apply. But in any case the collecting barrels should be thoroughly cleaned before re-use, an item that is again often regulated by the by-laws. The use of barrels or drums for collection and transport to the factory is inefficient compared with bulk handling, but the latter must be at least as satisfactory from the sanitary point of view, for example no dripping during transport and easy cleaning of all equipment. Bulk transport has now been introduced in some places.

FIGURE 10·1 Sorting of white fish offal

White fish offal is far from uniform in size or consistency. In addition, it is sometimes seriously contaminated with rubbish that finds its way into the offal barrels. When this includes large pieces of metal or stone it can cause serious damage to the fish meal machinery. Hence all white fish offal is sorted before processing. If large quantities of fish skins are present, it is preferable to send them to a fish glue factory, since they can cause trouble with sticking and balling-up in the fish meal plant. Exceptionally large and bony pieces of material may be unsuitable for feeding to certain types of dryer without preliminary chopping. All large, and as many smaller, foreign bodies as possible must be removed, an operation that cannot readily be mechanized, and, in fact, is always performed manually. It would be a valuable contribution to efficiency if the workers in filleting and similar plants could appreciate the problems and importance of the fish meal industry and co-operate to ensure that only fish offal goes into the offal barrels.

The supply of white fish offal is reasonably steady throughout the year. There is little incentive to accumulate large stocks of offal, even if local regulations permitted. Offal stored without preservative rapidly

FIGURE 10·2 Chopper (hacker) for reducing the size of white fish offal

putrefies and causes trouble with the Public Health Authorities. Further, the yield of meal subsequently obtained from it falls, due to losses in drainage liquors and in volatile material. The use of preservatives has been studied to some extent but, partly because the practical need is not great, has not been put into effect.

HERRING

Whole herring make up by far the greater part of the total oily raw material for the British fish meal industry. The surplus fish may be landed at a port far from a suitable factory, thus involving heavy transport costs. But in any case transport from the fish quay to the factory is usually necessary, since at few British herring meal factories can the fishing boats unload directly at the factory wharf, although this procedure is usual in some other parts of the world. Drums are generally used to transport the fish, but are typically emptied at the factory into a large storage bin or pit from which a mechanical conveyer feeds them continuously at a steady rate into the plant. These herrings, in contrast to white fish offal, provide a uniform raw material free from large foreign bodies and thus suitable for mechanical handling from the outset.

The storage bins or pits, whether above or below ground level, have facilities for drainage of blood and other liquors. There are often local restrictions on the permissible storage period, as for white fish offal. The longer the storage, and the greater the pressure from a large heap of fish, the more drainage liquor. There will always be some. Usually drainage is through slots in false inner walls as well as at floor level. Appreciable quantities of oil are present in the drainage liquors and this oil is recovered, for example in a trap or by centrifuging. The aqueous material, however, is often run to waste. Where it can be evaporated this is done, as otherwise it represents an appreciable loss in dissolved solids. This loss will be the more serious the staler the raw material, for example if it has been transported from a considerable distance.

Supplies of herring are not only markedly seasonal, particularly in any one area, but also fluctuate enormously from day to day during the season. Hence, in contrast to the position with white fish offal, there is a strong incentive to accumulate a buffer stock of raw material to enable a herring meal factory to operate at an economic and reasonably steady rate of throughput. The ideal would be to work like the white fish meal factory the year round, but this would involve both enormous storage capacity and an efficiency of preservation not yet contemplated. Even when local by-laws do not prevent the storage of unpreserved herring, the fall in yield of meal and in quality of oil would progressively offset the advantage of a steady processing rate.

Most fish meal factories are equipped with more than one complete production unit and considerable variations in supply can be absorbed without difficulty. However, at times of glut landings, processing capacity is liable to be swamped, necessitating uneconomic long hauls of fish to distant factories or to restriction of fishing. On such occasions storage of the surplus fish for a short period, say up to a week or two, would be most helpful. This can be achieved by the use of chemical preservatives.

Many chemicals can kill bacteria and prevent fish from rotting, but few of them would be acceptable in a product used in animal feeding stuffs. Although the regulations are not so strict as for human food, it is an offence against the Fertilizers and Feeding Stuffs Act to offer for sale a product containing any ingredient harmful to cattle or poultry. In fact, many common substances, such as ordinary salt, would be harmful if present at too high a level, so that the Act means, in practice, harmful at the concentration present in the product.

FIGURE 10·3 Storage pit for herring

Two preservatives have shown promise and are already used on a fairly large scale in Norway. These are sodium nitrite and formaldehyde. The amounts required vary enormously according to the seasonal temperature and the condition of the fish, and Norwegian practice cannot be applied without experiment and appropriate modification to British conditions. Formaldehyde toughens the fish as well as preserving it and may lead to difficulties in processing, although with fish originally very soft some toughening is helpful. Formaldehyde is far superior to nitrite in preventing infestation by blow-fly larvae, that is, maggots. Nitrite does not produce toughening, indeed the stored product becomes softer. Research suggests that satisfactory suppression of putrefaction and maggot infestation, with maintenance of satisfactory texture, for up to two weeks at summer temperatures can best be achieved by dipping the fish for one minute in a solution containing a mixture of nitrite and formaldehyde, each at a concentration of 1 %. The dipped fish may then be stored in the ordinary way if protected against re-infestation by flies. Simple spraying of the herring with stronger solutions of formaldehyde or formaldehyde-nitrite can do much to reduce nuisance and loss during short periods of storage. It is usually impossible to detect free formaldehyde in the final meal but after the above dipping treatment it does contain about 300 parts per million of nitrite. If whole meal instead of ordinary meal were made, the nitrite level might well be twice as high. Despite animal experiments showing an adequate safety margin, it would be desirable to blend any meal made from preserved herring, especially whole meal, with the normal product to reduce the nitrite content to 200 parts per million or less. This would accord with general practice in the food industry and with the Norwegian specification for herring meal.

COOKING AND PRESSING

COOKING

When fish are cooked and the protein denatured, much of the water they contain can be squeezed out under pressure. In contrast to this, the water of raw fish is firmly bound to the protein and very little indeed can be removed even under great pressure. When oily fish are cooked and pressed, a mixture of oil and water is squeezed out.

Since the primary operation in making fish meal is the removal of water, it might at first sight seem advantageous to cook and press all fish waste, including white fish offal, leaving far less water to be removed by evaporation. However, the water squeezed out carries with it both dissolved solids and finely divided solids in suspension, which can amount to about a fifth of the total solids of the original raw material. If the expressed watery liquor, generally known in the industry as *stickwater*, is simply run to waste, the loss of about 20% of the normal yield of meal will have to be set against any savings in fuel and plant. All oily fish is cooked and pressed before drying by evaporation, primarily to recover the oil instead of leaving it in the meal. The whole sequence of operations is shown in Figure 10·11.

The cooker consists of a long tube through which the fish are passed by a worm or similar conveyer. Until fairly recently heat was supplied only by means of a large number of independently controllable jets of low pressure steam, injected directly into the mass of fish at closely spaced intervals along the cooker. Condensed steam mixed with the fish, giving some 15–20% of extra water to be disposed of. The tendency nowadays is to supply part of the heat by means of a steam jacket round the cooker. It is possible to supply the whole of the heat in that manner, but the cooking operation is extremely critical and should be flexible according to the consistency of the fish. The provision of jets, as well as the jacket, permits great variation in heat input for a constant rate of throughput of fish.

If the fish are incompletely cooked, satisfactory expression of water and admixed oil cannot be obtained. If they are overcooked, the

FIGURE 10·4 Cooker in a herring meal plant

texture is too soft and mushy to permit ready straining of the expressed liquid through the fibres of the tissues during pressing. The proportion of dissolved and finely suspended solids in the press liquor is increased, hindering separation of the oil from it. The ideal is for fish to leave the cooker in the form of large pieces or whole fish, but it should be adequately cooked right through, for example the backbone should be opaque and not translucent. The toughening action of formaldehyde has been mentioned. This is sometimes utilized in the cooking and pressing of very soft raw material, when a strong solution of formaldehyde is allowed to drip on the fish as they pass from the storage container to the cooker. The skill and experience of the foreman is the best, if not the only, assurance of good performance of the cooker with varying raw material. Cooking time is about 15–20 minutes.

Some free water and oil accompany the cooked product and these are often removed before pressing by drainage between cooker and press, for example through a perforated extension to the cooker casing. Liquid so recovered is added to the main volume of press liquor.

PRESSING

The type of press universally used in Britain, and the commonest type in fish meal factories the world over, is the continuous screw press. The cooked fish is conveyed along a perforated tube, and subjected to steadily increasing pressure as it moves along. The water and oil squeezed out escape through the perforations and the final solid, known as *press cake,* emerges from the far end of the tube.

In the simplest type of screw press the fish is moved forward by a single rotating screw conveyer, rather like that in a mincing machine. However, unlike the conveyer in a mincing machine, that in a screw press incorporates one or more devices to ensure steadily increasing pressure. One is a taper on the shaft of the conveyer, so that it is thicker towards the press outlet, with a steadily diminishing free space between conveyer and press casing. Another is a decrease in the pitch of the screw towards the outlet, so that the product moves along more slowly and offers increasing resistance to the material following it. Typically both of these devices are used at once, that is a worm of diminishing pitch is mounted on a shaft of steadily increasing diameter. The exit end of the press is partly closed by an adjustable cone, affording another means of exerting pressure on the product.

A drawback of the single screw press is that the fish, particularly when it is unduly soft, may allow the screw to rotate in it without moving anything forward. The worm becomes filled with an adhering product which merely rotates with the shaft. The effect must be well known to all users of a domestic mincing machine. It has led to the development of the double screw press, which is now the most widely used type in this country. Two worm conveyers, each tapered and of diminishing pitch, rotate in opposite directions with the worms intermeshing. They are fitted within a common casing, appropriately modified from the tubular form of the single screw press. There are other ways of preventing passive rotation of fish with the shaft, even with a single screw, and in some countries, for example USA, other modifications to the simpler press may be seen. They do not offer any obvious advantage over the double screw press.

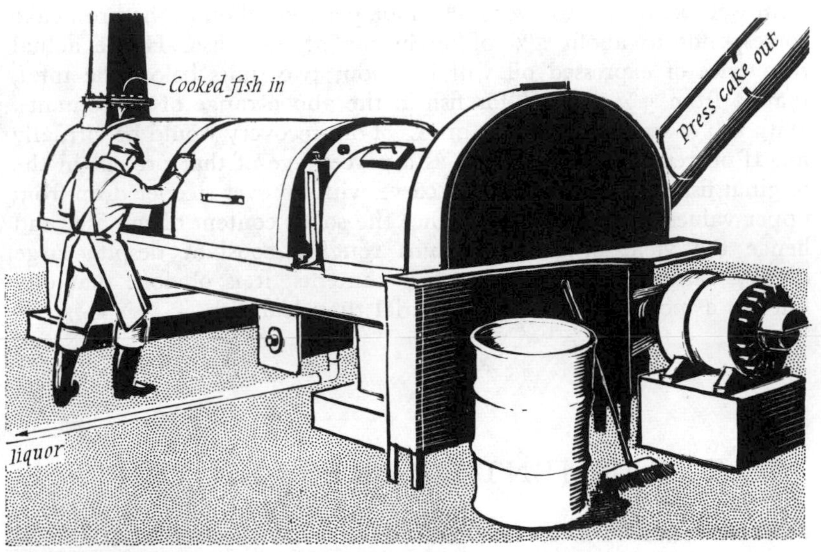

FIGURE 10·5 Continuous double-screw press in herring meal plant

The press casing must be able to withstand great pressure. The perforations may take the form either of holes or of longitudinal slots, being always tapered and wider towards the outside, to prevent blockage. The cooked fish should be pressed while still hot, so the press is often placed immediately below the cooker, being then fed from it by gravity.

Even under the best conditions the press cake still contains about 55% of water and some 4–5% of oil. Commonly the water content is nearer 60%. It is difficult to appreciate the significance of these figures without a little calculation. Let us assume that every 100 tons of original fish, for example herrings, contains 20 tons of solids, 15 tons of oil and 65 tons of water. Let us further assume that one-fifth of the original solids accompanies the stickwater. Then at 55% moisture we would get from 100 tons of fish 40 tons of press cake containing 22 tons of water, while at 60% moisture we would get 45 tons of cake containing 27 tons of water. In each case the cake would contain 16 tons of solids and about 2 tons of oil. Thus in the first case 66% and in the second case 58% of the original water has been squeezed out. This difference represents a reduction in load on the dryer of 5 tons of water for every 100 tons of original fish.

With efficient pressing, oil recovery depends almost entirely on the oiliness of the original fish. Herring vary seasonally in oil content from well below 5% to over 20%. Five per cent of oil in the press cake corresponds to about 2% of oil in the original fish. Hence actual recoveries of expressed oil will be about two units below the total, that is from 3% to 18% for fish in the above range of oil contents. With fish containing only 2% or less of oil, recovery would be virtually nil. If oil recovery is expressed as a percentage of that present in the original fish, the figures lie on a curve with an ever steeper drop from upper values of 90% or more. Since the solids content of the fish, and hence the yield of meal, remains roughly constant despite large inverse variations in oil and water contents, it is obvious that oily fish are a more valuable raw material than lean fish.

TREATMENT OF PRESS LIQUOR

PRELIMINARY TREATMENT

The press liquor is first freed from coarser pieces of solid material, either by a screen of suitable mesh or by centrifuging. To avoid choking of the screen it is inclined from the horizontal and continuously vibrated. A centrifuge is a device for increasing the effective gravitational force causing separation of two liquids or a liquid and a solid, when these are of different densities. If press liquor were merely allowed to stand, the coarser debris would eventually settle to the bottom and the oil would rise to the top. This process can be very much speeded up, with far sharper separation, by rapidly spinning the liquid. Centrifugal force thousands of times that of gravity can thus be brought into play. Many types of centrifuge are available, designed to perform this or that particular separation with maximum efficiency. For continuous removal of suspended solids from crude press liquor, a type of centrifuge is generally used that does not afford simultaneous effective separation of oil and water. The solids are flung to the outside of a rapidly spinning bowl and removed continuously by a spiral scraper. The combined liquid, oil plus water, is removed from the centre of the bowl. In Britain screening is far commoner than the use of such a centrifuge.

FIGURE 10·6 Centrifuge room in herring meal and oil plant

A vibrating screen will not remove the finer suspended solids. The type of centrifuge mentioned above would only remove the finest particles at the expense of speed of throughput. Subsequent treatment of the partially clarified press liquor in a centrifuge designed to separate the oil from the water removes the bulk of the remaining fine solids as a sludge which is recovered, either periodically or continuously according to design, from the outermost portion of the centrifuge bowl, for example continuously through holes fitted with suitably-sized nozzles. Solids recovered from press liquor are usually pressed in a special small screw press, known as a *foots press,* before adding them to the main flow of press cake. Liquor from the foots press is similarly added to the main flow of press liquor.

SEPARATION OF OIL AND STICKWATER

Efficient separation of oil and water at high speed, that is with high flow rate through the centrifuge, is facilitated by shortening the path that each substance must travel before its removal from the mixture. Without going into complicated details it may be said that this is achieved by fitting a series of overlapping cones within the centrifuge. The path for effective separation is only the short distance between the outside of one cone and the inside of that immediately above it. Oil travels up the outside surface of each cone, water down the inside surface, arrangements being made for their separate collection and discharge. It is common but by no means universal practice to pass the oil through another centrifuge, known as a *polisher,* before pumping it into storage tanks. The polisher is specially designed for efficient removal of the last traces of free water and solid debris from the oil, thereby ensuring stability during storage. Proteinaceous matter, together with moisture, can support the growth of moulds and bacteria, which in turn can cause deterioration of the oil.

EVAPORATION OF STICKWATER

The aqueous portion of the press liquor, that is the stickwater, contains both dissolved material and particularly fine solids in suspension. The latter can be removed by prolonged and powerful centrifugal action but it would be uneconomic to do this. The total solids, dissolved and suspended, usually amount to about 7% of the stickwater but under bad conditions, for example with very stale fish or overcooked fish, may be much higher. Until the Second World War stickwater was always run to waste. Commencing in the USA and then spreading to other countries, the practice has grown of evaporating the stickwater, usually to a thick syrup containing 40–50% of solids. This may either be marketed as such, under the name *condensed fish solubles,* or mixed with the press cake and so dried along with it to give *whole meal* or *full meal.* In Britain there is as yet very little production of condensed fish solubles but increasing production of whole meal. Sometimes, for one practical reason or another, only part of the stickwater is evaporated and added to the press cake.

Running the stickwater to waste means a loss of about 20% of the total solids of the fish, more under bad conditions. The Norwegians have shown that if no special evaporation equipment is available some of the unconcentrated stickwater can profitably be added back to the press cake, the proportion depending on the type of dryer and the operating conditions but sometimes amounting to 40% of the total stickwater. However, the evaporation of a liquid can be accomplished with less expenditure of fuel than is required for equivalent evaporation in a fish meal dryer. The ideal is to concentrate the stickwater as much as possible in the most efficient plant. A limit is set by the increasing viscosity of the product and its tendency to turn to a solid gel which would clog the plant and its pipe lines. This limit is typically 40–50% of solids varying with the kind of fish and the evaporating conditions.

Stickwater will rapidly putrefy if it is allowed to cool sufficiently and held in storage tanks without any precautions. In some factories it is evaporated almost as fast as it is produced. If it is then added to the press cake, and promptly dried, no special precautions are needed. If it is sold in the form of condensed fish solubles some preservative must be added. Typically this is sulphuric or other mineral acid, in amount necessary to lower the pH* of the product to about 4·5. This degree of acidity prevents growth of micro-organisms. If the dilute stickwater must be stored for some time before evaporation, it is usually preserved by similar addition of mineral acid to bring the pH to about 4·5. On concentration the pH will remain at about 4·5 and no further treatment is needed. Acidification of the dilute stickwater coagulates some of the fine suspended solids and these are sometimes recovered by further centrifuging before evaporation. However, the acidified liquid is corrosive and the centrifuge should be made of stainless steel. Corrosion of the evaporators is also serious. There is every advantage in evaporating the stickwater as fast as it is produced, particularly when whole meal is to be made.

Various types of stickwater evaporator are in use in the British industry. It would take too long to discuss them all. Instead a brief account will be given of the type most commonly found both here and in other countries and which is of high efficiency. This is the multiple effect evaporator, generally with three stages, that is triple effect. This makes use of the fact that the lower the external pressure, the lower is the boiling point of a liquid. Thus, in a series of evaporators operating at progressively diminishing pressures, the vapour from the first vessel can be used to boil the liquid in the second vessel and so on down the series. The only heat supplied from outside is that needed to boil the liquid in the first vessel.

Operating conditions in a multiple effect evaporator can be varied greatly according to individual requirements. Thus all stages may operate at pressures above atmospheric and the final steam be used somewhere else in the factory. In a fish meal factory this could have the double disadvantage that the relatively high temperatures might lead to loss of nutritive value in the product, and the steam would have a strong odour if released into the air, for example from an open

*The term pH is used by scientists when giving numerical values to denote degrees of acidity. Low values mean more strongly acid. Pure water has a value of 7 on this scale.

steam cooker. This would worsen the already unpleasant atmosphere of the factory. More usual is a triple effect evaporator in which the first stage is at a pressure above atmospheric, the second stage nearly at atmospheric pressure and the final stage under quite a high vacuum. Typical conditions would be a first stage pressure of 15–20 lb/in^2 gauge and a boiling point of about 250°F, a second stage at about atmospheric pressure and a boiling point of about 220°F, and a third stage under a vacuum of about 25 inches of mercury, with a boiling point of about 130°F. Sometimes the first stage is at about atmospheric pressure, or even below it, with progressively increasing vacuum in the subsequent stages.

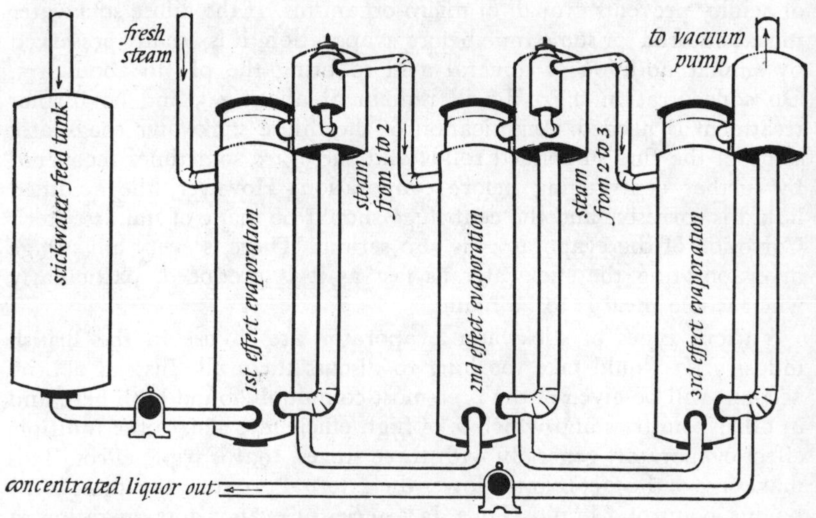

FIGURE 10·7 Triple-effect evaporator in herring meal and oil plant

These evaporators are continuous in operation, the dilute liquor entering the first stage at a controlled speed. In this it is pumped round between a tank holding the bulk of the contents and a heat exchanger consisting of a number of tubes, through which the liquid flows, surrounded by steam at a suitable temperature and pressure. A controlled flow of the partially concentrated liquor passes continuously into the second stage, where the same process is occurring, and from there into the third stage, from whence the final concentrate is continuously withdrawn.

Cooking and pressing of white fish or white fish offal, even although no oil will be obtained, seems an attractive alternative to a simple drying process. Indeed, it is practised by some white fish producers, particularly when they also handle a certain amount of oily fish and so must have a cooker and press available. Press cake does not have the same drying characteristics as wet fish or offal and, in particular, it can be handled in hot air dryers of high efficiency and relatively low capital cost. Most dryers of this type cannot handle wet white fish offal. Dryers are considered below but it may be mentioned here that the steam consumption of a triple effect evaporator is about ½ lb for every lb of water evaporated, whereas the overall steam consumption in a steam-jacketed white fish meal dryer is nearer three times that amount. A process in which cooking and pressing, with stickwater evaporation and feeding back of the condensed product, is coupled to a conventional steam-jacketed dryer should show a fuel saving of about one third. Since only about half the original load will be placed on the steam-jacketed dryers, either fewer or smaller units can be used. The capital cost of cooker, press and evaporator is generally less than the capital cost of the dryer capacity thus saved. If the steam-jacketed dryers are replaced by air dryers, even greater savings in fuel and capital cost should be achieved. It should be noted that cooking, pressing and evaporation is best suited to continuous operation if a uniform product is to be produced, and generally requires more supervision than the usual white fish meal manufacturing process. Scaling and corrosion of the evaporator can be very troublesome, varying in severity with constructional details and probably with the type of raw material.

DRYING AND DRYERS

Space will not permit description of all the types of dryer, with their various modifications, to be found in the British fish meal industry. Moreover, it is fairly common practice to combine dryers of different types, transferring the partly dried product from one to the other. This is usually done because the characteristics of each make it more suitable for a product of a particular consistency, or moisture content, or density. What will be done here is to consider briefly the structure and operation of a selection of dryer types, indicating some of their special features.

INDIRECT, STEAM-JACKETED DRYERS

Indirect dryers find their greatest use in the white fish meal industry. They are so constructed that they can handle a sticky product. White fish or white fish offal during drying passes through a stage in which it becomes very gluey. Not only will it readily cake into large masses but it will stick to any surfaces it touches. This property makes it unsuitable for any of the direct dryers described later, unless it is first cooked and pressed.

FIGURE 10·8 Indirect dryer for white fish meal

Dryers of all types are made in a range of sizes. In addition steam-jacketed dryers may operate under vacuum or atmospheric pressure, may be batch or continuous machines, or may involve any combination of these variables. For example the fish may first be partly dried under vacuum in a batch dryer and then finished under atmospheric pressure in a continuous dryer. But all such units incorporate a steam jacket and a stirrer. The shaft of the stirrer is hollow and heated by steam. The stirrer paddles approach as near as practicable to the dryer shell, to scrape off the sticky material and prevent a thick layer of scale from building up.

The commonest machine consists essentially of a long tube, with raw fish or offal entering one end and dried material leaving at the other. The stirrer shaft runs through the whole length and the paddles or *gate agitators* are arranged in a gentle spiral of about 4°/linear foot, moving the product through the dryer at a rate of about 1 foot a minute. Since a typical dryer will have a total length of some 150 feet or more, it is broken up into several, usually three or four, shorter tubes which are stacked one above the other and staggered to reduce space requirements to the minimum. The fish passes along each section in turn, falling by gravity from tube to tube. Vapour is withdrawn through ducts at various points.

A machine capable of drying about $1\frac{1}{3}$ tons of raw fish an hour will have three or four cylinders each about 44 feet long and of 3 feet internal diameter. Steam pressure used varies widely in different factories but is usually within the range 25–40 lb/in^2. Older machines are generally fitted with an additional preliminary cylinder of half the length and approximately half the diameter of the main tubes. This is called the *sterilizer*. Its purpose was ostensibly twofold, to cook and sterilize the fish, and to prevent overloading of the first main cylinder, since the fish swells on heating before appreciable drying has caused it to shrink again. The diameter of the sterilizer increased half way along it to accommodate this expansion. In fact, these dryers are very robust and the sterilizer tube is not fitted to new plants. The first section of the top main tube has taken over its cooking role. Total drying time in the equipment described is about two hours.

On older machines exhaust vapour ducts, fitted with individual dampers, are attached at intervals along each cylinder. These are connected to a common main duct leading to an exhaust fan and thence to a deodorization unit. Since no air is actually blown into the drier, and since the inlet end is largely blocked by a feeding worm and by wet fish, most of the air enters where the dried product emerges. Hence the main air stream and the fish are travelling in opposite directions. This is advantageous because the cold air, on entering the dryer and becoming hot, acquires a low relative humidity and so a high capacity for taking up further moisture just where it is most needed. This is where the product is already nearly dry and further drying is becoming increasingly slow, the vapour pressure of its remaining moisture steadily falling. Nearer the point of entry of wet fish, evaporation is quicker and easier. The water in the product exerts a higher vapour pressure and the by now much damper air stream is still able to carry off the water vapour at a reasonable rate. This suggests that there is no point in exhausting some of the air and vapour from most of the total length of the dryer but only from near the wet fish entry end. Experiments have confirmed this and newer plant has been modified accordingly. On older plant, dampers should be closed on most of the ducts.

The dimensions of the drying cylinders are probably far from ideal. In particular it is hard to see why the diameter should be the same from inlet to exit. The top tube is nearly full of material, the bottom tube nearly empty. Experiments are going on to improve the performance of this type of plant, for example by the introduction of extra heating surfaces, such as steam tubes, into the bottom cylinder. Contact with the hot surfaces is good in the top cylinder, but the bottom one is more of an air dryer than a contact dryer. It may pay to blow hot air into it rather than merely suck in cold air. A recent trend is the use of unit cylinders, 20–30 feet long and of larger diameter, namely 4 feet. These are so assembled as to give the desired total capacity. If it is found better for cylinder dimensions or internal arrangements to be different in different sections the unit cylinder idea may require reconsideration.

Scale formation on the surfaces of cylinders and shafts affects the efficiency of heat transfer and reduces capacity. This problem does not arise at the beginning or at the end of the drying sequence; it is somewhere in the middle that stickiness of the product reaches a maximum. Scale in the cylinder itself is limited in thickness by the clearance of the paddles but it becomes hard and polished by their action. Scale on the shaft can become very thick. Some of the scale breaks away by differential expansion or contraction when the plant is started up or shut down. It can be softened and loosened by admixture of a little oily raw material with the main feed. Much of it, however, must often be removed periodically with hammer and chisel. An advantage of the cook-and-press method is the elimination of scale formation in the dryer, although it then becomes a problem in the evaporator.

Roughly half of the white fish meal produced in Britain is made in fully continuous dryers of the above type. They also find a limited use in the drying of herring press cake. Most of the rest of British white fish meal is produced in batch-continuous units, in which about half of the water is first removed in the course of about three hours in a vacuum batch dryer. This has the usual steam jacket and shaft but the paddles are not designed to propel the fish along the cylinder, which in any case is short, say 15 feet long and of 5 feet diameter. The charge may be anything from 2–5 tons, according to the size of the unit, which is approximately half full at the start of a run. After drying to the desired extent the product is discharged into a buffer hopper, provided with heating and agitation, from which it passes at the desired rate into the continuous dryer. Often two batch dryers, working alternately, are coupled via such a hopper to one continuous dryer. Operating conditions vary widely and total processing time tends to be lengthy, sometimes as long as seven hours.

The batch dryer is described in the industry as a sterilizer but, since a vacuum of about 20–25 inches of mercury is applied from the start, the product cannot reach a high temperature there, even although steam at some 30–40 lb/in^2 is used on jacket and shaft. It presumably reaches a higher temperature in the buffer hopper, which is not under vacuum, and in the continuous drying cylinders.

Scale formation is not very serious in the vacuum batch dryers but may be heavy in the buffer hopper, where further drying occurs and which is often called the *after dryer*, and appreciable also in the first section of the subsequent continuous drying unit.

DIRECT DRYERS

Direct dryers are those that do not require an intermediary boiler between furnace and dryer. Not only is capital cost thus reduced but fuel efficiency should be higher. Moreover, the dimensions of a direct dryer in relation to its throughput are far smaller than for an indirect one. In particular the thickness and total weight of metal are much reduced. They are not universally used, however, because they demand a non-caking, non-sticking raw material, such as press cake. They are almost always employed in factories working mainly on oily fish, particularly if operations are on a relatively large scale.

The oldest dryer of this type, the flame dryer, which is still in widespread use in fish meal factories in many parts of the world, has almost disappeared from the corresponding British industry although still used here for drying many other commodities. It will, therefore, not be described in detail. Suffice it to say that flue gas and secondary air from an oil-fired furnace are passed in parallel flow with the press cake through a rotating cylinder which tumbles the drying material through the hot gas. A typical dryer can handle the press cake from 10 tons or more an hour of fresh fish. Parallel flow means that the very hot gases, at up to 1000°F or more, encounter the wettest fish and rapid evaporation keeps the product temperature reasonably low. When, towards the exit, the product is nearly dry and its temperature approaches that of the gas stream, the gas temperature has fallen to little above 200°F. Drying time is about 15 minutes.

Short though this time is compared with the two hours and upwards spent by the fish in a steam-jacketed dryer, it has been still further dramatically reduced in two types of dryer which are basically developments of the flame dryer; these are both important in the British industry.

PNEUMATIC, AIRLIFT DRYERS

Imagine the flame dryer, which is normally almost horizontal, turned into a vertical position with the flue gas and press cake inlet at the bottom. Rotation of the shell would now be useless, either to turn over the fish or to move it through the dryer. But if the gas velocity were sufficiently increased the product would become airborne and so be carried through the dryer. Moreover, as the product dried it would become progressively lighter. By having a tapering section, with progressively falling air speed, the dryer could also act as a classifier, individual pieces of press cake moving upwards at a rate governed largely by their moisture content. This would prevent scorching and reduce the average time spent by the fish in the dryer.

This principle has been utilized in an airlift dryer of American design which is now manufactured in Britain. It has proved perfectly satisfactory for drying herring press cake, without full return of condensed solubles, but less so for press cake from white fish or offal because the large pieces of bone which the latter contains are not readily airborne. Like most dryers it is made in a range of sizes, for example to handle the press cake from about 5 tons and 10 tons of fish an hour. Flue gas and extra air is drawn through by a powerful fan and controlled by a system of both fixed and adjustable vanes.

Operating conditions vary widely from factory to factory as with steam-jacketed dryers. Provided that overheating and scorching of the meal is avoided, higher inlet gas temperatures may be used with shorter drying times. The variables of temperature, air speed and rate of feed of press cake have usually been adjusted to suit the rest of the factory facilities in the light of local experience. Typical inlet gas temperatures are in the range 500–600°F and exit gas temperatures 200–230°F. The volume of gas drawn through the dryer is about 16 000 ft^3/min for the smaller plants and about double this for the larger ones. Average drying time is about 90 seconds.

With pneumatic dryers, involving high gas velocities, separation of exhaust gas and dried product must be effected in a cyclone, which is really a special form of centrifuge. It has no moving parts. Instead the rapidly moving stream of gas and solids is directed tangentially near the top of a tall conical vessel, tapering towards the base. The gas swirls around in this and the denser solid is flung against the casing, falling to the bottom from whence it is withdrawn. The gas is withdrawn centrally from the top of the cyclone. Some of this exhaust gas is usually re-cycled through the dryer, as part of the extra secondary air, with economy in fuel consumption. The proportion which can thus be re-used varies with the general operating conditions, in particular with the relative humidity of the exhaust gas.

DRYER-GRINDERS

The product leaving a fish meal dryer, especially a steam-jacketed dryer operating on white fish offal, contains some material too coarse

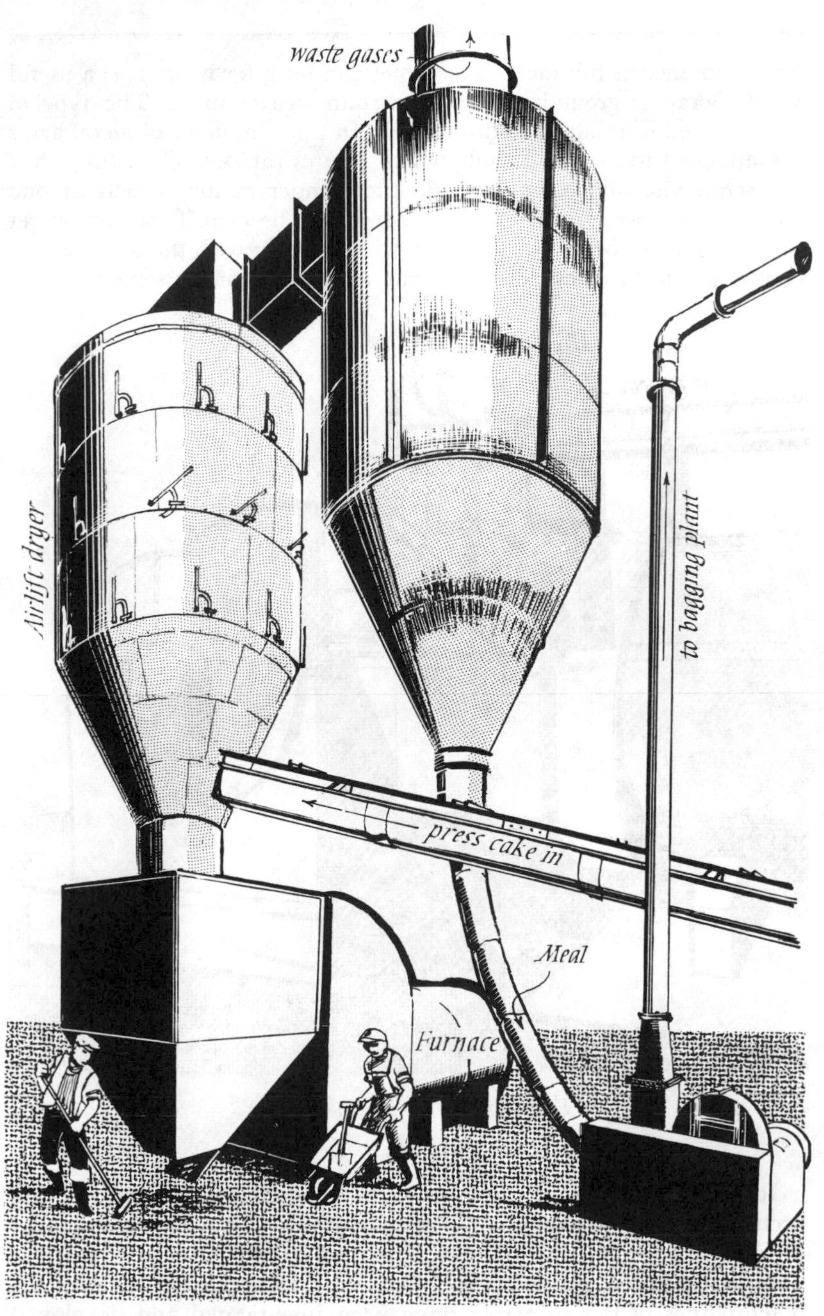

FIGURE 10·9 Pneumatic dryer

for acceptance as fish meal. The American term for it, *scrap,* is a useful word. Scrap is ground to pass a certain size of mesh. The type of grinder used is usually a hammer mill. In this a number of metal arms are attached to a central shaft which rotates rapidly. The arms whirl the scrap violently around inside the grinder casing, which at one section has a screen of suitable mesh size. The centrifugal action set up causes air to be drawn into the grinder along with the scrap at the centre and to be blown out with the meal through the screen.

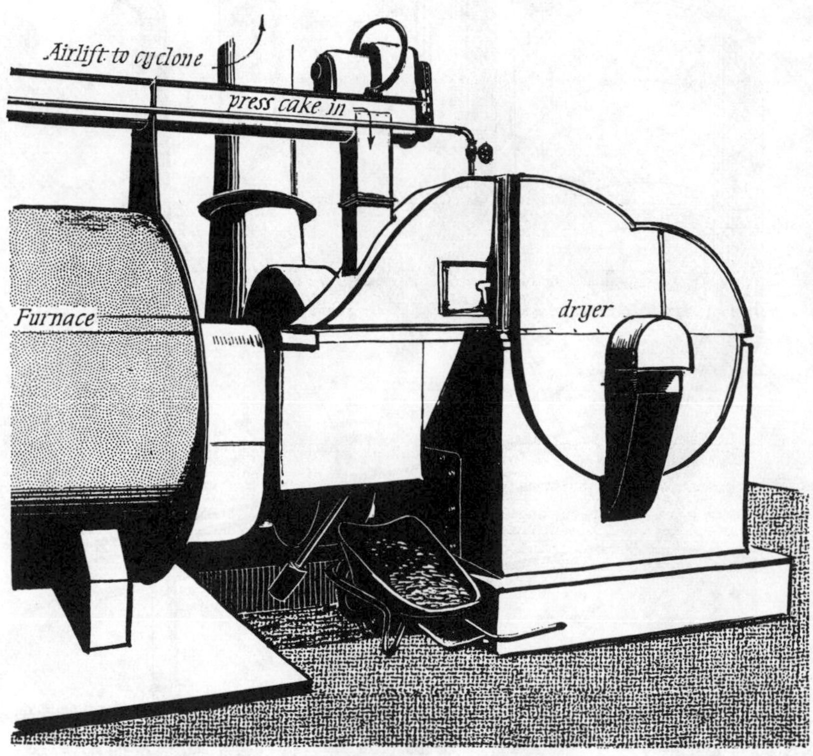

FIGURE 10·10 Dryer-grinder

It was always known that a certain amount of further drying occurred during grinding. It was, therefore, a logical development to blow hot air into such a machine and convert it into a dryer as well as a grinder. However, it is only in fairly recent years that the potentialities of this principle have been investigated and developed in the fish meal industry.

Two types of machine are to be found in the British industry. One is basically more of a mill than the other. It is not a hammer mill but a peg mill, in which the rotor is armed with pegs. These intermesh with stationary pegs, causing a tearing action. The air movement is very complex, with many eddies and vortices. It was originally developed for pulverizing all kinds of hard materials, for example coal. The other machine was designed primarily as a dryer for herring press cake, although a hammer mill in principle. Both machines are available in a variety of sizes.

The peg-mill machine is designed to evaporate 1 ton of water and the other machine 2¼ tons of water an hour. Inlet and outlet gas temperatures are the same on each, 750–800°F and 230–250°F respectively. There are, however, great differences in rotor speed and in gas flow rates. In the peg mill the rotor speed is relatively high, 1075 rev/min. In the other dryer it is only 230 rev/min. Gas flow through the peg mill amounts to about 6000 ft^3/min, that through the hammer mill about 24 000 ft^3/min. In each case average drying time is estimated at only a few seconds. The product is very light in colour and fluffy in texture.

Provided the problems of odour suppression are solved, the dryer-grinder might eventually replace other types of dryer in the herring meal industry. It is the most compact of all. Herring press cake is not a very abrasive material, so that the machine does not need to be particularly robust when designed for this purpose; it should have a long life. The product is not all ground sufficiently to obviate the need for the usual grinding plant but much of it can be sieved out without requiring further size reduction, so saving up to 50% of the normal grinding capacity and power consumption.

GRINDING AND BAGGING

The product of the dryers, for which the American term scrap will continue to be used for want of an English equivalent, must be cooled before it is ground. First, there is an appreciable risk of fire if it is ground warm, especially with oily meal in which oxidation of the oil will generate appreciable additional heat. Secondly, with white fish meal in particular, the warm product is somewhat sticky and, apart from requiring more power to grind it, would tend afterwards to cake badly in the sacks.

Scrap from white fish meal dryers is often delivered to the grinding room by worm conveyers. If these are lengthy they may allow sufficient cooling. In other cases it is common to feed the scrap to temporary storage heaps on the factory floor, where cooling can occur before grinding. Scrap from herring meal dryers is generally blown through pipelines to the grinding room; this affords adequate cooling.

A white fish meal factory normally uses a number of dryers, which cannot be expected all to turn out scrap of identical moisture content. Moreover, the raw material may be very uneven, for example some may be oilier than the rest, some may have a larger proportion of bone, and so on. Hence blending is general in this side of the industry. It is done in special mixing vessels before the scrap enters the grinders. In the herring meal industry it is usual to find a smaller number of larger dryers in use, operating automatically on a uniform raw material. Hence blending is less often required.

Before the scrap is ground it is passed over a magnetic device to remove any odd bits of iron such as nails. Scrap from pneumatic dryers or dryer-grinders is unlikely to contain large foreign bodies, such as stones or pieces of wood. Scrap from steam-jacketed dryers is commonly freed from such contaminants by shaking screens or similar devices. The grinder itself may be designed to reject large foreign bodies. The ground meal is separated in a cyclone from the air also passing through the grinders and filled directly into jute or multiwall paper sacks. The latter are being increasingly used.

The question of fire risk in fish meal stores is important. With white fish meal of low oil content it is considered safe for the sacks to be stacked closely immediately after filling. With oily fish meal the oil is actively oxidizing as the scrap leaves the dryer. Even although it has been cooled before grinding, oxidation has not stopped and will continue to generate a certain amount of heat. A multiwall paper sack, being less permeable to air, retards the rate of oil oxidation and so reduces the possible temperature rise in the sacks more than does a jute sack. Herring oil is not equally reactive to atmospheric oxygen at all seasons. It is chemically more unsaturated in the summer than in the winter and in general the more unsaturated an oil the quicker it will oxidize when exposed to the air. Herring oil is considerably less unsaturated than the oil of some species important in foreign fisheries and in these countries the precautions against fire in fish meal stores have to be more stringent. Here some herring meal factories take no special precautions apart from cooling the scrap before grinding. In other factories, the freshly filled sacks are stacked in an open formation, with free air circulation, for 24–48 hours before stacking them closely in the meal store. In some countries a special curing treatment lasting anything up to a week may be necessary before the scrap is ground and bagged. This is now being avoided by the addition of a chemical, known as an antioxidant, which even at very low level, say one part

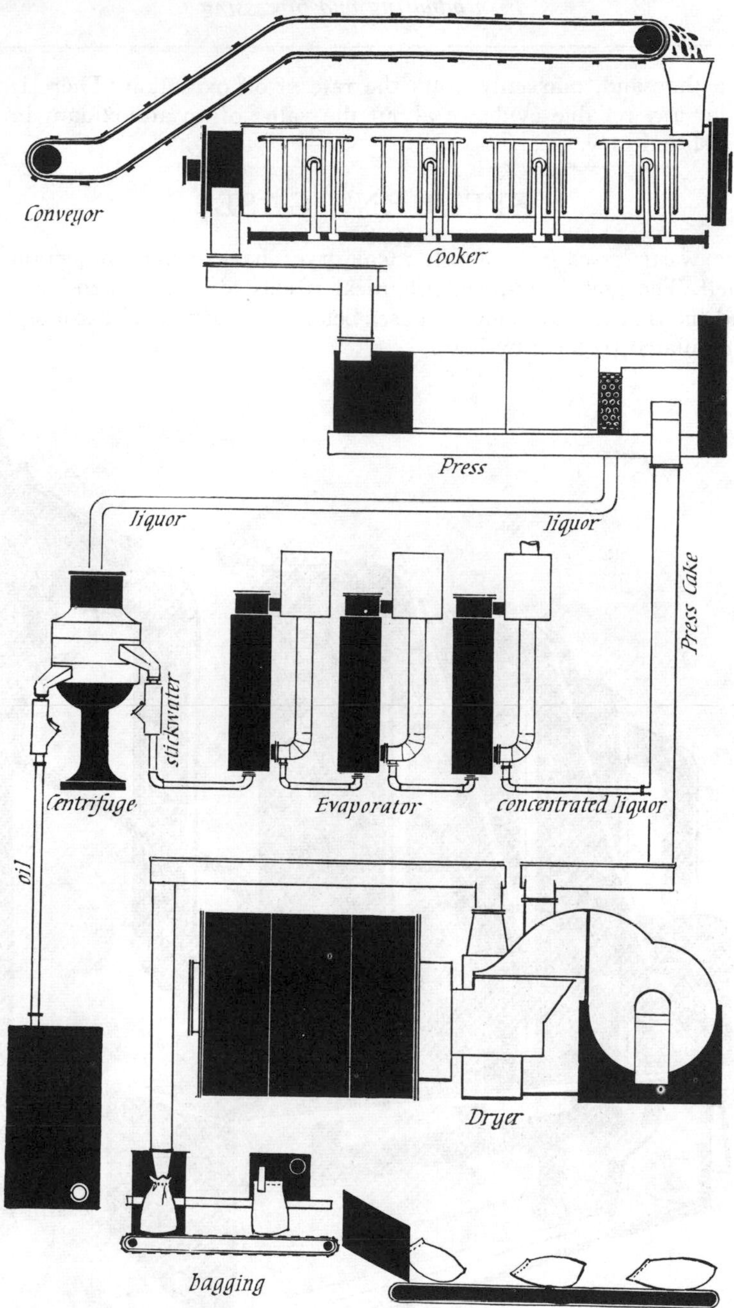

FIGURE 10·11 Sequence of main operations in making herring meal and oil

in a thousand, markedly slows the rate of oil oxidation. There is not so far any reliable evidence about the value of an antioxidant in the British industry.

EFFLUENT GASES

The waste gases from a fish meal dryer have a very objectionable smell. The manufacture of fish meal is classed as an offensive trade and the treatment of effluent gases before discharge to the atmosphere is regulated by local by-laws.

FIGURE 10·12
Scrubbing towers in a white fish meal plant

A steam-jacketed dryer discharges only a relatively small volume of waste gases. At all white fish meal factories these gases are collected into a common pipeline and led to a series of water scrubbing towers. Typically there are three of these. The gas enters at the bottom of the first, leaves at the top, is then piped to the bottom of the next, and so on. In each tower it meets a descending stream of water, either in the form of a series of cascades or as a spray. These scrubbers are not as efficient as those commonly employed in the chemical industry but they have the advantage of a low pressure drop, which means a big saving in fan power. The scrubbed gas still has a residual odour, which is usually treated by controlled admixture of chlorine gas or of a proprietary brand of *odour counterant.* Finally the treated gas may be discharged at a high level via the factory chimney, to ensure dispersal and dilution before it again reaches ground level.

The very large volumes of gas leaving a pneumatic dryer or a dryer-grinder present a difficult problem, which is currently the subject of active research both here and abroad. Ultimately it will be solved but at the time of writing it cannot be said that anyone has a complete answer. Adequate scrubbing at these volumes and flow rates, with only a low pressure drop, is one problem. But so is the nature of the effluent, which differs qualitatively as well as quantitatively from that leaving a steam-jacketed dryer. A great many very fine particles are present, removal of which by scrubbing is not simple. There are objectionable components which are not water-soluble, for example oily substances. The industry and the authorities are co-operating with the intention of alleviating this objectionable feature of the process.

COMPOSITION OF FISH MEAL

The moisture content of fish meal will vary somewhat, not only with conditions in the factory during its production, but with the humidity of the atmosphere if it is subsequently stored with fairly free access of air. Control of the moisture content of scrap leaving the dryer is largely a matter for the skill and experience of the foreman. However, some dryers, particularly hot air dryers, can be so instrumented that they will automatically deliver meal of a fairly constant moisture content, even if the feed rate or the moisture content of the feed varies considerably. Various instruments are used in the factory itself to determine the moisture content of any of the products as often as required for production control. More accurate values on the final meal are subsequently determined in a laboratory. Ideally the moisture content should be in the region of 10%. Appreciably higher values, say 15% or more, will permit the growth of moulds. Appreciably lower values, say 5% or less, involve the risk of overheating during manufacture, are wasteful of fuel and, in the case of herring meal, increase the risk of later overheating by oil oxidation.

Under the Fertilizers and Feeding Stuffs Act certain analytical data must be furnished when fish meal is sold. These are the content of albuminoids, phosphoric acid, oil and salt. All will vary inversely with the moisture content, which should be known. *Albuminoids* is a term for crude proteins, being defined as total nitrogen, other than ammonia or nitrate nitrogen, multiplied by a factor of 6·25. Phosphoric acid, mainly derived from the bones present, is stated in terms of phosphoric anhydride, P_2O_5. This does not mean that this particular compound actually occurs in the meal; it does not. It would be just as realistic to declare the result in terms of the element phosphorus itself. The same is true for albuminoids; it is actually nitrogen which is measured. But these are conventional notations of world-wide acceptance in trade. As would be expected, white fish meal, being mainly derived from offal, contains more phosphate and less oil than herring meal. Typical values, assuming a moisture content of 10%, would be fairly near the following:

white fish meal: albuminoids 65%, P_2O_5 8·5%, oil 4·5%, salt 1·5%;

herring meal: albuminoids 70%, P_2O_5 5%, oil 9%, salt 1·5%

The purchaser buys fish meal principally for its content of high grade protein. Some of the nitrogen, other than ammonia or nitrate nitrogen, is not of protein origin, although most of it is. A small and variable proportion represents what are collectively known as extractives, including such compounds as urea and trimethylamine oxide. If the proportion of these is unduly high, the conversion factor of 6·25 gives a falsely high value for the protein content. This would be the case, for instance, if a large proportion of dogfish, skate or shark had been present in the raw material. Occasionally relatively small quantities of meal are made from the processing waste of crustaceans such as crabs, lobsters and prawns. This contains quite a high proportion of non-protein nitrogen, derived from the substance *chitin* in the crustacean shell. Here the factor 6·25 is much too high.

A further problem is that of protein quality. For reasons not yet fully understood, some batches of fish meal are found in animal feeding experiments to be inferior to others, even when compared on the basis of a uniform protein content. This is not due to variation in the raw material so much as something having happened during or subsequent to manufacture. Although serious scorching can lower the quality of the protein, this is not the usual cause, which remains obscure. Luckily it is very rare to find low-grade protein in meal of British manufacture. Studies are now in progress in both this country and abroad designed to find a reliable and yet quick way of testing protein concentrates, including fish meal, for protein quality. The idea is that eventually they will be graded, that is a buyer would know not only how much nitrogen a particular fish meal contained, but how much of this nitrogen he would need to produce a desired balance in his mixed feeding rations.

Apart from protein and certain minerals, such as calcium and phosphorus, fish meal contains useful amounts of a whole range of vitamins. However, it is not bought on the basis of its vitamin content. It must be remembered that other components of the ration also contain useful incidental amounts of vitamins and it is simpler to evaluate fish meal on the basis of what it outstandingly contributes, high quality protein.

ALTERNATIVE PRODUCTS

In Britain, fish meal for use in animal feeding is virtually the sole product at present made from fish surplus to the requirements for human food and pet food. In some other parts of the world there is a production, either commercial or experimental, of alternative products. Since some of these are of considerable interest and possible future significance to the British fish meal industry, they will be briefly discussed here.

FISH FLOUR

The term *fish flour* is often applied to a fish meal prepared for human consumption. Such products are usually intended for use in areas where, mainly through poverty, the diet of most of the people is grossly deficient in good quality protein. The worst effects of such a diet are seen in young children and in pregnant and nursing women. In attempting to alleviate this deficiency by the introduction of special supplementary foods, local taste preferences and local cookery practice must be kept in mind.

First, let there be no doubt as to the value of fish meal in the diet of such people. Apart from what is known from animal nutrition, numerous studies have been made in various areas on protein-deficient human populations. Secondly, it is a supplement; it must be added to some other dish. Thirdly, it is a concentrate, one part of fish meal being roughly equivalent to five parts of fresh fish. Hence its price will seem high to the uninstructed buyer. Fourthly, the more it is refined, the more expensive it will become.

If the local preference is for a strongly flavoured product, for example for addition to soups or stews, the ideal would be a normal fish meal made under hygienic conditions from selected raw material. According to local preference it might be specially finely milled or screened from larger fragments of bone. If a tasteless and odourless product is needed, or one nearly so, quite elaborate extraction processes must be employed on the raw fish or on the finished meal. Not only must all oil be removed, but many trace constituents. Some such products, even if almost tasteless when freshly made, will develop a taste again if kept for some time. It is possible to mask some residual flavour by adding the fish flour to, for example, bread. This was recently done in South Africa, under a subsidized programme to improve the nutrition of the poorer section of the population.

It is extremely doubtful whether fish flour could be produced in a country with a high standard of living and exported on any large scale to some poverty-stricken area, for direct retail sale at a profit. A government subsidy would be needed. Further, a great deal of propaganda might be necessary to convince people that the fish flour was worth its cost to them, even if they only paid part of this cost. More hopeful would be a scheme whereby the development of local fisheries was encouraged and fish flour made locally as a means of preserving and distributing this fish. Such projects are being fostered in some areas by the local governments, often under the guidance of some suitable agency of the United Nations. This means that the more advanced countries export ideas, processes and perhaps machinery, rather than products. All this is not to say that limited amounts of fish flour might not be exported at a profit to certain areas. After all, drying is a well-known method of preserving fish. This is likely to apply only to areas where a good strong flavour is preferred, the flour being bought almost as a condiment rather than for its nutritional value.

In sponsored and subsidized schemes to alleviate protein deficiency, fish flour must compete with other products, such as skimmed milk powder, of which a considerable surplus is available in some parts of the world. For local production, and for a continuing self-supporting scheme, fish is a more promising raw material than milk. But it costs money to convert it into flour. Near the catching area it should be cheaper to eat the fresh fish, distributing the flour to more distant areas or storing it against seasonal shortages of fresh fish.

LIQUID PRODUCTS

Several processes have been developed to avoid the cost of the elaborate plant, fuel and so on, needed to dry the fish. Moreover, it is claimed that even under good conditions drying to fish meal will slightly lower the quality of the original protein.

The first of such processes to reach appreciable commercial production was the conversion of white fish offal into a liquid *silage* by acidification and self digestion. It is outstandingly a Danish development. The offal is simply minced and mixed with acid, which prevents putrefaction. If a mineral acid, such as hydrochloric or sulphuric, is used it is necessary to add sufficient to lower the pH to 2·0–2·5. This is very acid, the product is corrosive, and must be partially neutralized, for example with lime, before feeding it to livestock. By using a more expensive organic acid, typically formic acid, the pH need only be lowered to 4·0–4·5 for satisfactory preservation and it is not necessary to neutralize before feeding. After acidification a storage period of about two weeks results in almost complete liquefaction to a viscous, grey opaque mass. It will keep for a long time, certainly many months.

In a later development a silage more akin to the usual farm product was made, by admixture of molasses with the fish and inoculation with a particular bacterial culture. This ferments the molasses to produce an acid, lactic acid, which affords preservation. This bacterial process is far trickier to control than the simple acidification treatment and in unskilled hands can easily lead to putrid products.

There are several points to bear in mind when evaluating fish silage. It contains all the water of the original fish, that is about 80%. The price for a ton of fish silage may seem attractive but the price for a unit of protein will be very different. Approximately five times as much of silage as of fish meal will be required to balance a particular ration. This may add undesirable bulk and would certainly be a problem to the compound maker. All that water must be stored and transported. It cannot be distributed in bags but requires more expensive, and probably returnable, containers. It seems essentially more suitable for exploitation in areas where the farms are near to the fishing ports and where the farmer mixes his own rations rather than buy compounds as in Denmark.

It should further be noted that oily fish or offal is really unsuitable raw material. The cold silage is too viscous for efficient oil separation by centrifuging. If it is heated for this purpose, one of the sales advertising points, no heat damage, will be lost. Nevertheless, there have been commercial developments in production of silage from oily fish.

If no attention is paid to possible heat damage, it is practicable to concentrate a liquefied fish product in the same way as stickwater and to about the same final water content, 50%. An American development on a fair scale has been the production of such a liquid fish. The details of liquefaction have not been published. The advantage is the high thermal efficiency of a multiple effect evaporator.

Under the conditions obtaining in Britain it seems most unlikely that fish meal will be replaced by liquid products of high water content. There is a considerable international trade in fish meal and it is difficult to imagine a dilute, liquid product able to stand the transport costs involved, even if the users desired it.

11

Retailing

She wandered lonely as a cloud
To cry her wares o'er hill and dale.
While she herself was much too fresh
Her stock-in-trade was much too stale.

ANON

A THOROUGH knowledge of the proper way to handle fish is essential for the successful fishmonger or frier. He needs to know a great deal about many subjects; but he must be able to display his wares attractively, to know where he can buy the types of fish best suited to his customers, and to keep adequate records of his takings and expenditure. But this chapter is concerned only with the scientific aspects of fish handling. The care of the fishmonger's fish is so important, however, that it influences, or should influence, the whole design, layout and running of the shop.

There is no question but that the standard of retail handling of fish and fish products is not as high as it ought to be. Many shops are dirty, unhygienic and in a bad state of repair; what is sold to the public is sometimes of doubtful freshness (see Table 11·1). Some shops are very good and the quality of some fish that is sold is excellent, but there is no doubt that much harm is done to the industry as a whole by reason of the sale of poor quality fish. A considerable proportion of the very poor fish that is sold would in fact still be in edible condition by the time it reaches the consumer if it were properly handled.

TABLE 11·1

Percentage of samples of wet fish purchased in shops throughout England (1959) falling within certain eating quality

Number of samples: 750

Taste panel score		*% of samples*
3·9 and below	(very bad)	5%
4 to 4·9	(bad)	10%
5 to 5·9	(poor)	30%
6 to 6·9	(moderate)	40%
7 to 7·9	(good)	15%
above 8·0	(very good)	none

Scientists have been investigating various aspects of the industrial handling of fish for a number of years. The results of much of this work have over the years been given wide publicity in the trade press and elsewhere. It has been shown that the temperature of much of the fish, whether wet, smoked or frozen, reaching the public is far too high. When a fishmonger receives wet fish, for example, it is very often warmer than it ought to be and, under the usual conditions of handling in shops, it continues to warm even further until it eventually is sold. Temperatures of fish up to 80°F have been measured in shops on occasions. In fact, no differences in the temperature of wet fish in open-fronted and closed-fronted shops have been demonstrated, and this aspect of the design appears less important for temperature control than the general inadequacy of most of the measures usually adopted for keeping fish cool.

Smoked fish, which cannot be satisfactorily transported with ice, has been found to be in general warmer than wet fish. Frozen fish, which presents special problems of distribution and handling, has also often been found to be at too high a temperature. Frozen fish deteriorates rapidly as its temperature rises, even though it may still look and feel hard frozen.

Previous chapters have dealt with the details of the various processes used in the fish industry. This chapter attempts to bring together the various pieces of information that are of interest and use to the fishmonger so that some repetition has been unavoidable.

THE HANDLING OF THE COMMODITY

WET FISH

Fish goes bad mainly because of the activities of bacteria. Bacteria occur naturally on the skin, gills and in the intestines of a healthy fish but they do it no harm and they may even be beneficial. When the fish dies, however, the bacteria do not; they continue to grow and begin to feed upon the flesh of the fish itself. However much care is taken by the fisherman, some contamination of the fish with fish spoilage bacteria is inevitable.

The speed with which fish bacteria grow on fish flesh largely depends upon the temperature. Gutted cod, for example, if taken straight out of the sea and kept at 32°F by burying it in crushed ice, will remain fit to eat for 15–16 days. If it is stored at 42°F, it will keep for only $5\frac{1}{2}$ days and at 52°F it is objectionable after three days. The first essential in handling wet fish is therefore always to keep it as close to 32°F as possible, preferably by burying it in crushed ice.

The fishmonger does not, of course, receive his fish straight out of the sea. It takes four or five days to reach port from many of the distant water fishing grounds. The freshest fish landed from vessels fishing in those areas cannot, therefore, be less than four to five days old and, since trips last 20 days or so, the stalest fish may be 15–16 days caught. About 1% of the distant water catch is in fact condemned at landing as being unfit for human consumption. Therefore, fish reaching many fishmongers may become inedible before sale because the fish is not absolutely sea-fresh when it reaches the shop but has already been through the distribution network. It is best to hold stocks of wet fish for no more than 24 hours or so. As soon as boxes are delivered, they should be opened and the fish removed. Any that is not for immediate processing and display should be packed into a clean box with plenty of ice at the bottom and top and placed in a chill store.

There is a common belief that the best way to store wet fish is to put it into a chill store and in the past most fishmongers have used very little, if any, ice. The use of chill stores alone for cooling fish is bad practice. The appearance of the fish may rapidly be spoilt by drying of the surface, but there are other, and more serious, objections. The speed with which heat from the fish is removed by melting ice is very much faster than the speed with which it is removed by cold air. Even if the chill store thermostat is set at a temperature much lower than that of melting ice, 32°F, the speed with which the fish cools down is still very slow. A five-stone box of fillets that are all at 80°F, for example, may still be well above 40°F in the middle of the box after 24 hours when stored in a chill running at 32°F. If by any chance fish at about 32°F is put into a chill whose thermostat is set at 32°F there is a very real danger of slowly freezing the fish on the surface, since most chill room thermostats are made to turn on at a

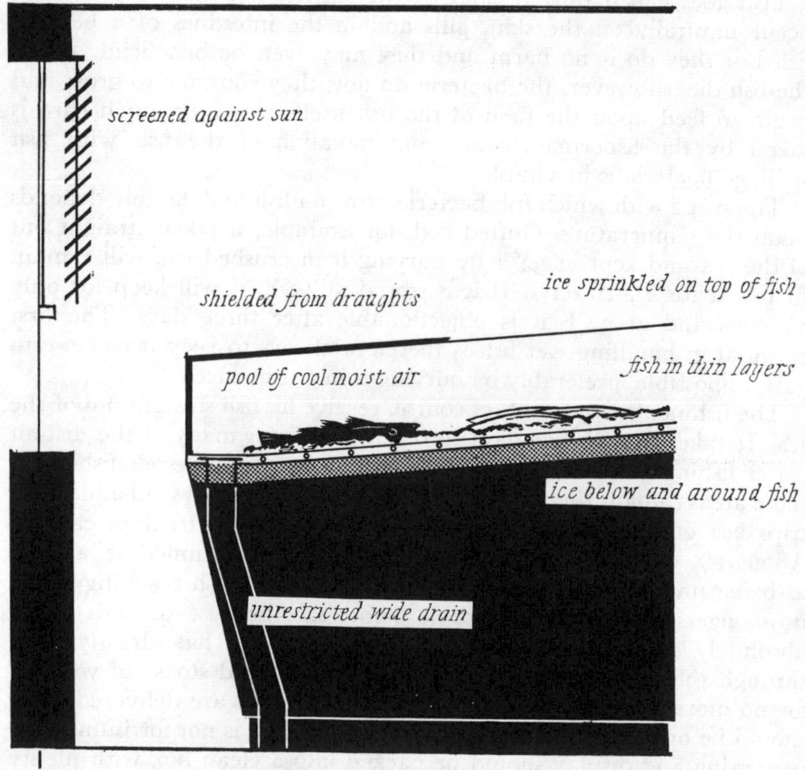

FIGURE 11·1 Fresh fish ideally displayed on ice

slightly higher temperature, and off at a slightly lower one, than the temperature actually indicated by the setting. In practice, a thermostat set at 32°F may turn on when the temperature rises to 34°F and off when it falls to 30°F. Many thermostats are in addition not accurately adjusted, so that they may keep the temperature a few degrees higher, or lower, than the setting indicates.

The best way to store wet fish is therefore to ice it well and then put it into a store set at about 34°F, when there should be no risk of freezing it. It is a good idea, however, to hang up two or three thermometers in different parts of the store to make sure that the setting is not too low. These thermometers should not be hung on or near the cooling grids, but near the fish itself.

Fish that is to be processed should not be allowed to lie in the shop or processing room. It should be removed from its box, washed and processed. The washing water should be kept cold by adding crushed

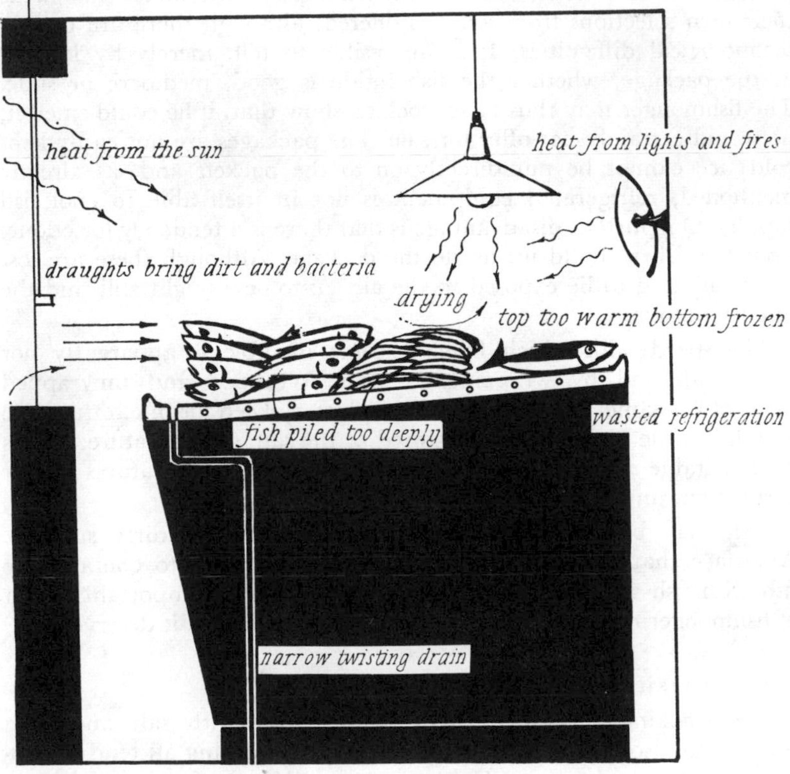

FIGURE 11·2 Bad practice in retail display

ice to it. Processed fish, whether filleted, cut in steaks or merely gutted, should be put at once into crushed ice. When it is to be displayed, fish should be laid in thin layers on a good bed of crushed ice. Customers should be served with fish from the slab. Some fishmongers lay out a display of uniced fish on the slab which they leave intact; this is not good practice because a thin layer of fish without ice can warm up very rapidly.

Refrigerated slabs and display cabinets should not be used for cooling fish. They will generally keep fish cold that is already cold, although even so some models dry the surface rather badly. Since fish is very rarely as cold as it ought to be when it reaches the fishmonger, it is much better practice always to use ice in conjunction with this equipment.

Of recent years there has been some interest in the trade in the possibility of pre-packing wet fish. The attractions of this method of presentation are considerable. For example, customers can make their own selections from what is offered, although there are certain technological difficulties. It is impossible to tell, merely by looking at the package, whether the fish inside is good, mediocre or stale. The fishmonger may thus have stock on show that, if he could smell it, he would certainly not offer for sale. The packages are not easily kept cold; ice cannot be put directly on to the packets and, as already mentioned, refrigerated equipment is not in itself able to cool fish rapidly. One further disadvantage is that there is a tendency for odours from the fish to build up inside the package. Although these are lost if it is allowed to lie exposed to the air, customers might still find the smell repulsive.

The speed with which fillets become inedible is apparently not greatly affected by wrapping as such. Wrapped and unwrapped fillets of the same age and temperature history keep for much the same length of time if they are both stored at the same temperature. Fillets in a suitable vacuum pack may keep at chill temperatures rather better than unwrapped fillets.

Fish may become contaminated by contact with dirty surfaces. A surface that looks clean may yet be able heavily to contaminate fish with fish spoilage bacteria. Shop hygiene is an important aspect of fishmongering that is not always given the attention it deserves.

SMOKED FISH

The smoking process involves treating fish with salt and with smoke and also drying it. Salting, smoking and drying all tend to preserve fish against spoilage. Consumers nowadays prefer their fish very lightly smoked and dried, however, and modern products therefore

TABLE 11·2

Storage life of smoked fish

Species	*Smoked product*	*Storage life*			
		room temperature 60°F		*chill temperature 32°F*	
		in first class condition	*remains edible*	*in first class condition*	*remains edible*
Cod	single fillets cold smoked	2–3 days	4–6 days	4–6 days	8–10 days
Haddock	single fillets cold smoked	2–3 days	4–6 days	4–6 days	8–10 days
	block fillets cold smoked (golden cutlets)	1–2 days	2½–3 days	4 days	6 days
	finnans cold smoked	2–3 days	4–6 days	4–6 days	10–14 days
	pales cold smoked	1–2 days	2½–3 days	4 days	6–7 days
	smokies hot smoked	1–2 days	2½–3 days	3–4 days	5–6 days
Herring	kippers and kipper fillets: cold smoked unwrapped	2–3 days	5–6 days	4–6 days	10–14 days
	wrapped	1–2 days	3 days	3 days	3–4 days
	bloaters cold smoked	1–2 days	2–3 days	3–4 days	5–6 days
	buckling hot smoked	1–2 days	2–3 days	3–4 days	5–6 days
Salmon	fillets cold smoked	2–3 days	4–5 days	4 days	10 days
Trout	whole gutted hot smoked	3 days	7 days	6 days	10 days

do not remain edible for very long. The most that can be expected from products available commercially is a shelf life of a day or two at 60°F. Kippers, if made from really fresh herring and smoked reasonably well, will remain edible at 60°F for a maximum of about a week. Smoked white fish will become inedible after three to four days at 60°F.

Smoked fish should not, therefore, be kept in the shop for more than three or four days. It is better to order small quantities at intervals of two or three days than to order larger quantities less frequently.

Smoked fish cannot be iced to keep it cool. Some producers add a few knobs of dry ice or solid carbon dioxide in an attempt to keep the temperature down during distribution. Others freeze their products before despatch so that they thaw during distribution and arrive at their destination ready for sale but still cold. Both methods, if properly used, can help to maintain quality during distribution. It is as well to examine boxes of smoked fish as soon as they are delivered.

Smoked fish warms very quickly almost to the temperature of the air and therefore a minimum quantity should be displayed and the rest should be put into a chill store. It will cool most rapidly if it is spread out on trays. Since smoked fish freezes at a lower temperature than wet fish, these trays may be placed quite close to the cooling

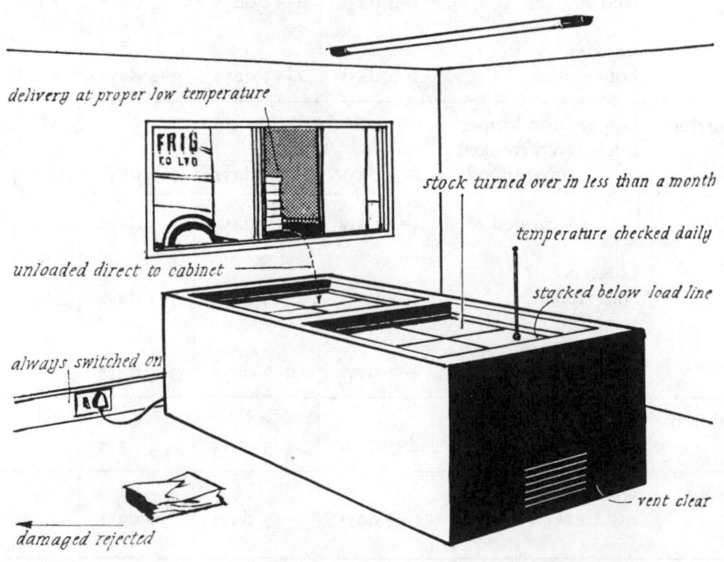

FIGURE 11·3 Correct use of a zero cabinet

grids even in a store running at 30°F. It is in the presentation and display of smoked fish that the refrigerated slab and refrigerated display cabinet are probably most useful.

Smoked fish readily develops moulds. Moulds are especially liable to appear during the warm and humid weather that often occurs in August and September. Wrapped kippers are more susceptible than unwrapped. Fishmongers should therefore examine stocks every day; it is as well to open a few packets of ready-wrapped products every morning in order to make sure that the contents are still in good condition. Table 11·2 gives the approximate storage life of various smoked products.

FROZEN FISH

A considerable proportion of the fish frozen in Britain is in the form of catering packs. Many fishmongers at certain times sell the fish from these thawed blocks; it is common practice for suppliers to send the frozen blocks packed in ordinary boxes so that they are thawed and ready for sale by the time they reach the shops. Although this practice cannot be viewed entirely with favour there are no special points for the fishmonger to watch in the handling of thawed fish. It should be treated as if it were ordinary wet fish.

Much frozen fish is, however, packed in small cartons and stored in so-called *zero cabinets* in shops. There are a number of points to

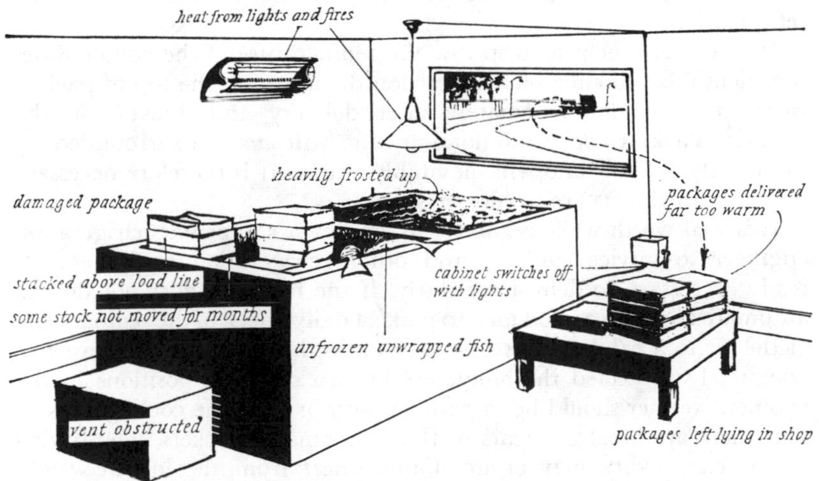

FIGURE 11·4 Misuse of a zero cabinet

watch in handling this material which are not always observed in practice.

Frozen fish will not remain indefinitely in good condition. Its quality deteriorates during storage, even though bacteria on the flesh stop growing at temperatures around 15°F. Chemical changes occur in the protein of the flesh; the thawed product when cooked becomes tough and stringy and unlike fresh fish. In addition, the flavour becomes unpleasant; the fat of oily fish such as herring, salmon and trout will combine with the oxygen of the air to give rancid flavours which can be highly objectionable. The speed with which these changes occur depends upon the storage temperature of the commodity. Even under ideal conditions, however, when really fresh fish has been rapidly frozen and stored at minus 20°F or lower, the product will slowly deteriorate. The expected storage life of well-frozen fish stored at three different temperatures is given on page 150, Chapter 7.

The zero cabinet is built to hold stocks of frozen fish only for short periods. It is not made to freeze foodstuffs and any attempt to use it for this purpose can only result in damage to the commodities already stored in it. Even if a cabinet is able to maintain a temperature of 0°F, and those that conform to British Standard 3053: 1958 should be able to do so, packages should not remain in it for longer than a month. Frozen fish is not always as cold as it ought to be when it is delivered to shops. Deliveries of stocks to shops should be from insulated vans; many firms use solid carbon dioxide to prevent heat that leaks into the insulated container from warming the product. Increasing numbers of road vehicles are fitted with mechanical refrigeration.

It would also help to keep low the temperature of the commodities if a plentiful sprinkling of solid carbon dioxide over the top of packets were to be made immediately after delivery and stacking in the cabinet. Packets above the loading line will not be surrounded by sufficiently cold air and will inevitably warm. It is therefore necessary to see that cabinets are never overfilled.

It is well worth while regularly to employ a competent refrigeration engineer to service cabinets and other refrigeration equipment. It is also necessary to defrost regularly, if the machine does not do this automatically. It is a good idea to make a daily check of the temperature of the air in a cabinet. The best way to do this is to place at least two adequately protected thermometers in two different positions in the cabinet. Neither should be in contact with or near the cooling coils.

If there are breaks or tears in the wrapping of packets, then drying of the commodity may occur. Quite apart from the loss in weight involved, the final thawed product is dried and unattractive. This condition is known as freezer burn.

Spotlights should not be fitted close to, or shining into, a frozen food cabinet, which should always be sited away from the direct rays of the sun. Draughts will destroy the well of cold air upon which the efficient functioning of the cabinet depends.

COOKED FISH PRODUCTS

Wet fish becomes quite inedible before it is likely to cause bacterial food poisoning. Fish spoilage bacteria do not themselves cause food poisoning. When cooked fish products are made, however, there may be contamination in the factory, during distribution or in the shop, with other types of bacteria that, if allowed to grow and increase in numbers, can cause food poisoning. Food-poisoning bacteria do not grow, however, at chill temperatures. Fish cakes and similar cooked products, therefore, if obtained from a reputable supplier and kept in the shop at not more than 40°F for not more than three days are unlikely to be a danger to health. It is essential to stress the importance of keeping the temperature below 40°F.

CLEANLINESS

GENERAL

The word clean is often used in everyday speech to mean absence of visible dirt. An article that appears to be clean in this sense may, however, still carry bacteria capable of causing food poisoning. In this section 'clean' means free of these bacteria as well as visibly clean. Clean food, which is free of any risk to health, can only be ensured by hygienic handling. *Hygiene* is the system of rules devised to prevent contamination. The Food Hygiene Regulations 1955 lay down the minimum requirements for the maintenance of shop hygiene.

Cleanliness is important if a shop is to look attractive. It is also important in preventing contamination of foodstuffs not only by bacteria that cause food poisoning, but also by those organisms that spoil the commodity by making it distasteful. In practice it is impossible to separate the various aspects of cleanliness.

PERSONAL HYGIENE

It is very necessary that a high standard of personal hygiene be maintained. No smoking or spitting on any part of the premises should be permitted. There should be water closets in good condition and washing facilities should be adequate. There should always be an abundant supply of hot water, soap and paper towels. Hands should be washed immediately after using the lavatory. Fingernails should be kept short and clean. Suitable protective clothing and head gear should always be worn and it should be kept clean. It should also be worn only within the premises.

Cuts and abrasions should be dressed immediately; wound dressings should always be clean. Boils, pimples and septic spots should always be kept covered; bacteria from infected spots are a common cause of food poisoning. Coughing and sneezing may also be responsible for infecting food with food-poisoning bacteria; it is essential always to use a handkerchief. A medical check on the health of people handling food should be carried out regularly. This is just as important in a small fish shop as it is in a large food factory. Many of these points are, in fact, incorporated in the Food Hygiene Regulations, but they must be observed all the time, and not merely when the Public Health Inspector is in the vicinity.

SHOP DESIGN

One important characteristic of a fish shop is that it shall look attractive. This aspect of shop design cannot be considered here; but it must be remembered that any shop, however attractive it may be, that is difficult to keep clean is *badly designed.*

Probably the biggest controversy amongst fishmongers is whether or not the closed-fronted shop is better than the open-fronted one. In the survey of fish temperatures carried out a few years ago, very little difference in the temperature of fish in the two types of shop was found. One possible explanation for this is that closed-fronted shops are usually well ventilated. From the general considerations of hygiene, the closed-fronted shop is certainly to be preferred to the open-fronted one.

It is a cardinal principle of clean food handling that all surfaces should be easily cleaned and readily accessible.

It is essential that all floors and walls should be constructed of suitable impervious material. Concrete floors are not usually very satisfactory. They absorb fish juice and it is difficult to prevent them smelling of stale fish. They are also liable to crack. Asphalt floors, to which granite chips have been added in order to render them less slippery when wet, are widely used and appear to be satisfactory. Tiled floors, also, are satisfactory although the joints between each tile must be kept in good condition. Walls should be coved to the floor. Tiles are probably the best form of wall finish.

Whatever the nature of the flooring material, it is essential that there should be a good drainage slope, if possible from the front to the back of the shop. Drains carrying water likely to contain pieces of fish can easily become blocked. It is therefore worth while to make sure that drains are adequate and that they are well trapped. One common source of fish smell is the rebate of drain covers. It should be possible easily to remove covers and to clean out the rebate.

Ceilings should be finished with some impervious material; at least they should be gloss-painted. Where it is possible to do so, self-closing doors should be fitted. The use of wood in the construction of the shop and its fittings should as far as possible be avoided. Counters, benches and tables should be made either of tubular metal, and be easily movable, or be built to the floor, coved and tiled. Wood absorbs fish juices and is impossible to clean and keep clean. Wooden counters are also often heavy and difficult to move; the fishmonger therefore has difficulty in cleaning beneath them. An adequate supply of piped hot water for washing walls and floors should be readily available in the shop.

When new shops are built or old ones modernized, it is commonly found that not nearly enough attention has been given to the design and layout of the preparation room and yard. The design and condition of these two is usually a very good guide to the general standard of working of the shop.

The finish of floors, walls and ceilings of preparation rooms should be as good as that of the shop itself. All equipment should be easily cleaned. There should be no inaccessible corners. Any equipment that is no longer used should be removed. There should be a separate room for storage of dry materials such as wrapping paper and cartons. Adequate provision should be made for offal receptacles, which should always be kept covered when not in use.

The yard should be covered with concrete, tar-macadam, or some similar material; earth or brick yards are not suitable. A tap and hose should be readily available, and the yard should be properly drained. Gully and channel drains should be freely accessible. Offal bins should be covered in a suitable out-building, but if they must stand outside, then a proper concrete plinth should be provided. Fish boxes should be scrubbed as soon as they are emptied and should then be stored in an out-building. There should be access to the yard from a carriage road behind the shop; it should not be necessary to carry bins of offal or boxes of fish through the shop.

PESTS

The main pests associated with fishmongers' shops are flies. Other pests are occasionally encountered; rats and mice may be found in old buildings. Measures for the eradication of these vermin are best carried out with the advice of the local Health Inspector.

Shops are occasionally infested with cellar slugs. These pests live in surroundings that are permanently moist; they lay their eggs in cracks and crevices and feed on scraps of fish. They cannot exist in a shop that is properly cleaned.

Infestations with flies are more difficult to control. The flies that are commonly found in shops are of two main kinds, blowflies, that is blue bottles and green bottles, and houseflies. Blowflies are rather large, heavy-bodied creatures that buzz loudly when in flight. Houseflies are smaller and remain in flight for long periods. They tend to circle round and round and, if there are a large number present, they may appear as a cloud in one corner of the shop. They settle on objects such as lampshades, or upon ceilings, and leave the familiar and objectionable black marks that are caused by their excrement and vomit. The numbers of houseflies in a shop may increase from day to day in summer unless measures are taken to keep them down.

Houseflies lay their eggs in rubbish, such as kitchen refuse, rotting vegetable matter or in dung. Creamy-white maggots, tapering to a point at the head and blunt at the rear, hatch out within a few hours if the weather is warm and grow rapidly. Under warm conditions when there is plenty of food available, they may reach a length of ½ inch or so within three days. They turn into a brown *puparium*, similar to the chrysalis of butterflies and moths. Finally, after a resting period when nothing appears to be happening, the adult fly bursts out of the end of the puparium, its wings expand and dry and it flies away in search of food and a mate. The whole cycle may take as little as nine days in warm weather, but it may take longer in cold weather. Houseflies tend to remain around the same spot and, if suitable rubbish dumps are available, will remain and breed on them from generation to generation.

Blowflies have quite different habits. They lay their eggs on any meat, fish, dead animals and even rotten eggs. The staler the material the more attracted they are to it. The life cycle is rather like that of the housefly with one or two important differences. The maggots or gentles are larger and, when full-grown, leave their food and bury themselves in dry soil, cracks in walls and similar situations. When they emerge as adult flies they may fly relatively long distances in search of food and a suitable place to lay eggs.

Houseflies are mainly attracted to fish shops because there they can find food and a fairly humid atmosphere. Blowflies are attracted to shops mainly because fish is a suitable material in which to lay their eggs. Most of the blowflies entering shops are female ones laden with ripe eggs. It is the blowfly, crawling over the surface of fish and, if permitted to do so, laying eggs in fissures and cracks in the food that customers find so revolting.

It is these differences between the habits of the two kinds of flies that it is important to know if control is to be effective. The first essential is, of course, that the shop shall be kept clean at all times. Flies, blowflies especially, are attracted by the odours of stale fish; any measures that keep down these odours help to reduce the fly problem. Methods for the control of flies must be applied regularly, even when there is apparently no problem. Carelessness on one day is only shown nine or ten days later.

Strong-smelling solutions do not necessarily destroy odour, although they may prevent *humans* from detecting an underlying smell of stale fish. Every possible source of stale fish odour must be satisfactorily dealt with. Offal buckets and similar containers should be fitted with lids and emptied and scrubbed out with a suitable disinfectant cleaning fluid at least once a day. Offal bins should be covered and stored under cover; they should be emptied daily if possible and should never remain on the premises for more than two days. If they must stand in the open they should stand on a plinth, and walls and fences in their vicinity should be sprayed fortnightly with DDT dispersible powder or emulsion (1 lb of 50% dispersible powder to a gallon of water for every 1000 square feet of surface). The best type of offal bin is one with a heavy lid and deep lip. Bins should be well sprayed with DDT emulsion or dispersible powder every week. A bucket-type pump is the most suitable for this purpose. 2 oz of 50% dispersible powder in 1 pint of water, or the equivalent quantity of emulsion, is recommended.

Insecticides should not be used in the shop while food is on display. Haphazard spraying when the shop is open is not likely to be very effective and, apart from this, there is a danger of tainting fish with the spray chemicals. Some fly sprays contain poisonous chemicals. Dead and expiring flies on the slab are also likely to be even more revolting to customers than live bluebottles buzzing about in the window.

Premises should be sprayed every night, after the fish has been removed and the shop closed up. An insecticidal aerosol dispenser, vapour dispenser or an efficient hand sprayer may be used. The most satisfactory spray is one containing pyrethrum, a substance with quick 'knock-down' properties, and DDT or some similar compound which eventually kills the insects. It is important to choose a hand spray that gives a fine spray with no coarse droplets. After spraying, all corpses should be swept up and all surfaces, especially those likely to come into contact with food, thoroughly scrubbed and finally hosed down.

Although persistent insecticidal films, for example insecticidal lacquers, are sometimes recommended by suppliers, their use cannot be advised for food shops, because of the risk of partially paralysed flies falling on to the fish.

Plastic containers, refrigerated showcases and similar equipment where fish can be kept covered will help to reduce odours that attract flies. Fresh fish adequately chilled does not attract flies.

Fish boxes, as soon as they are emptied, should be scrubbed out with detergent solution, rinsed with hypochlorite solution, stacked on end and allowed to dry. Non-returnable fish boxes should not be stored at all, but should at once be cleaned and removed from the premises or broken up and burnt. Returnable boxes should be stacked under cover, if possible well away from the shop. Wooden boxes should be rinsed with rather strong hypochlorite solution containing ½–1 pint of commercial hypochlorite solution in 8 gallons of water. Aluminium boxes require much less, since they are non-absorbent; 1–2 fluid oz of commercial solution in 8 gallons of water is sufficient.

Finally if, after the greatest care has been taken, flies are still a problem, the health department of the local government authority should be consulted.

CLEANING

Detergents are used to remove dirt; disinfectants are used for killing bacteria. It is desirable to disinfect immediately after removing the dirt.

Some surfaces, many plastics, tiles and metal for example, are much easier to keep clean than porous materials like brick and wood. There should never be bare brick surfaces in a shop; wood should be used as little as possible. All surfaces in the shop should be of smooth and impervious material that can be easily cleaned.

All trays and other similar equipment should be washed and scrubbed in warm water with detergent immediately after use and rinsed thoroughly in hot water. Detergent should be used in the concentration recommended by the manufacturer; too little will be ineffective, too much may be corrosive. After rinsing, equipment should be sterilized with a *suitable* chemical disinfectant and then given a final rinse. Certain types are very strong-smelling and are quite unsuitable for use where food may become tainted. Chlorine, in the form of hypochlorite, is one of the best all-round disinfectants.

Walls, floors, counters and slabs should also be scrubbed daily after the shop has closed. If pieces of fish and slime are allowed to remain they will go bad and attract flies; they will also dry on the surfaces from which they cannot easily be removed. A suitable detergent should be used for scrubbing down these surfaces which should then be well hosed with water, and afterwards swilled down with buckets of disinfectant solution; this should be allowed to lie for the period recommended by the manufacturer. Finally every disinfected surface should be thoroughly hosed down. Many disinfectants are also corrosive and should not therefore be allowed to remain on surfaces for longer than the recommended period. They should also be used only in the strengths recommended.

12

Fish as living animals

> The fish, when he's exposed to air,
> Displays no trace of *savoir-faire*,
> But in the sea regains his balance
> And exploits all his manly talents.
> The chastest of the vertebrates,
> He never even sees his mates,
> But when they've finished, he appears
> And O.K.'s all their bright ideas.
>
> OGDEN NASH

FISH technology is concerned with the treatment of dead fish, but it should always be remembered that before fish can be dead it must be alive. Fish is not merely a commodity to be bought and sold; once upon a time, before ever it reached the quay, it was alive in the sea, fighting fiercely for food and, indeed, for life itself. What happened to it in life may influence its quality, how it keeps and its suitability for processing.

It is not intended to give here either a complete account of the biology of fishes or to describe the characteristics of British food fishes. To do this would require a volume or two to itself and it has already been well done. It is desirable, however, that everybody who has to handle fish should know something about the way they live, how they are constructed and how the types that exist in the seas, rivers and lakes of today came to be what they are.

WHAT IS A FISH?

Although everyone has a good idea what a fish looks like, many people might find it difficult to write down a description that would fit all types of fish from a shark to a plaice and yet would not include also whales and porpoises, which are warm-blooded mammals and not fish at all.

Fishes, whatever their shape, all have certain characteristics in common; for example, they all have backbones, they all have gills and they are all cold-blooded. Most of them also have, in addition to fins along the back and under the tail, two pairs of fins, rather as a man has two pairs of limbs, and a large vertical tail fin and not, as in whales, a horizontal one.

This description is, of course, very general; it has to be, in order to include such diverse creatures as sea horses and cod, dogfish and lung fish. Nevertheless, scientists have found that, in spite of the considerable variation in the appearance of different species of fish, they all have the same basic plan.

Cod may be taken as an example of a typical commercial bony fish. It is torpedo-shaped, is covered with a transparent slimy skin beneath which lie row upon row of scales from head to tail rather like tiles on a roof (see Figure 12·1). It has three vertical fins along the back, the *dorsal fins*, and two beneath the tail behind the vent, the *ventral fins*. In addition, it has a pair of *pectoral fins* and a pair of *pelvic fins*. Strange as it may seem, the main function of all these fins in the vast majority of fishes is merely to act as stabilizers and brakes. A few species move through the water by rapidly waggling their dorsal fin, as the sea horse does, but this is unusual and it is the *tail* which usually propels the fish. The tail is usually taken as the whole of the body behind the vent, and not as it is popularly understood, just the *tail fin*. Perhaps a third or even more of the weight of a cod fillet is therefore tail muscle!

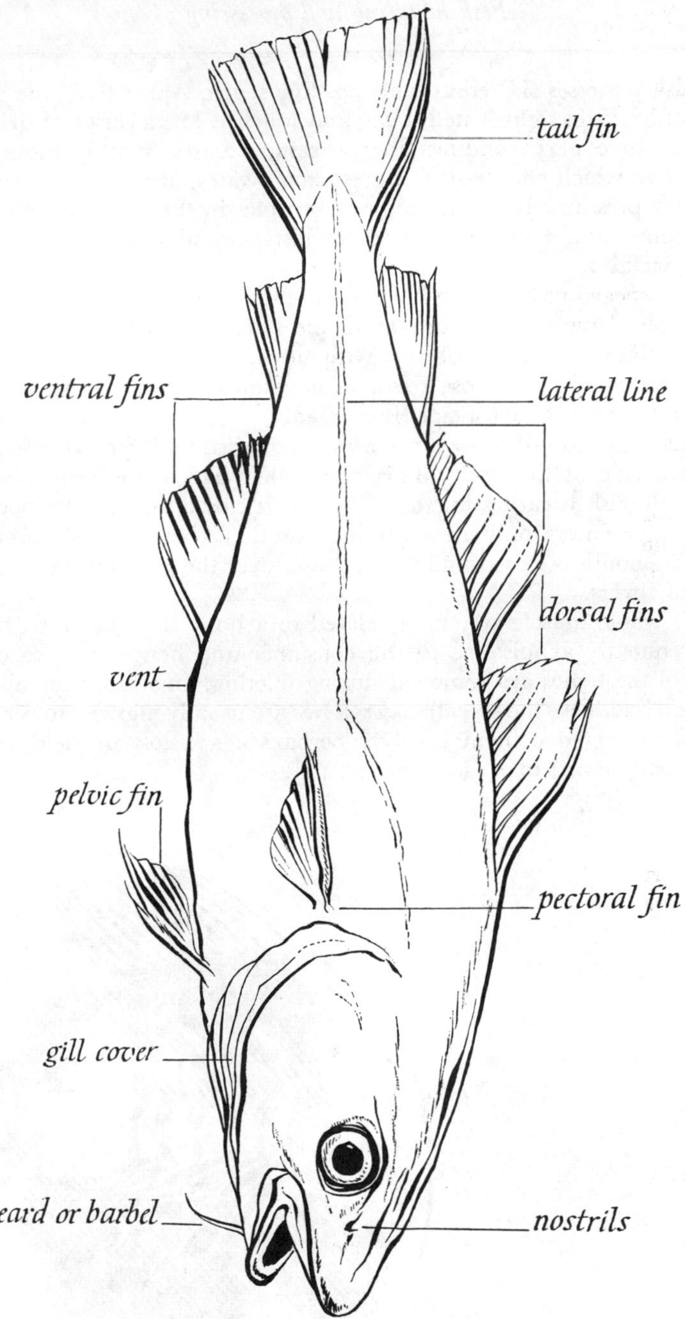

FIGURE 12·1 Diagram of a cod

A fish possesses six senses, and possibly seven. Apart from the usual ones of hearing, sight, smell, taste and touch it has a series of delicate and sensitive nerve endings in the skin, situated mainly along the *lateral line*, which enables it to detect small water currents and changes of water pressure. It can detect small ripples in the water which arise from some distant source such as, for instance, another fish swimming in its vicinity.

Some species of bony fish have a sense that is virtually an additional one. They have sensitive organs in the skin and on the fins and are able to 'taste' or 'smell' objects without having actually to eat them. Cod and many of its close relatives have an especially sensitive *beard* or *barbel* that is in fact a smelling organ.

Bony fish, too, all possess a very characteristic *gill cover* or *operculum* on each side of the body. This acts rather like a non-return valve. When the fish breathes in, the gill cover is closed against the body so that water enters only through the mouth. When the fish breathes out, the mouth is closed and water passes over the gills and out behind the gill covers.

The whole muscle system is related to a bony skeleton. Fish bones are frequently a nuisance to the consumer and hence the processor. Most of the bones are removed during filleting. In white fish, such as cod and haddock, the so-called *dorsal ribs* are usually allowed to remain, because to remove them involves perhaps a 5% loss in yield and a peculiarly shaped fillet (see Figure 12·2).

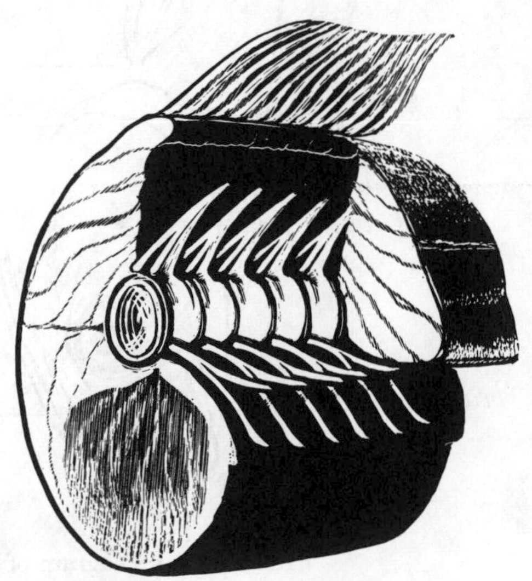

FIGURE 12·2 Steak of cod dissected to show dorsal and ventral ribs, and backbone

The herring has even more bones than most white fish. It possesses not only *dorsal* and *ventral ribs* but fine hair-like *intramuscular bones*. It is impossible to remove *all* the bones from herring and yet produce an attractive fillet. Short needle-sharp bones that support the dorsal fins are also sometimes encountered and can be troublesome.

Sharks, dogfish, skates and rays, which do not have bony skeletons but have *cartilage* instead, the so-called *cartilaginous fishes*, differ slightly from the bony fishes such as cod and herring; they breathe in water through a special hole just behind each eye, instead of through the mouth, and they breathe it out again through a separate series of gill slits, usually five, which lie on each side of the head.

The structure of the mouth of fish differs from species to species. For example, hake has rather long jaws armed with sharp teeth, cod has similar, though shorter, teeth and the bottom jaw is shorter and stouter, while the Arctic catfish has powerful jaws with conical teeth in front and flat crushing teeth on each side which are used for smashing the molluscs on which they feed. The angler fish, sometimes called the monk fish, has huge teeth that can lie pointing back into the mouth, or can swing out to point straight up or down; when an angler fish bites its prey the teeth swing forward and act like barbs. The mouths of most fish with skeletons of cartilage are not, as in the bony fish, at the end of the snout but lie some way beneath it. The teeth are also constructed differently; the precise differences need not be considered here. What is interesting is that there is as much diversity in the teeth of the cartilaginous fishes as there is in the bony ones, from the flat, crushing plates of some of the skates and rays to the sawlike teeth of some sharks and dogfish.

A fish does not usually masticate its food, but swallows it without chewing it. Some have teeth in their throats that are used for breaking up lumps of food, but none chew their food in the way that mammals do. Food passes straight into the stomach which is often a U-shaped structure capable of an incredible amount of stretching. Some bony fish, however, notably the roach, possess no stomach at all, and in this case the food, which is largely aquatic plants, passes from the mouth into the intestine. When food is available, a fish will frequently gorge itself; when it is not, it starves. Fish can survive periods of starvation lasting many months; indeed, in some seas, they are forced to make an annual fast either because food is not available or because they cannot hunt during the long Arctic night. Very many species starve when they are preparing to spawn. The roe of salmon, for example, grows to such a size that the gut becomes compressed and it is doubtful whether the fish could eat very much even if it had any appetite to do so. Types that live in very deep parts of the ocean may also be forced to go for long periods between meals, because of the difficulty of finding food at any time. Some of these deep-oceanic forms can swallow fish almost as big as themselves, so elastic are the walls of their stomachs and so large their gape.

Starvation is one possible cause of soft fish, that is, fish that easily tears and sags and is difficult to hang up for smoking or even satisfactorily to market fresh.

The stomach wall contains microscopic glands that pour out digestive enzymes as soon as food is eaten. This fact explains why the so-called feedy herring rapidly becomes soft and broken; enzymes in the stomach, and also the intestines, after the death of the fish, will digest any protein with which they may come into contact, including the stomach and intestines themselves. In a fish that is feeding heavily, large quantities of enzyme are produced.

Enzymes are manufactured in the microscopic glands in the lining not only of the stomach but of the intestine and, in the bony fish, of the so-called *pyloric caeca* (see Figure 12·3). These are peculiar hollow finger-like structures attached to the intestine near the bottom of the stomach and which are not found in other vertebrate groups, even, indeed, in the cartilaginous fish. Their number differs from species to species; the angler fish has only one, while the mackerel has nearly two hundred.

It has been claimed that the characteristic flavour of both bloaters and salted herring is partly due to the presence of gut enzymes. Bloaters are, of course, prepared from ungutted herring and, in herring cured in the traditional manner, great care was taken to ensure that a portion of the intestine was left in position after gutting.

Bile, which is produced in the fish liver, enters the intestine just behind the stomach through a fine tube. Also near this point enters the duct from the *pancreas,* an organ that produces some of the digestive juices. The entire intestine of all fish is short, perhaps at the most twice the length of the body. It is a characteristic of vertebrates that the flesh eaters have short intestines compared with those of the vegetarians. Very few fish are vegetarians.

The apparent length of the gut of the cartilaginous fishes is even further abbreviated by the presence of an interesting portion called a *spiral intestine.* The main purpose of the gut is to digest food and absorb it through its walls into the body. The spiral intestine is a structure that reduces the apparent length without decreasing the absorptive area; if the normal intestine is looked upon as a lift shaft, then the spiral intestine is a shaft in which the lift has been replaced by a spiral staircase.

The other main organs that should also be mentioned here are the liver, kidney, swim bladder and reproductive organs (*testes* or *ovaries*).

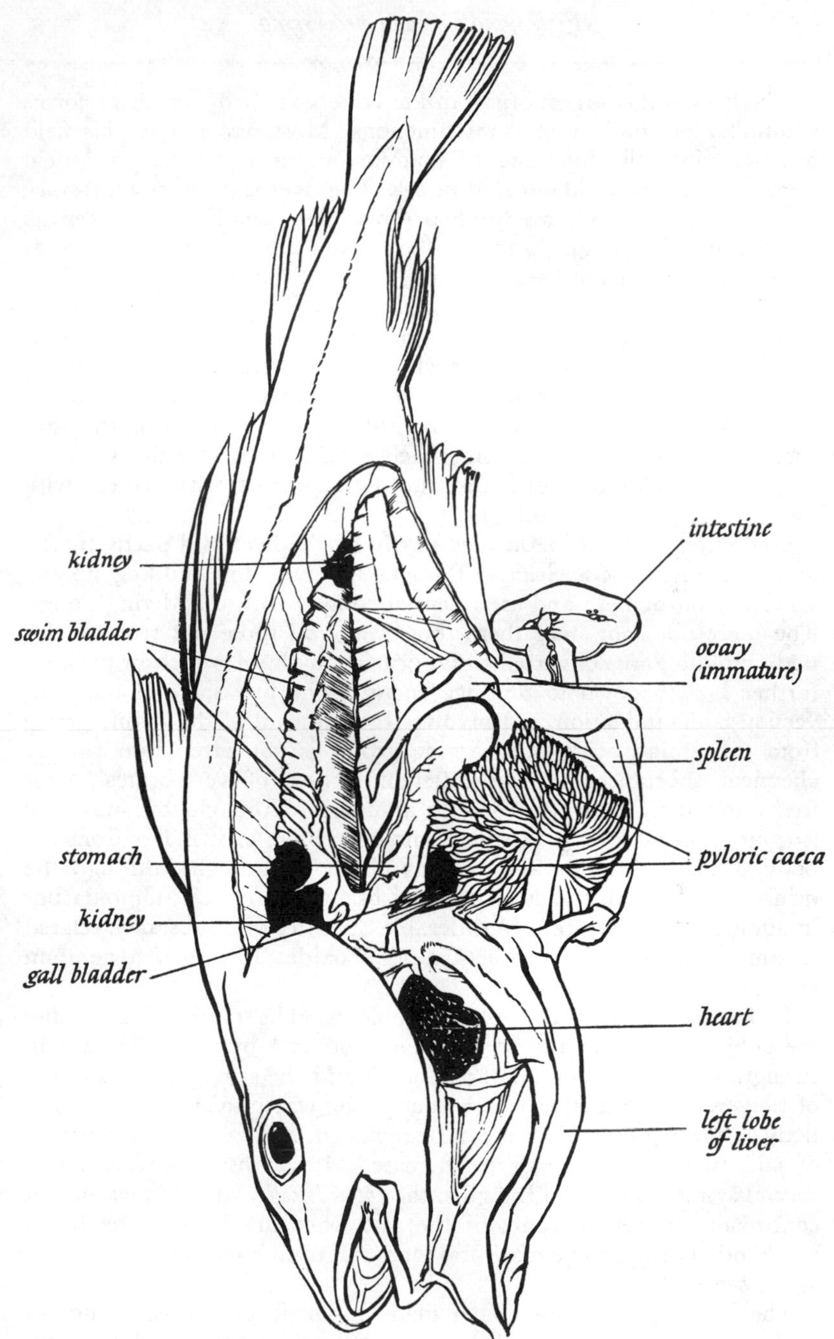

FIGURE 12·3 Diagram of a dissected cod

The liver is the largest organ in the vertebrate body and it performs a number of vitally important functions. Most research in this field has been into the functions of mammalian liver, and a great deal remains to be learnt about that of fish. The liver is primarily a factory and storehouse; foods are broken down into simpler substances by the enzymes in the gut, and are then taken into the bloodstream. It is one function of the liver to build these substances up again into a suitable form for storage. Sugars, for example, are built up into a type of starch, animal starch or *glycogen,* which is stored until it is required. Fat also, in some fish such as cod, halibut and many sharks, is laid down in large amounts. So much so, indeed, that if the fat of a cod liver is dissolved away, very little structure remains. In other species, the so-called fatty fish, much of the fat is laid down in the flesh and sometimes round the gut. Fat contents vary widely with species, season and condition.

One important task performed partly by the liver and partly by the kidney is that of *de-amination.* Proteins are used for building up the substance of the flesh and are essential components of all living things. The body does not store them, however, and those not required for body building are converted to fat or carbohydrate, which in turn are further broken down to produce energy. The first stage of this conversion is de-amination and involves the removal of the amine group from the amino acids. The amine group is converted into various chemical substances in the different classes of vertebrates. Most freshwater bony fishes turn it into ammonia, although this substance is quite poisonous even in small amounts. Most of it is lost from the body through the gills and not, as might be thought, through the urine from the kidneys. Marine bony fish also produce ammonia but in addition they excrete a considerable quantity of a substance related to ammonia known as trimethylamine oxide. It is less toxic than ammonia itself.

Living in the sea presents its problems. The marine bony fishes are subjected continually to a strain produced by the difference in strength of the salt solution of the blood and the stronger salt solution of the sea. There is always a tendency for water to be lost from the body at a rate greater than it can be replaced, so that the concentration of salts in the blood tends to increase. Marine fish have evolved a special system of cells on the gills, the *chloride cells,* whose function it is continually to remove part of the salt from the blood. The blood would otherwise become more and more salt until it was as concentrated as the sea itself.

The cartilaginous fishes have made a virtue of necessity and, as well as producing trimethylamine oxide and ammonia, make urea, which is the same waste product as that made by mammals, including man. Urea can be tolerated in quite high concentrations and the sharks, dogfishes, skates and rays have as much as 2·5% in the blood and tissues. The effect of this large quantity of soluble substance is to reverse the tendency present in the bony fish to lose water to the sea. Fishes of this group, therefore, tend to gain water from the sea, and excess water is much more easily removed than excess salt. Not only the liver and kidney can make urea but all the tissues of the body except the brain and blood. Urea is lost to the sea mainly through the gills but also through the kidneys.

This difference in the functioning of the bony and cartilaginous fishes appears to be one extending back a long way in time. The fishy ancestors of present-day forms probably began their existence in fresh water and subsequently colonized the sea. These two groups have solved the problem of how to live in what is virtually an impure 10° brine in two different ways.

The kidneys of fish lie beneath the backbone. As a general rule, the kidneys of animals are not kidney-shaped. Those of fish are long, reddish-brown structures which may easily be overlooked even by the man who is handling fish every day of his life. In herring and many other fish they look rather like clots of stale blood and in cod they are completely hidden by the white thick-walled swim bladder. Kidney tissue is rather soft and spongy and, as in other animals, soon begins to turn stale after death. Staining of the surrounding flesh by products from the decomposing kidney is a characteristic of, for example, cod after about the tenth day in ice. When a finnan haddock is prepared, it is the dark kidney tissue lying along the backbone that is scrubbed out. The brushes in a herring splitting machine also clean out the kidney.

The swim bladder is an organ found only in some bony fish. It is not found in the cartilaginous ones, nor is it found in, for example, flat fishes like plaice and halibut. It is primarily a buoyancy organ and believed to have evolved from a primitive type of lung.

The kidneys and the gut pass waste matter to the outside through a common opening, the vent. The products of reproduction, eggs or sperms, are also voided at the same point. There is an important difference between the cartilaginous fishes and the bony ones, however. The cartilaginous fishes produce a few large eggs containing a lot of yolk and with horny shells. The empty shells are the familiar *mermaid's purses* so often seen on the shore. Fertilization occurs internally and a thick horny shell is secreted by a special gland before the egg is laid.

It is interesting to note that the females of many species of cartilaginous fish possess only one ovary. A similar situation is found in birds. The eggs produced in both instances are so large that there is not sufficient room in the abdomen for the development of both ovaries. The traces of the second one are, however, sometimes to be found. Since fertilization is internal, the males of this class of fish therefore deposit sperms in the egg duct of the female by means of two *claspers*. In some species the fertilized eggs are retained in the egg duct until hatching and in a few a type of *placenta* is formed, similar in function to that used to nourish the human baby in the womb.

Live young are also produced by a number of bony fish, both freshwater and marine. Tropical aquarium enthusiasts will know of the way in which a guppy gets fatter and fatter and one day appears surrounded by a swarm of babies. The redfish, also called soldier, 'bream' or bergylt among other names, also produces live young. The majority of bony fish, however, produce large numbers of small spherical eggs which are fertilized in the sea by the male, which squirts the *semen*, the fluid containing *sperms*, over them as they are laid. The process is an inefficient one; not all the eggs are fertilized, and the ones that are have to overcome numerous hazards before they can hatch into baby fish. Even at this stage the infant mortality is enormous. The wastage is even more apparent when it is realized that a good plump female cod may produce three or four million or even more eggs during one spawning season. Assuming that the numbers of breeding cod, for example, remain constant from year to year, it will be seen that on average only two of the many millions of eggs produced by every pair during their lifetime reach maturity and live long enough to replace their parents.

There is an important market for ripening *ovaries*, or hard roe, of cod; the *testes*, or soft roe, chitterlings as they are called by fishermen, are generally discarded although there is a small market for them. Occasionally an *hermaphrodite* cod is found; in these, half the reproductive organ is ovary and half testis, although whether these both function at the same time is to be doubted. Although both the ovaries and testes of herring are eaten, only soft roe is normally sold commercially.

Fish are sensitive to temperature. The body temperature of birds and mammals is rather finely controlled; in ill health this control can be upset, and the body temperature may rise or fall beyond the normal limits. The body temperature of fish is not controlled, and it is therefore usually very close to that of the surrounding water. Muscular activity does produce some heat, however, and the temperature of a cod, for example, that has been threshing about in the cod-end may be a degree or two higher than that of the surrounding water.

There is a range of temperatures in which any particular species of fish can live. At the two extremes of this range, individuals have some difficulty in surviving and beyond these extremes they cannot exist at all, except for very brief periods.

WHERE DID FISH COME FROM?

Scientists have discovered a lot about the ancestral fish by studying their *fossils* found in rocks. Rocks are of various ages and the newest, most recent rocks lie on top of the older ones. The rock of ages is, more often than not, built of the sands of time. Many rocks were originally laid down in rivers or the sea as sand or sediment and animals that died were overlaid with a protective layer of mud and their hard parts preserved as fossils. The age of rocks, and hence that of the fossils in them, can be estimated to within a few million years or so.

Fish-like vertebrate creatures first begin to appear in rocks about four hundred million years old and they were even at this remote time highly organized, indicating that they already had a long and respectable history behind them. They were nevertheless nothing like fish as they are known today. For one thing, they did not have jaws; their mouths were probably used merely for sucking up minute creatures from the bottom of the lakes and rivers in which they lived. For another, their entire bodies were heavily armoured with large bony plates. Truly, any fisherman would be greatly puzzled if he were to trawl a catch of such peculiar and outlandish creatures (see Figure 12·4). A study of the way in which the vertebrate egg develops suggests that at a date even more remote than four hundred million years ago, the vertebrates and the *echinoderms*, the starfishes and sea urchins, sprang from the same group.

These first stages in the evolution of living organisms are not, however, known with any certainty and can only be guessed. Every

FIGURE 12·4 Two fossil fish as they may have appeared when alive

year sees some addition to, or modification of, the current theories of how life first arose and then evolved into plants, animals, bacteria and viruses, the various forms of life now existing.

Evolution, as the term is used by biologists, means the process by which living things change from generation to generation so that, after a long period of time, forms that are markedly different from their ancestors may arise. These forms have evolved in such a way that they are better adapted to the particular environment in which they live. Evolution does not necessarily imply progress; numerous examples could be given of both animals and plants that have evolved into degenerate forms. The tapeworm is in many respects a simpler animal that its ancestors; it has no gut, for instance, although its relatives still possess one.

Even groups that do progress in their evolution from simpler to more complicated types are not immune from obscurity and total destruction. Specialization has its own special dangers and an organism that is highly specialized is almost doomed to eventual extinction because, when the environment changes, as it slowly does, it may be incapable of adapting itself to changed circumstances. The evolutionary history, where it is known, of every single group of animals is littered with the corpses of those that specialized too much and that were finally buried in the mud and sediment of time.

By studying both the existing fish and the fossils of extinct ones scientists have contrived to work out their general relationship to one another. There are many, many gaps in detail, but this is hardly surprising when it is realized that both the chances of a particular fish being preserved as a fossil in the first place and then of its being discovered by someone who knows its value are both very small indeed, much smaller than the chances of winning a first prize on the pools. What has been worked out amounts to a family tree of all known fish, whether fossil or still living. It is known as a system of classification.

Systems of classification have been produced for all living things; the general rules are the same whatever forms of life are being considered. Although the fish are particularly mentioned below, it should be remembered that similar considerations apply in classifying, for example, bacteria.

WHY CLASSIFY LIVING THINGS?

When one fish merchant tells another that he has just purchased a certain number of boxes of plaice, there is little likelihood of them misunderstanding each other. If a merchant offers frozen plaice to an importer in Australia there may be some doubt in the mind of the Australian about precisely what British plaice is like and whether it is sufficiently similar to local Australian species of flatfish to be acceptable to his customers. If, instead of plaice, bream were offered there might be even more doubt. There are a number of species of bream landed in Britain; in addition the redfish, soldier, or Norway haddock sometimes masquerades as bream. This last species is also known elsewhere as rosefish or ocean perch and is sometimes exported as snapper. Although it may be argued that such a lack of precision is permissible, or even desirable, in trade, it would be intolerable in science. If a scientist reports that he has studied a certain animal, then there must be no doubt in anyone else's mind about precisely what animal it is. It would be quite inadequate to give merely the common local name.

Classification aims at giving precision to the naming of living things. It furthermore shows how different species are related to each other. There are very many reasons why this is useful; for example, if it is shown by a scientist in one part of the world that all the animals of a particular group contain a certain vitamin in large quantities in the oil in the liver it immediately suggests that this may also be true for other, but closely related, species occurring in other places and so stimulates further investigation.

Exporters in the British fish industry are often confronted by problems that are really ones of classification. Is a particular British species suitable for export to a certain country? Is it sufficiently like what is eaten in that country to be acceptable? What is the local name in the country overseas for the nearest equivalent to the British species? Local names are notoriously misleading; even in different ports in Britain itself the same fish may have quite different names. Certainly in many instances time and money could be saved if the scientific name of a local fish, as well as its common name, were known.

The system of classification now used universally by scientists was invented by Karl Nilsson Linné, or Linnaeus to use the more usual Latinized form of his name. Linnaeus was born in 1707, the son of a curate in a small Swedish town, and at an early age showed great interest in the living things around him. He lived at a time when men were looking for order in nature and he, by describing, naming and classifying all the plants and animals known to him was destined, during his lifetime, to achieve a fame comparable to that of Sir Isaac Newton (1642–1727) who had already formulated the natural physical laws that hold the universe together.

The system may perhaps be explained by a parallel example. If a visitor from some remote part of Tibet where motor traffic is unknown were suddenly to arrive in Britain he could still learn a great deal about transport even though he could not speak English. After a little study he would notice that there were basic differences in the type of engine in different road vehicles propelled by steam, electricity, petrol or diesel fuel. He would then observe that there were certain differences between vehicles propelled by the same type of engine. In the petrol driven ones, for example, he could not fail to be struck by the fact that some had refinements and additions lacking in others. Some are more streamlined and efficient than others; some have preselective gears, some synchromesh gears and some merely crash gears. There are differences in carburation and ignition and methods of suspension. He would find that some cars were very similar to others, perhaps differing merely in minor matters such as the shape and size of the rear windows. He would also notice, if he were very observant, that some, such as shooting brakes, were basically very similar to other forms that had bodies of a different shape.

If he so wished he could classify all these vehicles down to particular models, although even here he would find different varieties according to colour of body or upholstery or whether or not they were standard or *de luxe* models. He would have to find places in his classification for certain forms, like steam cars and steam lorries, that are now extinct, fossils in fact, preserved by a few enthusiasts up and down the country, and for the various steam rollers, almost extinct, but still being used by a few impoverished rural district councils.

Linnaeus applied the type of reasoning indicated in this example to the organic world. The basic unit of his system is the *species*. Although no two living things belonging to one species are exactly alike, even when they are twins, they do share a great many features in common, far more than they share with any other living thing of a different species. Members of the same species can freely interbreed; generally, different species cannot interbreed or, if they do, their offspring are sterile. The species is represented in the simile by a particular model of vehicle; there are differences in colour and upholstery, but basically all cars of one model and year are the same. Species are grouped together into *genera* (singular *genus*) equivalent, perhaps, to the models of slightly different design produced by one manufacturer from year to year.

Every species, plant, animal or bacterium, is given a *specific name* that consists of two words, first the name of the genus and then one distinguishing the particular species in question from all others. For example, the genus to which cod, haddock, whiting and coalfish belong is *Gadus*; cod itself has the specific name *Gadus callarias*, haddock *Gadus aeglefinus*, whiting *Gadus merlangus* and coalfish *Gadus virens*. Also in the genus *Gadus* are classified a number of species less familiar on the slab such as the bib, *G. luscus*, poor cod, *G. minutus*, and pollack

G. pollachius. Where, as here, there is no doubt about which genus is intended, it is usual merely to give its initial letter.

Similar genera are grouped together into families, families into orders, orders into classes and classes into *phyla* (singular *phylum*). All members of a phylum have certain major features in common. It is sometimes found convenient in certain groups to introduce further divisions, such as sub-phylum, sub-class and sub-order.

Linnaeus did not produce his system all at once in a complete form. He kept on altering and improving it during his lifetime and it has been revised and altered very much since then. Nevertheless, the rules that he suggested are still followed. Since Linnaeus' time, very many living things have been described; bacteria, indeed, were not discovered until over a hundred years after Linnaeus first published his system.

When a worker names an organism that he believes to be new to science, the first thing that he does is to describe it in such detail that it cannot be confused even with closely related species. A specimen of the new species is deposited at a recognized museum or institution, such as the British Museum (Natural History), where it is available to specialists for study and comparison. The name that is given, provided it has not been used for some other form, then becomes the accepted scientific name.

The full significance of classification did not become apparent until Charles Darwin published his famous book, *The Origin of Species*, in 1859. Modern ideas of the evolution of organisms are largely due to Darwin, who showed how every living thing, including man himself, had evolved from simpler forms. It was only then that it became clear that a system of classification is in fact a family tree indicating how living things are related to each other.

THE CLASSIFICATION OF FISH

The fish that exist today are thus seen to be the products of a very long period of evolution during which many peculiar forms have arisen that have been in the long run unable to compete successfully in the constant bloody battle for life that occurs in nature. Some of those forms that still exist are less successful than others; John Dory and red mullet, for example, are less successful than cod and herring. It may yet be that red mullet or John Dory will give rise to a highly successful form that will at some future time become as common as cod but they may alternatively represent the end of an unsuccessful evolutionary trend like mammoths and flying reptiles or pterodactyls.

Modern fish can be grouped into three main classes which are no more closely related to each other than, for example, birds and mammals.

CLASS: CYCLOSTOMATA

This class contains a large number of fossil forms; only a few rather peculiar and not very common forms belonging to this class survive today. These are the hagfishes and lampreys, relatives of some of the

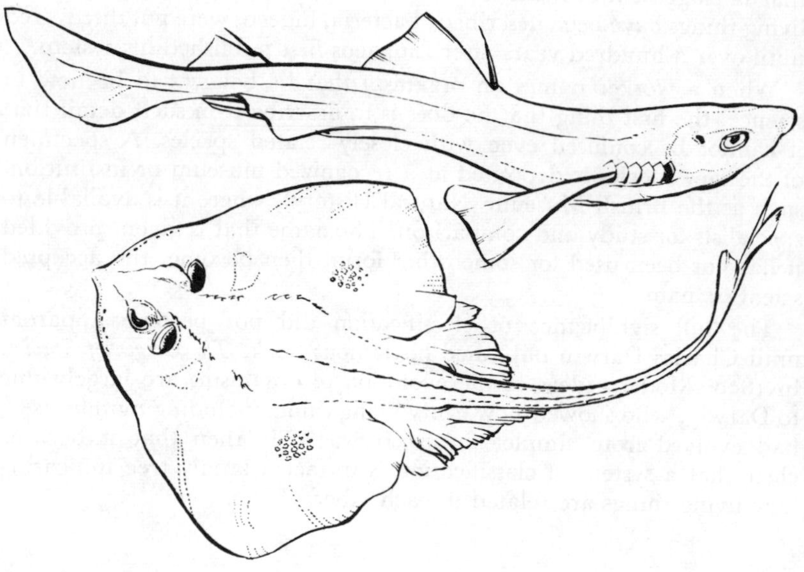

FIGURE 12·5 Dogfish and skate, common representatives of the cartilaginous fishes, the Selachii

very first primitive creatures. These modern forms are in some ways simplified, do not have a heavy bony armour covering them and are rather like eels in appearance. They are almost parasites and live on flesh of other fish; they have no jaws, only a round, sucker-like mouth, and attach themselves to their prey which they then proceed to rasp away with their tongues studded with horny teeth. In the Middle Ages the lamprey, especially the Severn lamprey, was considered a delicacy and, indeed, Henry I '. . . Eating Lamprys at the Town of

St. Dennis he surfeited on them, and after a short Sickneys Dyed', but they are no longer of commercial importance.

CLASS: SELACHII

This class, which is sometimes referred to as Elasmobranchii, includes all the sharks, dogfishes, skates and rays found throughout the world (see Figure 12·5). The skeletons of modern representatives of the group are, as already mentioned, of cartilage and never of bone although fossil forms were bony.

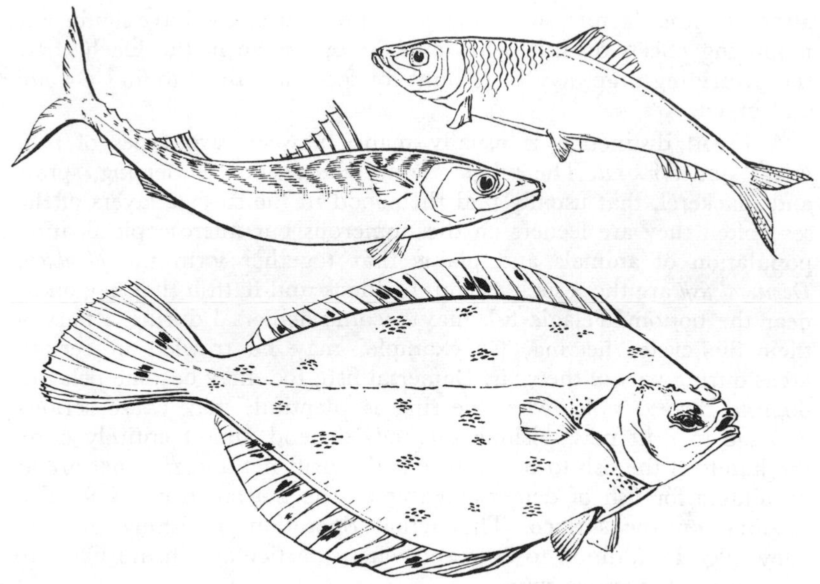

FIGURE 12·6 Herring, mackerel and plaice, common representatives of the bony fishes, the Pisces

CLASS: PISCES

These are the bony fishes, the group that contains the majority of the commercially important species, such as cod, haddock, herring and halibut (see Figure 12·6). Included in this group is also a number of scientifically important species, although too rare to be of commercial significance; such forms, for example, as the lung fishes, that breathe by a lung as well as gills, and the Coelacanth which, until 1938, was thought to have been extinct for at least fifty million years.

A complete classification of the fishes is not required here; an attempt has been made, however, to show what scientific names mean and why they are used in preference to the more familiar common name.

THE NATURAL HISTORY OF FISH

The behaviour of fish, as of other animals including man himself, is mainly conditioned by the desire to find food and to reproduce, although other factors, such as water temperature, also have significant modifying effects on fish behaviour. At one stage in the life history, the overriding urge may be to hunt for food, at others to find a mate and breed.

A broad distinction is usually made between two types of fish, *pelagic* and *demersal*. The *pelagic fish* are those, such as herring, sprats and mackerel, that usually find their food in the surface layers of the sea; often they are feeders on the numerous but microscopic floating population of animals and plants that together form the *plankton*. *Demersal fish* are those such as cod, haddock and flatfish that live on or near the bottom. Pelagic fish may become demersal during a part of their life cycle; herring, for example, may be trawled in certain areas during part of the year. Demersal fish, too, may become pelagic; dogfish, for example, when herring is plentiful, may cause serious damage to drift nets. Fishing methods depend almost entirely upon the habits of the fish to be captured; the drift net and ring net are as unsuitable for fish of demersal habit as the bottom trawl is for fish shoaling on the surface. The actual operation of fishing involves knowledge in addition to how and where particular fish are likely to congregate at various times.

It is obviously not possible, in a short introduction of this nature, to deal in detail even with the main species of commercial importance in Britain. Consequently two species only will be discussed, cod or *Gadus callarias*, and herring or *Clupea harengus*, as being illustrative of demersal and pelagic types, respectively.

COD

Cod is the most important British food fish. It is caught mainly by trawling, although some is also seined and a small quantity captured by long line. It is a widely distributed species and is found as far north as the Arctic Seas and as far south as the Bay of Biscay; it is also common on the other side of the Atlantic. It occurs in water ranging from above 50°F to about 30°F; there is evidence that temperatures of 35°F to 40°F suit it best and it cannot remain indefinitely at temperatures below 32°F.

Cod usually spawn in the spring in various areas, for example in the North Sea, Faroes, Iceland and the Lofoten Islands off Norway. In most of these areas, the spawning fish gather together from regions in the vicinity. At the Lofotens, however, the mature cod have often travelled long distances, even from the eastern Barents Sea and from Bear Island and Spitzbergen. The young and immature cod in these distant areas, as spawning time approaches, appear to carry out a dummy run in the direction of the Lofoten Islands. They start off but turn back before they get there. The closer they are to maturity the greater the distance they travel. The behaviour at this time of both mature and immature cod is controlled by a small gland beneath the brain, the *pituitary*.

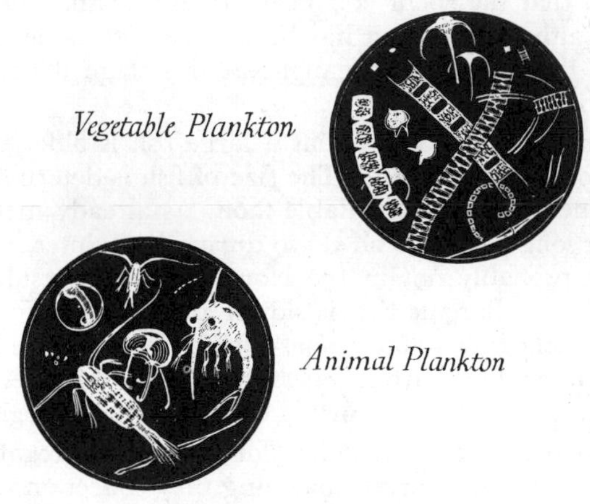
Vegetable Plankton

Animal Plankton

When spawning, mature cod concentrate in the spawning regions which are generally quite small in area. Large catches are often made on some of these grounds such as the Lofotens in February and March. The fertilized eggs rise to the surface, develop and hatch out as larvae of perhaps ¼ inch in length. They feed voraciously on the minute floating animal population of the surface water, the so-called *animal plankton,* and after perhaps 12 weeks or so, when they are slightly more than ¾ inch long, they seek the bottom and begin to feed upon crustacea such as small crabs and shrimps. At 18–24 months of age they begin actively to hunt for prey. Fish, especially herring, becomes an important part of their diet. The precise time taken for the egg to develop depends upon water temperature.

Although in many spawning areas the swirl of the surface water currents tends to retain the young larvae in the same region, certain regions receive young fish from spawning areas a considerable distance away. West Spitzbergen cod, for example, is largely spawned off north-west Iceland and Bear Island cod off the Lofotens.

Spawning, coming as it does after the rigours of winter, puts a considerable strain on the organism. Body reserves are drawn upon in order to provide for the development and ripening of the ovaries and testes; a similar situation occurs in all animals. The old saying 'a tooth for a child' is based on the fact that the body reserves of calcium are drawn upon to such an extent in an expectant mother that the cement holding the teeth in the jawbone may become so softened that they may drop out. At spawning time both fat and protein reserves are drawn upon, and tissue may consequently become more and more watery. Feeding is interrupted during this phase and it is only after its conclusion that the spent cod begin to eat again. There follows a period when although feeding has begun the flesh is nevertheless soft and it seems that it is in fish caught at this stage that spoilage can occur at its most rapid rate.

Although it is often thought that a large fish is older than a small one, this is not necessarily true. The size of fish is determined both by age and by the amount of available food. As already mentioned, fish can starve for long periods; cod do so during the winter in the Barents Sea although probably not in the North Sea. The result is that the growth rate in the Barents Sea is slower than in the North Sea, and maturity is reached at a later age. North Sea fish five to six years old may be larger than Barents Sea fish of twice this age. The rate of growth becomes less after maturity, when spawning begins.

There is consequently a tendency for cod to congregate at certain times during their lives; during spawning they collect on the spawning grounds which are usually of small area and they may also be found in considerable numbers when shoals of fish, such as herring or sand eel, are plentiful. Fishermen, as far as they can, take advantage of these facts as is shown by the greater landings from certain areas at certain times of the year.

Initial, or biological, quality is influenced by whether or not the fish is mature, when it spawned, whether it has recently starved or what it has been eating. Size and age may also affect initial quality; size is certainly important in deciding keeping quality.

HERRING

British fishermen catch herring mainly by drift nets, although ring net vessels also land a fair quantity. Little trawled herring is landed, although this method of capture is commonly used on the Continent. Herring are a widely distributed species found all round the British coasts, in the Baltic, off Norway, round Iceland and in the Barents Sea and on the other side of the Atlantic. A closely similar species occurs in the Pacific.

Herring can be caught somewhere near Britain at any time of the year. In January to March, herring can be caught in the Minch, around the Firth of Forth and the Clyde. In April and May there are small fisheries off the north-east coast and off Shetland which become more important during June to August, when herring may also be caught on the west coast and off North Shields. From September to December there is a small fishery in the Clyde area and another in the Minch but the main emphasis is on the more southerly regions off Scarborough and Whitby, culminating in the East Anglian fishery of October and November.

It is impossible in a short introduction of this nature to deal in detail with the various races of herring and their spawning behaviour. It must be sufficient to say here that there are distinct differences between the types of fish caught in the different areas. Broadly, it is possible to distinguish between spring and autumn spawners. They may occur together in catches at some times of the year.

Herring may be classified according to the stage in the spawning cycle. *Immature* fish are those that have not yet become adult; *full* and *filling* are preparing for spawning, and the roes are well developed; *spawning* fish are those in which the spawn or milt is actually running; *spents* are those in poor condition following spawning and the *recovering spents* are those that, although not full and filling, are yet recovering from their former condition.

The most convenient measure of the suitability of any particular sample of herring for processing is the fat content, and this is closely related to the biological condition. Herring, as already mentioned, feed upon the animal plankton; the animal plankton increases rapidly in amount in the spring and the herring begin to feed voraciously on it. The fat content rises rapidly; there is an initial period when the fat does not appear to be completely absorbed by the tissue and is very liquid. This is the spring oily herring that is unsuitable both for salt curing and kippering.

After the spring abundance of animal plankton, there is a fairly rapid falling off in the quantity present in the sea and the herring consequently lays down fat more slowly after the first rapid increase in the spring. By about July, fat content has reached its peak and begins to fall, and if the fish is an autumn spawner the fall is rapid. The fat content of the spring spawners remains relatively high until the following January or February.

Spawning, as in cod, occurs in certain well-defined areas, where the fish must congregate in enormous numbers. Unlike cod, herring produce eggs that remain for the early part of their development on the sea floor. The sea bed in a certain area may be thickly carpeted with eggs; some fish, notably haddock, will gorge themselves on the spawn. When the baby fish hatches out it, like the baby cod, enters the plankton. Here it feeds, grows and changes shape until at about 1¾ inches long, it appears as a perfectly formed little herring. In the North Sea, shoals of these appear off the coast, especially round the estuaries and, after six months or so, scatter over the North Sea. They become sexually mature when they are from three to five years old. It is only when they are approaching maturity that they join the main shoals.

13

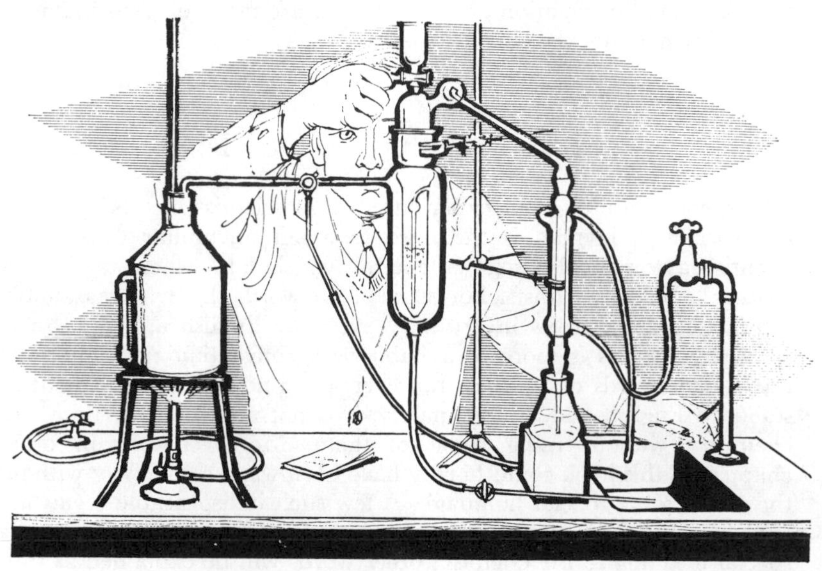

What fish are made of

As the following Treatise is chiefly designed for Persons not bred up in the Profession of Physick, it is necessary to give a general Notion of the Meaning of some Chymical Words that frequently occur in it.

JOHN ARBUTHNOT, MD, 1732

A RESEARCH worker attempting to improve a product obtained by processing fish, for example by smoking, finds a constant need for detailed knowledge of the chemistry of fish tissues. The practical man in the industry does not require so much intimate detail. Nevertheless, without at least a superficial knowledge of the chemical composition of fish he cannot really understand what is going on during a fish processing operation. He may know that certain variations in treatment lead to improvement or impairment of quality, but he cannot be expected to know *why* they do. Without such knowledge he cannot fully understand a book like this, or best apply in his daily work what he has gleaned by reading it. The purpose of this chapter is to give, in as non-technical a way as possible, what may be regarded as the basic essential information about what fish are made of. It is intended for the layman, not the chemist.

TERMS TO BE USED

One great difficulty often experienced by a scientist in talking about his work to a layman is the need to avoid unfamiliar terms. The scientist automatically uses a special vocabulary when talking or writing to fellow scientists. For some of the words there are reasonably satisfactory alternatives in ordinary everyday English and the use of specialized words is more of a habit or a convention than anything else. Other words or phrases, however, have no simple equivalent in standard English and, unfortunately, it is not possible to write about chemistry without using some of them. Moreover, certain other chapters in this book could hardly have been written concisely without the use of some special language. A few such indispensable terms are explained here. They are sometimes common English words with a special meaning to the chemist; other words will be explained as they arise later in this chapter.

ELEMENT

An element is one of the ultimate materials into which any substance can be broken down by chemical treatment. The physicist has shown that even the chemist's element is composed of still simpler units such as *electrons* and *protons*. But for most chemical purposes it is sufficient to regard such substances as iron, sulphur or oxygen as elements which cannot be broken down into anything simpler or changed into any other element. About 90 elements go to make up all the substances we know on Earth. Some of these are very rare, others very plentiful. Only a few need concern us here. The chief elements out of which all the complex organization of fish tissues is built are carbon, hydrogen, oxygen, nitrogen, calcium, phosphorus and sulphur. Probably all these names will be familiar to the reader, as also some of the substances themselves. The chemist has a list of internationally agreed symbols to represent the elements. They are a kind of shorthand. Some of them are the initial letter, or the initial plus another letter, of the English name, for example, for the seven elements listed above. Sometimes, however, the letters are taken from a Latin or Greek name, for example, the symbol for iron is Fe, from the Latin *ferrum*.

COMPOUNDS AND MIXTURES

Although elements can occur as such, most of them are more often met with as compounds. A *compound* contains two or more elements joined together by special bonds, so that they cannot usually be separated by simple means. Moreover, in a true chemical compound these elements are always present in the same proportions. In a *mixture* the proportions can vary in an infinite gradation.

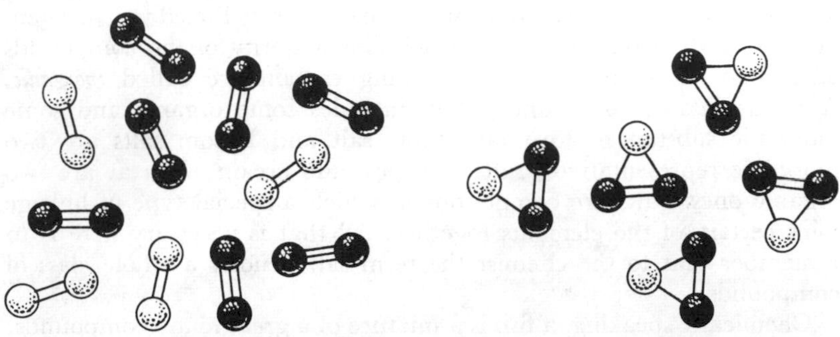

FIGURE 13·1 The black circles represent nitrogen atoms and the white circles, oxygen atoms. The left hand diagram shows a mixture of oxygen and nitrogen molecules, as in air. The right hand diagram shows molecules of the chemical substance, nitrous oxide

The difference between a compound and a mixture can be illustrated with the elements oxygen and nitrogen. As a mixture they form the major ingredients of the atmosphere. When this mixture is breathed our body has no difficulty in separating the oxygen it needs from the nitrogen, which is so much inert residue to be breathed out again. But if oxygen and nitrogen are combined chemically, the resulting gas is very different from air. One particular combination is the colourless anaesthetic gas used by dentists. Another combination gives a brown-coloured and highly poisonous gas. The body cannot separate these gases into oxygen and nitrogen again and so they cannot support life.

Elements may be combined in different ways, or in different proportions; they then give rise to different compounds. Two entirely different combinations of oxygen and nitrogen have been mentioned. The anaesthetic gas contains two parts of nitrogen combined with one part of oxygen. The brown gas contains one part of nitrogen combined with two parts of oxygen. The chemist would show this in his special shorthand by using the formulae N_2O and NO_2 respectively (see Chemical Formulae on page 299). The total number of known compounds containing solely the three elements carbon, hydrogen and oxygen runs into many thousands. With the addition of other elements, for example nitrogen, the list reaches almost astronomical proportions. In these compounds not only do the relative proportions of the various elements vary, but also the pattern in which they are combined. Some examples of this are given later.

The multifarious compounds containing carbon form the subject matter of what is called *organic chemistry*. Like many other statements in this chapter, this is a simplification. Strictly, organic compounds are those in which at least some of the carbon is also linked to hydrogen, but this is the case on the overwhelming majority of the compounds of carbon. Compounds not containing carbon are called *inorganic*. A third class of compounds, *salts*, includes some organic and some inorganic substances. Ordinary table salt and Epsom salts are two inorganic representatives; salts of lemon and cream of tartar are two organic ones. Salts are compounds in which a special type of linkage joins certain of the elements together. All that is necessary here is to remember that to the chemist the term salt denotes a whole class of compounds.

Chemically speaking, a fish is a mixture of a great many compounds, some present in large proportions, some in smaller proportions, and some in mere traces. All three classes of compound, organic, inorganic and salt, are included.

ATOMS AND MOLECULES

An *atom* is the smallest possible particle, or unit, of any particular element. Each atom is a definite assembly of the physicist's fundamental electrical particles, a grouping of electrons, protons and neutrons. If an atom is split up, as happens, for instance, when an atomic bomb explodes, it becomes an atom, or atoms, of a different element, or elements. This sort of thing does not happen in ordinary, everyday chemistry, although certain atomic disintegrations are going on naturally all around, and a special branch of chemistry was concerned with them long before atomic bombs were invented. For present purposes atoms of elements can be regarded as the ultimate building stones out of which all matter is made. Some slight idea of of the actual size of an atom may be gained by noting that in 1 oz of iron, for instance, there are about 3×10^{23} atoms. That is 3 followed by 23 noughts!

A *molecule* is the smallest possible unit of a compound. Actually elements often exist as molecules instead of atoms, these molecules being composed of several atoms of the same element which are quite firmly linked as a definite compound. But the molecules of such an element can be broken down into atoms of that element without loss of its identity. A molecule of a compound containing two or more elements cannot be broken down without complete loss of identity of the compound.

Thus the term atom can only relate to elements, while the term molecule can relate to compounds composed of one or more elements. For present purposes compounds will be regarded as containing two or more elements, as defined above. As an example, one molecule of water contains two atoms of hydrogen combined with one atom of oxygen.

CHEMICAL FORMULAE

Reference has already been made to the use of letters to represent elements and of formulae to represent compounds. A formula not only tells the chemist what type of compound he is dealing with, but gives him a pretty good idea of its probable chemical properties and behaviour. Formulae are a great help in understanding what goes on during some particular chemical change or reaction. The formula of water is often written H_2O. This means that one molecule of water contains two atoms of hydrogen and one atom of oxygen, but it does not give as much information as the alternative formula H–O–H. This shows that each hydrogen atom is separately joined to the oxygen atom. Information of this type becomes increasingly vital the more complex the molecule. Take such a relatively simple substance, for instance, as alcohol. The short formula, what the chemist would call the empirical formula, is C_2H_6O, that is two atoms of carbon, six atoms of hydrogen and one of oxygen make up a molecule. But two entirely different compounds have this same assortment of atoms. One of these is alcohol, the other is a compound called dimethyl ether, closely related to, but not identical with, the ether used as an anaesthetic. The more detailed formulae, what the chemist calls structural formulae, for these two substances are quite different.

$$\begin{array}{ccccccccc} & & H & & H & & & & \\ & & | & & | & & & & \\ H & — & C & — & C & — & O & — & H \\ & & | & & | & & & & \\ & & H & & H & & & & \end{array} \qquad\qquad \begin{array}{ccccccccc} & & H & & & & H & & \\ & & | & & & & | & & \\ H & — & C & — & O & — & C & — & H \\ & & | & & & & | & & \\ & & H & & & & H & & \end{array}$$

alcohol — dimethyl ether

The lines indicate which atoms are joined to which. The chemist often uses contracted forms of structural formulae, to save space, for example, $CH_3.CH_2OH$ for alcohol and $(CH_3)_2O$ for dimethyl ether.

CHEMICAL REACTIONS AND EQUATIONS

Chemical reactions, that is, chemical changes, are the basis of life. In breathing, in the digestion of food, chemical reactions of many kinds are involved. Equally they are involved when a match is struck or the engine of a car is started. It is impossible to get away from the interplay of chemical reactions. Many are exceedingly complex but, for present purposes, it will suffice to consider a simple example.

Dogfish and skate contain appreciable amounts of a substance called *urea,* which after death takes part in a chemical reaction by which ammonia is produced, with its characteristic smell. The substances that interact are urea and water, and the products are ammonia and carbon dioxide. This reaction can be shown by an *equation*:

$$
\begin{array}{ccccccc}
\text{H—N—H} & & \text{H} & & \text{H—N—H} & & \\
| & & | & & | & & \\
\text{C}=\text{O} & + & \text{O} & \longrightarrow & \text{H} & + & \text{O}=\text{C}=\text{O} \\
| & & | & & \text{and} & & \\
\text{H—N—H} & & \text{H} & & \text{H} & & \\
 & & & & | & & \\
 & & & & \text{H—N—H} & & \\
\text{urea} & & \text{water} & & \text{ammonia} & & \\
 & & & & \text{(2 molecules)} & & \text{carbon dioxide}
\end{array}
$$

The reason for the two lines joining oxygen to carbon need not be considered here. There are certain chemical rules to be followed. Suffice it to say that in any formula each carbon atom must have a total of four lines, each nitrogen atom three lines, each oxygen two lines and each hydrogen atom one line. The shortened version of the above equation would be:

$$CO.(NH_2)_2 + H_2O \longrightarrow 2NH_3 + CO_2$$

Many chemical reactions are reversible, under appropriate conditions, and this is one of them. At high pressures and temperatures, ammonia and carbon dioxide react to give urea and water.

One thing to note about chemical reactions is that they always take place more rapidly the higher the temperature. As a rough generalization it may be said that a rise of 18 deg F will about double the speed of any chemical reaction; or conversely, a fall of 18 deg F will halve it. This is why fish keeps fresh longer the lower the temperature. In frozen fish the effect of temperature on the rate of at least some chemical changes is even more marked than in unfrozen fish, for example a rise of only about 5 deg F will double the rate; hence very low storage temperatures are of benefit.

ENZYMES AND CATALYSTS

If urea and water are mixed and kept at ordinary temperatures there will be no detectable production of ammonia over a long period. How then does this reaction occur so rapidly in fish that is becoming stale? The answer is that it is enormously speeded up by the presence of a stimulating agent secreted by some of the bacteria on the fish. This particular agent, known as *urease,* is one of a whole class of substances known as *enzymes.*

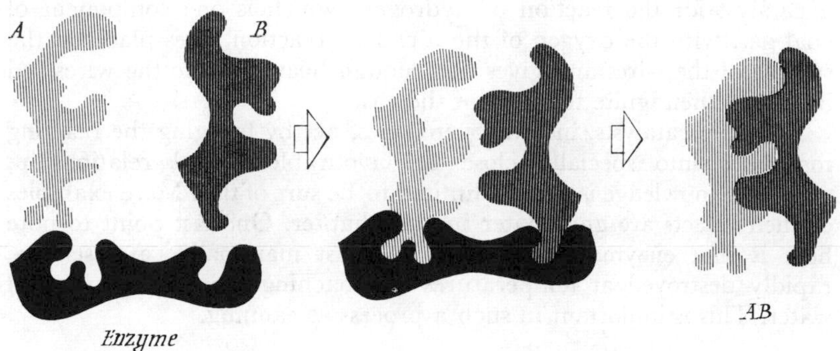

FIGURE 13·2 A and B are molecules of two chemical substances, brought into the position favourable for rapid chemical interaction by the enzyme molecule

The great majority of the chemical reactions that are continually going on in all living tissues, plant and animal alike, will only take place at measurable speed at ordinary temperatures in the presence of the appropriate enzyme. Any living tissue produces and contains a great variety of enzymes, each stimulating some particular chemical reaction. Their activities are perfectly co-ordinated to the requirements of the organism in question. But enzymes themselves are not alive. They are compounds, or particular associations of compounds. Many of them can be separated from the tissue and still show their characteristic promoting action, for example in a laboratory test tube. When a fish dies, the delicate balance of its various enzyme activities is disorganized and they then contribute to the complex series of changes grouped under the term *spoilage.* They promote a kind of self-digestion, known as *autolysis.* When bacteria invade a dead fish and convert it into a progressively staler and finally putrid product, they do so by means of their enzymes.

Enzymes may be considered as a special class of a wider group of substances known as *catalysts*. A catalyst, like an enzyme, enormously increases the speed of some particular chemical reaction simply by its presence. It does not directly enter into the reaction that it promotes, and it is not used up by it. Hence a relatively small amount of catalyst can effect the rapid interaction of relatively large amounts of material. Catalysts of all types are widely used in the chemical industry. One example of *catalysis*, the action as distinct from the agent, may be familiar. Automatic lighters are fitted to some gas appliances such as gas fires, and they can be purchased as separate items for general use. Their essential feature is a bundle of fine platinum wires. Platinum is a catalyst for the reaction of hydrogen, which is one component of coal gas, with the oxygen of the air. This reaction takes place on the surface of the wires and gives out enough heat to make the wires red hot; they then ignite the bulk of the gas.

Probably catalysts, including enzymes, act by bringing the reacting molecules into specially close or favourable spatial relationship, although knowledge is still too limited to be sure of this. More examples of their effects are given later in this chapter. One last point to note here is that enzymes, in contrast to most man-made catalysts, are rapidly destroyed at temperatures approaching the boiling point of water. This is important in such a process as canning.

PROTEIN

The term protein will be found occurring repeatedly throughout this book. There is a whole class of substances known as proteins, each of which is an extremely complex chemical compound, that is one molecule contains a large number of atoms. All proteins contain carbon, hydrogen, oxygen and nitrogen as main elements. Sulphur is almost always present in small amount, and sometimes other elements.

For practical purposes it is not very helpful to think of such a complex compound as a protein in terms of its component elements. It is simpler, and customary, to think of it in terms of a number of building stones, or fragments. In any protein the building stones are compounds which the chemist calls *amino acids*. When a protein is eaten by an animal, it is broken down by the enzymes of the digestive juices into its component amino acids. These simpler substances are then readily absorbed through the wall of the intestine and are subsequently rebuilt by the animal, again by enzyme action, into other proteins.

Some 20 different amino acids go to make up a typical protein. In the protein these amino acids are not present as such, but are joined chemically to one another in a definite pattern. The same amino acids joined in a different pattern would give a different protein. The essential structure of a protein can be illustrated by considering two typical amino acids, *glycine* and *serine*. These have the formulae shown below:

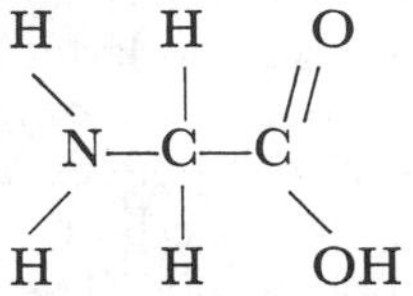

$$\begin{array}{ccccc} H & & H & & O \\ & \diagdown & | & & /\!/ \\ & N & - C - & C & \\ & \diagup & | & & \diagdown \\ H & & H & & OH \end{array} \qquad \text{or } NH_2CH_2COOH$$

Glycine

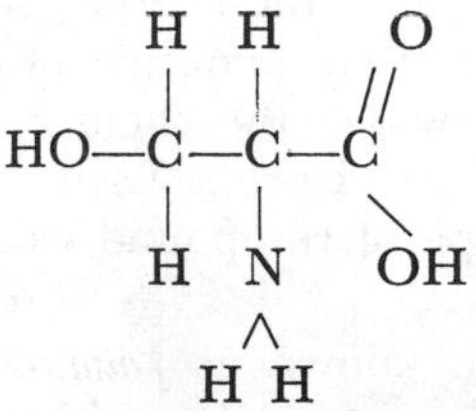

$$\begin{array}{ccccc} & H & H & & O \\ & | & | & & /\!/ \\ HO - & C - & C - & C & \\ & | & | & & \diagdown \\ & H & N & & OH \\ & & \wedge & & \\ & & H\ H & & \end{array} \qquad \text{or } CH_2OH.CH(NH_2).\ COOH$$

Serine

The atom group NH_2 is the *amino* part of the structure and the COOH group, called a *carboxyl* group, confers the properties of an acid. The carboxyl group of one amino acid can combine with the amino group of the other, to produce a larger molecule plus water; this is shown in the following equation. The reaction is reversible in the presence of the appropriate enzyme, as mentioned above, and this is indicated by using two arrows.

$$\underset{\text{serine}}{CH_2OH.CH(NH_2).COOH} + \underset{\text{glycine}}{NH_2CH_2COOH}$$

$$\downarrow\uparrow$$

$$\underset{\text{serylglycine}}{CH_2OH.CH(NH_2).CO.\ NH.\ CH_2COOH} + \underset{\text{water}}{H_2O}$$

Note that the new compound still has an amino group and a carboxyl group, either of which can react with the appropriate portion of still another molecule of any amino acid, leading to ever bigger molecules. Many hundreds of amino acid molecules linked in this way go to make up a molecule of even a simple protein, and many thousands in the case of some very important proteins, for example the chief ones of fish flesh.

FAT

Everyone knows what fat looks like, but how many know anything about its chemistry? Yet to the food processor the latter is the more important. Fat contains only carbon, hydrogen and oxygen. Just as with protein the term covers a whole class of compounds, a mixture of which is present in any natural fat. Some fats are liquid at ordinary temperatures and it is then usual to call them oils. The term *fats* is less ambiguous chemically.

Like a protein, a fat is too complex a compound to think about in terms of its component elements. Once again the chemist thinks of it in terms of building stones or fragments. One molecule of any fat is made up of one molecule of glycerine, which the chemist prefers to call *glycerol*, for reasons which need not be discussed, combined chemically with three molecules of a type of compound called a *fatty acid.*

A typical fatty acid is the substance known as *palmitic acid,* the structure of which has a long *chain* of 16 carbon atoms with a carboxyl group at one end:

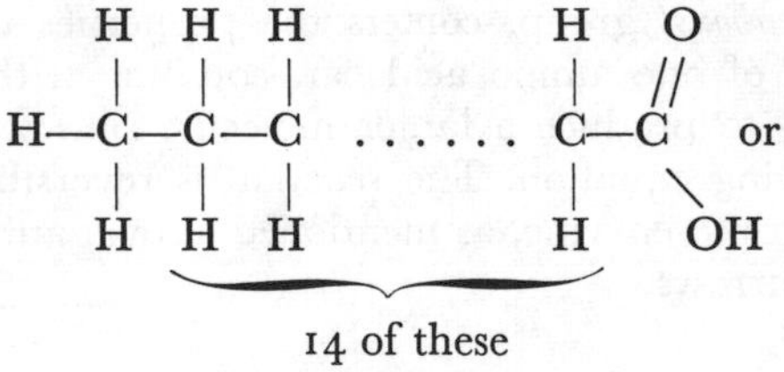

```
   H  H  H           H  O
   |  |  |           |  //
H—C—C—C . . . . . . C—C      or CH3.(CH2)14.COOH
   |  |  |           |  \
   H  H  H           H   OH     palmitic acid
   \_____________________/
          14 of these
```

$$CH_3.(CH_2)_{14}.COOH$$

palmitic acid

In different fatty acids the length of the carbon chain varies from as little as four atoms to as many as 24 or more. In most fish the range is from 14 to 22. Curiously, acids with an odd number of carbon atoms seldom occur in more than traces in a natural fat. There is another important way in which fatty acids may differ chemically besides variation in chain length. Some of them are what the chemist calls *unsaturated.* This means that they contain fewer hydrogen atoms than the type shown above, which is a *saturated* compound. Those carbon atoms that are attached to only one hydrogen atom each, instead of two, are then shown diagrammatically as being linked together by two lines instead of one; it should be remembered that carbon must have a total of four lines. The chemist speaks of this linkage as a *double bond.* A common fatty acid of this type is that called *oleic* acid:

```
    H  H          H   H   H   H              H    O
    |  |          |   |   |   |              |   //
 H—C—C  . . . . .  C—C═C—C . . . . .  C——C
    |  |          |       |              |    \
    H  H          H       H              H    OH
   \_____________/        \________________/
     7 of these              7 of these
```

or

$$CH_3.(CH_2)_7.CH{=}CH.(CH_2)_7.COOH$$

oleic acid

Contrary to what might be imagined, a double bond between carbon atoms is chemically a point of weakness, not strength. Unsaturated compounds are more reactive than saturated ones, that is they are less stable. Some fatty acids contain more than one double bond, and such *highly unsaturated* acids are particularly plentiful in fish fats. This is why fish fats are particularly liable to react with atmospheric oxygen, for example in cold-stored fatty fish, so producing rancid flavours.

The connection between glycerine, fatty acids and fat is illustrated below. The letter R stands for the long carbon chain with its attached hydrogen atoms. Remember that in any fat a whole assortment of different fatty acids is available to choose from, for example up to 40 in a fish fat.

```
CH2.OH          HOOC. R          CH2.O.CO. R          H2O
 |                                 |
CH.OH     +     HOOC. R ———→     CH.O.CO.  R     +    H2O
 |                      ←———       |
CH2.OH          HOOC. R          CH2.O.CO. R          H2O

1 molecule      3 molecules      1 molecule       3 molecules
   of              of               of               of
glycerine       fatty acid          fat             water
```

The reaction is reversible, as shown by the arrows, and in the body is promoted in either direction as necessary by appropriate enzymes. The dotted line shows which atoms go to form water.

Herring oil is used in the manufacture of edible fats, such as margarine and cooking fats. This involves turning a liquid fat into a solid one. Fish fats are liquid because they contain a high proportion of unsaturated fatty acids, and if these are converted chemically into saturated ones the fat will become a hard solid at ordinary temperatures. At some intermediate stage, with only some of the total number of double bonds removed, the fat will have the desired soft consistency. The double bonds can be removed by direct reaction with hydrogen gas, provided a suitable catalyst is used. In the fat industry the catalyst used is metallic nickel. The process as a whole is called *hydrogenation* and it removes the characteristic fishy taste and smell of the fish fat as well as altering its consistency.

SOLUTION

The final term to be considered in some detail is the word *solution*, which refers to a more complicated phenomenon than may be supposed. In many cases one compound has so much affinity, or attraction, for another that they can mutually dissolve in each other. Usually we thinks of solids dissolving in liquids; for example salt or sugar will dissolve in water. Actually we can have solutions of gases in liquids, of liquids in other liquids, and of solids in other solids. Consideration is limited here to some examples of liquid solutions important in the composition and technology of fish. The substance that dissolves is called the *solute* and the liquid in which it dissolves is called the *solvent*.

Water is the best general solvent, that is it can dissolve a wider range of compounds than any other known liquid. However, many substances not appreciably soluble in water are easily dissolved by some other solvent, for example fat in ether. It is particularly with organic compounds that organic solvents are necessary. Similarity in structure leads to affinity and hence solubility.

Fish flesh is largely composed of a stiff solution of protein in water. This stiff and virtually solid solution is called a *gel* and in it the water molecules are so firmly held by the protein particles that they cannot move about.

Chemical reactions involving two or more molecules can only take place when the molecules approach each other closely. This happens readily in liquids or gases, but not so readily in solids. Hence the chemist dissolves solids if he wishes them to take part in some quick chemical reaction.

Speed of solution is always increased by higher temperatures. But the effect of temperature on *solubility*, that is how strong a solution can be made, is not so straightforward. In most cases, apart from solutions of gases in liquids, solubility is greater at higher temperatures. Thus 100 parts of water at 32°F can dissolve 179 parts of sugar, whereas at 212°F they can dissolve 487 parts. Sometimes this effect of temperature is only slight, an important example being the case of common salt in water. The amounts of salt that can be dissolved in 100 parts of water at 32°F and 212°F, to give a full strength brine in each case, are 35·7 and 39·0 parts respectively. In a few cases solubility is actually less in hot than in cold solvent, for example with slaked lime, calcium hydroxide, in water. If lime water is heated it becomes turbid, due to some of the lime coming out of solution.

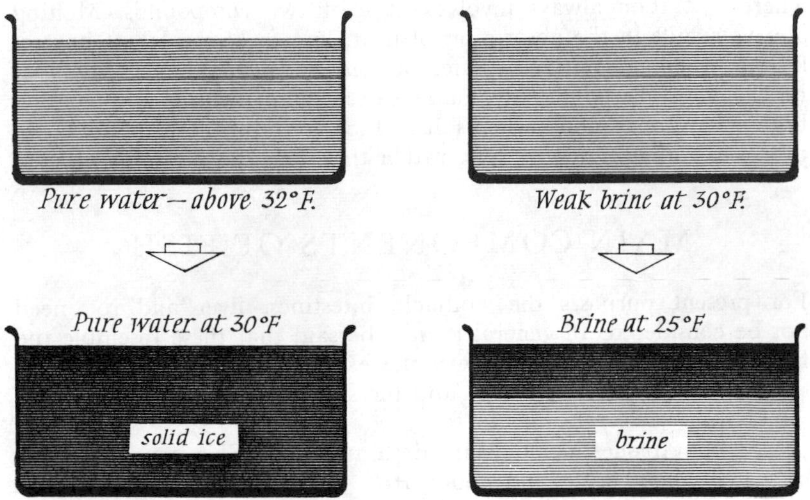

FIGURE 13·3 The difference between freezing pure water and weak brine

A solute always lowers the freezing point of the solvent in which it dissolves. The presence of numerous dissolved substances in fish tissues means that their freezing point is a few degrees lower than 32°F. However, it should be noted that when a weak solution of any solute is progressively cooled, the first effect is to freeze out pure solvent, leaving a more concentrated solution. Hence, as the temperature of fish is lowered beyond the point at which freezing commences, about 30°F, more and more of the total water is converted into ice. This is quite different from the freezing of pure water, where at any temperature below 32°F the entire mass must be ice.

The position in fish tissues is particularly complicated since not only are there many solutes, each exerting its separate effect, but also much of the water is bound in some manner by the protein. It is not possible to calculate, from our knowledge of the chemical composition of fish, how much of its water is frozen at any particular temperature; it can be determined experimentally by various physical measurements. At ordinary cold storage temperatures there is still some liquid water in fish, which probably explains why certain types of chemical change can still go on in it.

There is sometimes popular confusion of the terms *solution* and *melting*. Strictly solution means the dissolving of one substance in another, due to mutual affinity of the molecules, whereas melting means a change from the solid state to the liquid state. Melting can occur with a single pure substance, for example ice becoming water, whereas solution always involves at least two compounds. Melting always results in the absorption of heat, known as the latent heat of fusion, whereas solution is often accompanied by the evolution of heat. There are admittedly many cases where melting and solution both occur, for example when salt and ice are mixed, but they are two processes going on side by side, rather than one and the same process.

MAIN COMPONENTS OF FISH

For present purposes the stomach, intestines, liver and roe need not be considered. In general it may be said that these resemble the flesh for main components, with the exception of the liver of some species. Fish of the cod family and the sharks and rays have livers of which from 50 to 75% may be fat.

The body tissues include skin, flesh and bone. Skin consists mainly of water, about 80%, and about 16% protein. Bone contains much mineral matter, principally calcium phosphate, which amounts to about 14% of the total bone material; the rest is mainly water, about 75%, and protein, about 9%. Skin and bone are eaten to some extent, particularly in the case of canned fish, where even the larger bones are softened by the high temperature cooking. Apart from this they are raw material for important by-product industries such as the manufacture of fish meal and fish glue.

The really important tissue is, of course, the flesh. This is made up predominantly of numerous tiny cells. The chief ones are the muscle fibres, which are held together by smaller proportions of what is called *connective tissue*. These cells are surrounded by fluid, the *extracellular fluid*. In addition to these three components, that is muscle fibres, connective tissue and extracellular fluid, the flesh contains such structures as blood vessels and nerve fibres; on a weight basis these are not important.

FAT IN FISH FLESH

Fat is the form in which surplus body fuel is stored in any animal. Fish are no exception. If a fish eats more food than it requires to satisfy its immediate needs for energy and for growth, it will deposit the surplus as fat. Animals can readily convert *carbohydrates*, that is starchy foods or sugars, into fat for this purpose, and with less efficiency can use also surplus protein to make fat. Fish do not eat much carbohydrate, but their food always contains plenty of protein and fat. All three types of food, carbohydrates, fat and protein, can be used as fuel to meet energy requirements. When the total food intake is less than required for energy expenditure and growth, some of the tissue fat is used to satisfy the energy requirements. Fat cannot substitute for protein required for growth but it can spare protein that might otherwise have to be used to supply energy.

Fish vary enormously in food consumption with the season and the area where they live. Apart from seasonal or geographical variations in abundance of food, there is a direct effect of temperature on the appetite of the fish. Lower temperatures lead to lowered food consumption but so can abnormally high temperatures. Lower temperatures also lead to diminished physical activity, so the overall effect is a complex one. Over and above the effect of external variations there is an annual change within the fish itself caused by the incidence of spawning. When the roe is developing a fish eats far less than normal, despite *increased* nutritional requirements; the result is a marked consumption of the body reserves, especially of fat.

Hence it should not come as a surprise to learn that the total fat content of any particular species of fish is very variable. When in addition different species of fish are compared, it is found that they may deposit the bulk of their fat in quite different tissues. The very oily livers of the fish of the cod family, for instance, have been mentioned. Virtually the total fat of these fish is stored in the liver, and the flesh contains mere traces. In this case there is no margin for significant fluctuations in the fat content of the flesh, although marked variations occur in the amount of fat in the liver. Some fish deposit part of the total fat in the body tissues and part in the liver or other organs. Thus the halibut deposits most of it in the head and body tissues, but appreciable amounts do go into the liver. In the conger eel it is split mainly between the body and the membrane lining the abdominal cavity, the *peritoneum*, with relatively small amounts in the liver. In the herring almost all of the fat is present in the head and body tissues.

Fat distribution is far from uniform within the head and body tissues as a whole. Typically there is a thin layer of nearly pure fatty tissue, which is known as adipose tissue, being connective tissue heavily loaded with fat, immediately below the skin. This recalls the subcutaneous fat so usual in mammals, but it is a very much thinner layer in fish. Fat is present in muscle tissue as microscopic droplets within both fibres and extracellular fluid. Muscle from various areas of the same fish may show appreciable variations in fat content.

For practical purposes it is the total fat content of the flesh which matters. For the herring this can be as low as 1% in spring, following the combined effects of winter and spawning. The minimum is more usually about 3 to 5% and is reached about April. Then in May and June there is a marked increase in abundance of food, coupled with

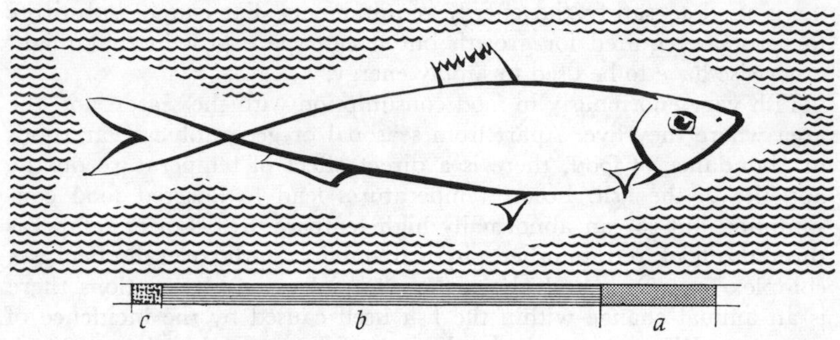

FIGURE 13·4 Starved herring: (*a*) shows the proportion of protein in the whole fish, (*b*) the proportion of water, and (*c*) the proportion of fat

rising water temperatures. Within a few weeks the fat content of the herring may have risen to over 20%. Subsequently, during summer and autumn, there will be a gradual fall in fat content, associated with less voracious feeding. With the onset of spawning, the fall becomes much more rapid. Some herring spawn much earlier than others. The biologist talks of autumn spawners and spring spawners; these will reach very low fat contents at different times. Who then can say what is an average fat content for herring? The best one can do is to say, for instance, that summer herring may be expected to contain about 20% of fat, autumn herring about 10 to 15%, winter herring about 5 to 10% and spring herring less than 5%. It is impossible to be more precise; conditions from one year to the next may be different.

A similar pattern, and a roughly comparable range of fat contents, is found in the flesh of many other fish, for example sprats, sardines, pilchards and mackerel. Sometimes fish actually starve to death at spawning time; the salmon is an example. The flesh of salmon entering our rivers in the early months of the year, known as *fresh run* fish, contains about 13% of fat. As they journey up river to the spawning grounds the fish do not feed and by the spawning season, about November, the fat content has fallen to about 5%. Following spawning the fish are not eaten but it may be of interest to note that their fat content goes on falling and when they die it is usually well below 1%.

It is rather remarkable that, in healthy fish flesh, fat and water together total about 80%. This value does not vary much and what variation there is cannot be correlated with fatness or leanness of the fish. This means that when a fish deposits fat in its flesh, this fat

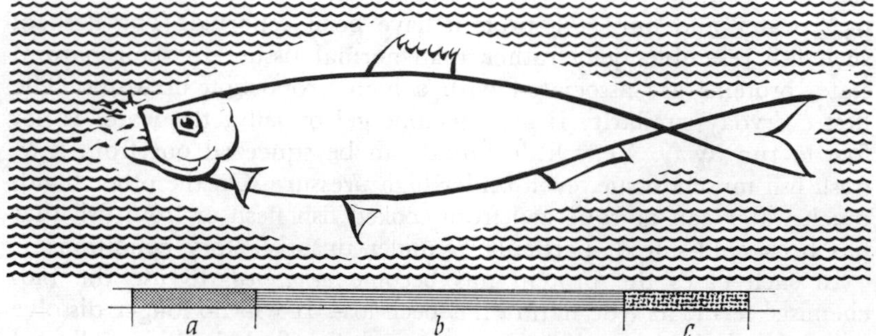

FIGURE 13·5 Well-fed herring: (*a*), (*b*) and (*c*) have the same meaning as in Figure 13·4. Notice that fat has increased at the expense of water, while protein remains the same

replaces an equivalent amount of water. Similarly, when the fish draws on its reserves of flesh fat, water replaces the fat withdrawn. This is distinctly different from the position in mammals, where very little fat is really deposited in the muscles, but rather as adipose tissue interspersed with the muscles. It is also different from the position with fat in other tissues of a fish, such as the liver or peritoneum. But so far as fish flesh is concerned, fatness does not automatically mean increased bulk.

There are various ways of determining the fat content of a tissue. The most widely used method is to extract a known weight of the finely-minced material with some suitable solvent that is later evaporated and the residue of fat weighed.

PROTEINS IN FISH FLESH

It has been stated earlier that the term protein refers to a whole class of compounds. Most tissues contain a mixture of proteins, and fish flesh is no exception. The chief proteins in the flesh are called *myosin* and *actin*, which may really be combined in the muscle as *actomyosin*. There are various proteins grouped under the name *albumins*; the enzymes of the flesh are to be found among these. The name albumin has given rise to a word *albuminoid*, often used instead of protein in the animal feeding stuffs trade, but this is an unsatisfactory term to the scientist since it means albumin-like substance, whereas the main bulk of the protein in many feeding stuffs is made up of other types of protein.

The protein content of healthy fish flesh is about 16 to 18%. Under conditions of prolonged partial starvation, particularly when this follows the drain of spawning, fish may become so depleted in protein that they contain well over 80% of water; for example, cases of about 95% water and only 3% protein have been reported. The flesh of such fish resembles jelly rather than normal tissue. In their natural state, proteins are associated with a high proportion of water. The best everyday similarity is to a gelatine gel or jelly; the water is not free to run away. Very little liquid can be squeezed out from raw, fresh fish muscle tissue, even under high pressure. On the other hand, much water can be squeezed from cooked fish flesh or from fish that has been cold-stored, particularly under unsatisfactory conditions.

In such cases the protein has become *denatured*, to use the biochemists' term; its true nature has been lost. It will no longer dissolve in weak salt solutions, for instance, and the flesh itself is dull and opaque instead of glossy and translucent. Loss of solubility in salt solutions is responsible for failure to obtain a nice surface gloss on a smoked fish made from a badly cold-stored article. The gloss is due to a dried film of extracted protein. From the nutritional point of view, as distinct from the food processor's problems, protein denaturation of this kind does not matter. It does not interfere with its breakdown by the digestive enzymes.

It should be noted that animals, including man, have only a very limited power to convert one amino acid into another. Certain amino acids must be obtained as such from the food. The precise requirements vary with the species of animal but in every case a list could be compiled of what are called the essential amino acids; this means essential in the animal's food. Other amino acids as well will be essential to the animal but it can make them by interconversions among its total supply of amino acids.

It so happens that the proteins of many cereals contain a mixture of amino acids with only small proportions of certain ones essential for most animals, and very large quantities of such proteins would have to be eaten to meet the animal's requirements of those particular acids. As the cereal products in question are mainly starchy in nature, they could not be eaten at this level. But the total proteins of fish flesh supply an assortment of amino acids admirably adapted to the general requirements of higher animals. Moreover they can supplement the proteins of cereals, so that the animal can make efficient use of the cereal proteins. People living on a good mixed diet are unlikely to suffer from protein deficiency and for them fish may be considered on its merits as a tasty item contributing to the general variety of their meals. But poor people in various parts of the world do suffer from protein deficiency. In such cases fish may offer the cheapest source of high grade protein, capable of removing much ill health. In the diets of farm animals fish, in the form of fish meal, is widely used to supplement cereal proteins.

The protein content of a tissue is usually determined analytically by measuring the content of nitrogen and multiplying by a factor of 6·25. This assumes that an average protein contains 16·0% of nitrogen. Actually the nitrogen content of different proteins can vary appreciably, for instance the proteins of fish milt are unusually rich in nitrogen. In addition, the tissue contains small amounts of nitrogen-containing compounds other than proteins. So the value obtained by this method is only an approximate one, but it is good enough for most purposes. How the chemist determines the nitrogen content of the sample is outside the scope of this book.

MINOR COMPONENTS OF FISH

There are three groups of minor components of fish tissues, which affect the quality or the dietary value of fish; very brief consideration will now be given to these.

EXTRACTIVES AND VOLATILES

The term *extractives* is applied rather loosely to a group of otherwise unrelated substances that share the properties of solubility in water and of usually possessing some kind of taste. When a tissue is extracted with water most of its flavour can be removed. If the water extract is concentrated by evaporation it acquires a very strong flavour. Familiar examples are the various brands of meat extracts. *Volatiles* overlap with extractives; they are those compounds, whether water-soluble or not, that are evolved in small quantities in the gaseous form from foods and so contribute to their characteristic odours as well as tastes.

Not all the substances removed by the water are called extractives, even if they also contribute to the taste. Thus some inorganic salts will be present. Inorganic salts are not called extractives but organic salts are included with them.

The extractives of fish tissues include a great range of compounds. The following groups may be noted as of particular interest for present purposes: nitrogenous bases and related compounds, free amino acids, and sugars and related compounds. Volatiles will not be dealt with here.

(1) *Nitrogenous bases.* Nitrogenous bases may be considered as compounds related chemically to the well-known substance ammonia. If the chemical relationship is close, they smell very like ammonia. But it should be noted that both ammonia itself and these bases do *not* have any smell when they are combined as salts, or in certain other ways. Two compounds to be considered are the substances known respectively as trimethylamine oxide and urea. These are colourless solids, the former almost without smell, the latter completely so. However, when certain types of bacteria begin to attack a fish their enzymes convert trimethylamine oxide into trimethylamine itself and they produce ammonia from urea.

Both trimethylamine oxide and urea are especially plentiful in sharks, dogfish, rays and skates. This is why such fish so rapidly develop a particularly strong ammoniacal odour as they become stale. In other marine species trimethylamine oxide is still fairly abundant, but urea tends to be present in little more than traces. Freshwater fish contain much less trimethylamine oxide than do marine fish but even a little free trimethylamine can produce a strong odour. Indeed,

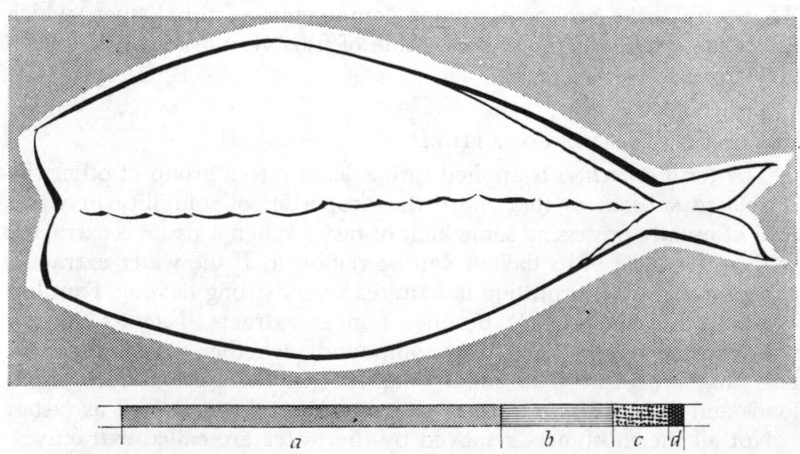

FIGURE 13·6 Components of fish flesh: (*a*), (*b*) and (*c*) represent the proportions of water, protein and fat respectively and (*d*) represents the minor components, such as extractives and salts

while trimethylamine in high concentrations smells almost exactly like ammonia, in very low concentrations it smells like stale fish. It probably accounts, in fact, for much of the characteristic smell of stale fish. As fish pass from the stale to the putrid condition, however, other types of compound derived from protein breakdown become important constituents of the total odour source.

Numerous other nitrogenous bases, and compounds derived from them, occur in fish, usually in very small proportions. They contribute to the total characteristic taste of a species.

(2) *Free amino acids.* Amino acids have already been referred to as the compounds out of which proteins are built. It may be of interest to note that the word *amino* indicates to the chemist a relationship to ammonia. However, because of the presence in their structure of an acidic unit as well, the chemist does not usually group them with the nitrogenous bases, even although in some of them the basic properties are stronger than the acidic ones. It is really a matter of convenience in classification.

Although the great bulk of the total amino acids of a fish are present in combined form in its various proteins, small amounts of many of them occur free. In addition some are present combined into much simpler compounds than proteins, involving only small numbers, two or three, of amino acid molecules in each molecule of the compound. These particular compounds are called *peptides*, and for present purposes can be considered along with the free amino acids. Serylglycine is an example of a peptide, although not one occurring in fresh fish. Other compounds forming important proportions of the total free amino acids are the substances called *taurine* and *creatine*. These, especially taurine, are not amino acids of the typical pattern and do not occur in proteins. However, taurine is closely related to another amino acid, called *cysteine*, which does occur in proteins. Both cysteine and taurine contain the element sulphur. Creatine is largely present in combination with phosphoric acid and plays an important role in the mechanism of muscular movement.

From the food processor's point of view, free amino acids are important components of the characteristic flavour of a product. They are also involved in a type of deterioration which may be troublesome during the storage of dehydrated and canned products, including fish. This is the development of a brown colour and an associated off flavour. The chemical reaction responsible involves the reaction of a certain type of sugar with an amino group, of which the free amino acids are an important source.

(3) *Sugars.* Just as with the word salt, the chemist has appropriated the word *sugar* to denote a whole class of compounds. Ordinary cane

sugar he calls *sucrose*, and the ending *-ose* is used for all sugars and related compounds; a well-known example is *glucose*.

While the characteristic taste of a sugar is always sweet, there are enormous variations in the intensity of this. Thus glucose has much less taste than sucrose, while a sugar found in milk, and called *lactose*, is nearly tasteless. Like amino acids, which can link up chemically with one another giving peptides and ultimately proteins, sugars can link up into bigger and bigger units. One very large such unit is the compound *cellulose*, which is the main structural material of plants; the paper on which this is printed is made of cellulose. A smaller unit is starch. Cellulose and starch, both built up from many units of glucose, are tasteless. Ordinary sugar is made up of one unit of glucose and one of a sugar called *fructose*, found in many fruits.

The principal free sugar of fresh fish is glucose, which as already noted is not very sweet and at the level of its occurrence in fish can hardly be expected to affect its taste, unless some other substance enhances its flavour. Information is still very scarce on the sugars of different species, and some fish, such as catfish, do have a sweetish flavour. The main significance of glucose in fish to the processor is in regard to the browning reaction mentioned under amino acids. Another sugar, called *ribose*, occurs in live fish attached to a complex nitrogenous substance and to *phosphate*; after death autolysis frees the sugar. Ribose is a particularly reactive browning agent.

MINERALS

To the layman the term mineral means something found in the ground, something mined. To the chemist it means something inorganic. To the dietician it covers a large range of elements, rather arbitrarily selected but typically those that will remain in the ash when a tissue is burned. For present purposes it will be restricted to certain elements essential to animals, including man. A satisfactory diet must include adequate amounts of these elements and they must be present in a form or a compound that the animal can use. For instance, phosphorus is such an element but it cannot be used directly in the free form. It is available to the animal when in the form of phosphoric acid, typically combined with some other substance, organic or inorganic, as a phosphate.

The list would include the following metals; potassium, sodium, calcium, magnesium, iron, copper, manganese, zinc and cobalt. It would also include the following non-metallic elements; phosphorus, sulphur, chlorine, iodine, to say nothing of such elements as carbon, hydrogen, oxygen and nitrogen which are not grouped with the minerals for present purposes. It is possible, even probable, that certain other elements are required in very tiny amounts. All the elements listed above are found in fish flesh.

Certain of these elements are so plentiful in most diets that the amount of them in fish is not likely to matter much one way or the other. In general it may be said that fish flesh resembles meat in its content of useful minerals. It approximates so closely to the flesh of higher animals that it is, in fact, a well-balanced supply of what they need. The element sulphur, for instance, is mainly required in the form of a particular sulphur-containing amino acid, called *methionine*, and fish flesh is a good source of this. In respect of one mineral marine fish, in common with other marine organisms, are an outstanding dietary source; this is the element iodine. A deficiency of iodine leads to the disease goitre, which often occurs among people living far from the sea where the crops are not affected by fine sea spray. Nowadays small amounts of iodine are often added to table salt in some countries, including Britain, but where this is not done marine fish may sometimes be the only good source of iodine in the local diet.

When fish bones are also eaten, for example the very small bones in a herring, or the softened bones of canned fish, they afford a rich source of calcium and phosphorus in a favourable combination. Fish are the only animals of which some of the bones are likely to be eaten.

VITAMINS

The term *vitamins* covers a very varied assortment of compounds. The only properties they have in common are that in very small amounts indeed they are necessary for health or even life, and that they must be obtained ready made from the diet. The latter statement is not quite true, for example vitamin D can be formed in the skin by the action of the sun's rays. But this is not something man does himself; the same effect could be produced in a test tube. It seems to be a matter of convention which substances with such properties shall be called vitamins and which shall not. One guiding principle is undoubtedly the scale on which they are required. The chemical structures of the various vitamins are not relevant to the present book.

Some 10 or more vitamins are recognized as necessary for man. All of them are found in fish but their distribution in various tissues is very uneven. Thus in many species, for example the cod family, almost the whole of the fish's store of vitamins A and D is found in the liver. This is why cod liver oil is a valuable pharmaceutical and veterinary product. In some other fish, such as herring, much vitamin D is present in the body tissues. In the freshwater eel, most of its store of both vitamins A and D is found in the body tissues. Again vitamin C in an animal is concentrated in the adrenal gland and this is not a part of the fish that is usually eaten. Fish roe, particularly the hard roe, is generally a much richer source of two of the B vitamins, B_1 or thiamin, and B_2 or riboflavin, than is fish flesh. Fish liver, in addition to its large content of vitamins A and D, is rich in B vitamins, such as thiamin, riboflavin and vitamin B_{12}, although these do not pass into the oil when it is prepared. In places where fish liver is eaten, and there are many such places, it is a rich source of all these vitamins.

Fish in general contain much higher proportions of vitamins A and D than do mammals, although there are exceptions. The sharks and rays are almost devoid of vitamin D. Some young fish contain very little vitamin A. On the other hand the levels may be extremely high, for example for both vitamins A and D in halibut liver, or for vitamin A in that of some sharks. It is useless to attempt to give average values. Apart from variations from one species to another there are enormous individual and seasonal variations within a species.

Generally speaking, it may be said that fish flesh resembles lean meat in its content of vitamins, but that some species have flesh far superior to meat in their content of vitamin A or vitamin D, or both.

Vitamins occur in far too small amounts to have any effect on flavour or texture. Some of them are rather sensitive to heat, or to atmospheric oxidation, and the food scientist should attempt to ensure the least possible damage to them during processing.

14

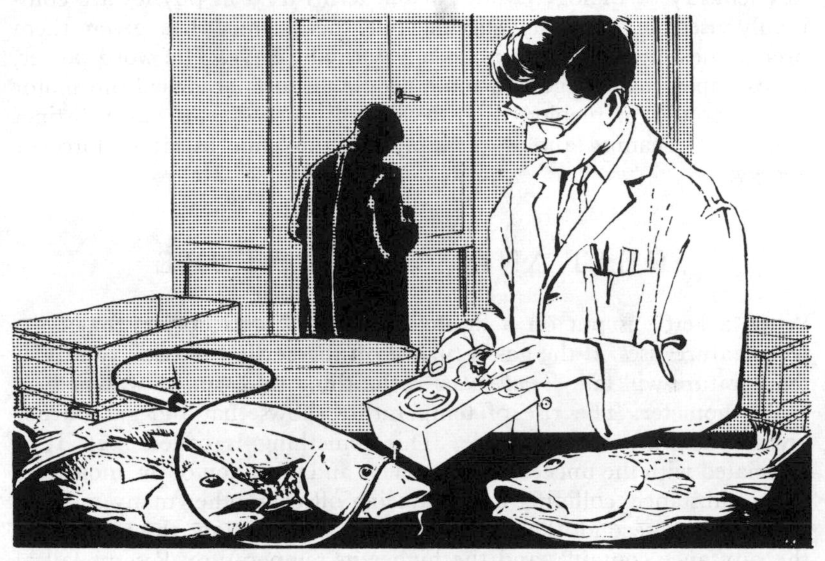

Fish and physics

This is a bit irregular
And not entirely true
But Chemistry is what things are
And Physics what they do.

D. L. NICOL

WHAT PHYSICS IS ABOUT

PHYSICS is the branch of science concerned with such things as heat and power, the force of gravity, and the behaviour of electric currents. Chemistry deals with what the world is made of, its bricks and mortar. Physics deals with the rules or the laws followed in the universe. Some knowledge of physics, particularly that part of it concerned with heat and temperature, is essential for understanding many operations in the fish trade.

As mentioned in the previous chapter, one great problem when explaining any scientific idea to a layman is to find the right words, and this problem is even more difficult in physics or engineering than in chemistry or biology. Many of the terms used in physics are commonly used in everyday speech, but the scientist has given them precise meanings which they do not normally have. The word power, for example, may suggest to some people all sorts of ideas from motor cars to politics or theology. To the scientist, however, it has a distinct and exact meaning and is certainly not the same as either force or energy.

HEAT AND TEMPERATURE

When a kettle is put on a lighted gas ring it gets hotter, that is its temperature rises. If the gas is turned off the kettle will cool, that is its temperature will fall. *Temperature* is the thing that can be watched on a thermometer. The rise of temperature shows that the gas burner gives something to the kettle. This something is called *heat*. It is associated with the unceasing movement and vibration of the molecules of the substance, colliding and bouncing off each other many millions of times in a second. The more energetic the motion, the more heat the substance contains, and the higher its temperature. Rise or fall of temperature is the effect that is noticed, and gain or loss of heat is the cause. In ordinary speech the words heat and temperature are roughly interchangeable but in science their meanings are quite separate and distinct, and must not be confused.

Heat and cold are also commonly judged by sensation, so that the tropics are thought of as hot and the Arctic as cold. Liquid air, however, is a lot colder still, and would boil in the heat at the North Pole. And even liquid oxygen contains some heat, because its molecules, although sluggish, are still in motion. Nevertheless, at a temperature somewhat lower than that of liquid oxygen, and known as *absolute zero*, the molecules stop moving altogether. Nothing can ever be colder than this. There is no known upper limit of temperature and the interiors of stars are unimaginably hot.

Physics is concerned with the precise measurement of causes and effects. It is not enough merely to say that heat flows into the kettle and its temperature rises. The question arises 'How much heat?' and 'How far does the temperature rise?' To answer these questions, it is necessary to measure temperature and express it as a number; that is, a scale of temperature must be devised. To make such a scale, two fixed and easily recognized temperatures are chosen and the range between them divided into equal steps called degrees.

Convenient fixed temperatures to use are the melting point of ice, and the boiling point of water. In the *Centigrade* (hundred step) *scale* these two points are known as 0 degrees (0°C) and 100 degrees (100°C) respectively. When most substances are heated they get bigger, that is they expand in all directions by an amount depending on the temperature and the actual material. If a gas, such as air, cannot expand because it is confined in a closed container, its pressure increases.

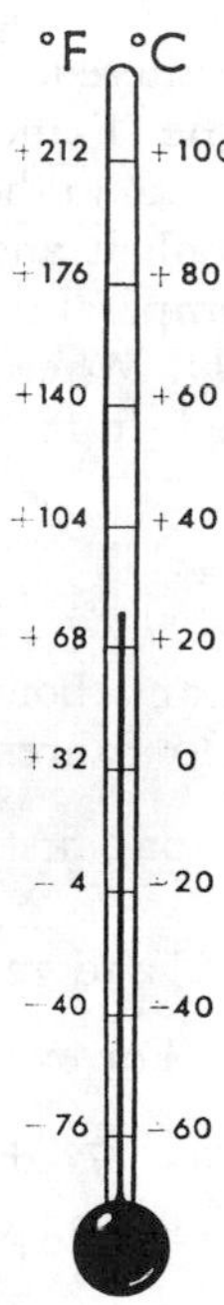

FIGURE 14·1 Fahrenheit and Centigrade temperature scales

This property of heat expansion is used in various sorts of thermometer. The most familiar type measures the expansion or contraction of mercury, or coloured alcohol, in a glass bulb connected to a tube with a fine bore. The finer the bore, the greater the movement of liquid in the tube. The bulb is first placed in melting ice and the resulting level of the liquid in the tube is marked. The procedure is then repeated for boiling water. If the tube has the same inside diameter throughout, the distance between the two marks can then be divided into 100 equal lengths, each of which is a degree. The expansion of solids and gases is used in other types of thermometer.

The Centigrade scale is used in scientific work in all countries and in commercial practice in Europe and many parts of Asia. In the *Fahrenheit system*, which is used in commercial practice in Britain and the USA, the melting point of ice is 32 degrees (32°F) and the boiling point of water 212 degrees (212°F), and the difference between subdivided into 180 equal steps. The Fahrenheit system, so named after its inventor, is slightly the older of the two and was proposed in 1720. The Centigrade system was introduced in 1745. In continental countries it is known by the name of its inventor, Celsius, the abbreviation, °C, remaining unaltered.

It is unfortunate that both the °F and °C systems should be in use, but it is not difficult to convert one to the other.

Figure 14·1 shows the Fahrenheit and Centigrade equivalents of a number of temperatures. Temperatures in between the intervals given can be estimated roughly. Where this is not accurate enough, temperatures can be converted from one to the other directly as follows:

Method I

Between the freezing point and boiling point of water there are 100 Centigrade degrees and 180 Fahrenheit degrees; thus to convert from Centigrade to Fahrenheit one first multiplies by $\frac{180}{100}$, that is by $\frac{9}{5}$. However, freezing point is 0°C and 32°F; therefore when going from °C to °F add 32 to the answer. For example, to convert 40°C to °F:

$$40 \times \frac{9}{5} = 72; \quad 72 + 32 = 104$$

$$40°\text{C} = 104°\text{F}$$

To convert from °F to °C the process is reversed. First subtract 32 and then multiply by $\frac{5}{9}$.

For example, to convert 50°F to °C:

$$50 - 32 = 18; \quad 18 \times \frac{5}{9} = 10$$

$$50°\text{F} = 10°\text{C}$$

Method II

To go from °F to °C. Add 40 and multiply the answer by $\frac{5}{9}$. Then subtract 40.

For example, to convert 50°F to °C:

$$50+40=90;\ 90\times\frac{5}{9}=50;\ 50-40=10$$

that is, 50°F = 10°C.

To go from °C to °F. Add 40 and multiply the answer by $\frac{9}{5}$. Then subtract 40.

Thus, to convert 40°C to °F:

$$40+40=80;\ 80\times\frac{9}{5}=144;\ 144-40=104;$$

that is, 40°C = 104°F.

Temperatures below that of melting ice and above that of boiling water can be measured by extending the scale at both ends. For example, minus 30°C represents a temperature 30 steps on the Centigrade scale colder than the melting point of ice, and 250°C is a temperature 150 steps hotter than that of boiling water. At temperatures below the freezing point of water, including some cold storage temperatures, the relationship between the Fahrenheit and Centigrade scales is sometimes confusing, for a temperature may be positive in Fahrenheit and negative in Centigrade. It will be seen from Figure 14·1 that minus 40° is the same on both scales. As this book is for British readers, only Fahrenheit temperatures will be used.

The unit of heat, known as the *British thermal unit*, abbreviated Btu, is the quantity of heat that will raise the temperature of 1 lb of water, that is four-fifths of a pint, by 1 deg F.

More heat is needed to warm 1 lb of water than to warm a pound of almost any other substance to the same extent. For example, to raise the temperature of 10 lb of mineral oil, or ice, by 1 deg F would require 5 Btu; 10 lb of most metals, including aluminium, would require only 1 Btu. Putting this in another way, Table 14·1 gives the heat-holding capacity or *specific heat* of a number of common substances relative to water which is taken as 1·0.

TABLE 14·1

Specific heats of some common substances

Btu required to raise 1 lb by 1 deg F

Mineral oils	—	about 0·5	
Most metals	—	„ 0·1	
Ice	—	„ 0·5	
Wet fish	—	„ 0·96	(usually taken as 1)
Frozen fish	—	„ 0·4	
Air	—	„ 0·25	(1 lb of air at 50°F occupies about 13 ft³)

When any substance gains or loses heat the amount of heat gained or lost can be worked out by multiplying:

$$\underset{(\text{lb})}{\text{Weight}} \times \underset{(\text{deg F rise or fall})}{\text{change in temperature}} \times \underset{(\text{Btu/lb deg F})}{\text{specific heat}}$$

That is to say:

$$\underset{(\text{deg F})}{\text{Temperature change}} = \frac{\text{Heat gained or lost (Btu)}}{\underset{(\text{lb})}{\text{Weight}} \times \underset{(\text{Btu/lb deg F})}{\text{specific heat}}}$$

Thus to heat 1 gallon (10 lb) of water from 40°F to 80°F requires

$$\underset{\text{lb}}{10} \times \underset{\text{deg F}}{(80 - 40)} \times \underset{\text{sp. ht.}}{1} = 400 \text{ Btu}$$

Similarly, to cool 10 stones (140 lb) of frozen fish from minus 20°F to minus 25°F means removing:

$$\underset{\text{lb}}{140} \times \underset{\text{deg F}}{(25 - 20)} \times \underset{\text{sp. ht.}}{0{\cdot}4} = 280 \text{ Btu}$$

There are cases where heat may be gained or lost without there being any temperature change at all. The temperature of a kettle of boiling water will remain steady at 212°F, regardless of how much heat is supplied, until all the water has boiled away. A mixture of ice and water will remain at 32°F until all the ice has melted. The thing to notice in these two examples is that what is known as a *change of state* is occurring, that is a change from liquid to gas or solid to liquid. In one case water is turning into steam and in the other ice is turning into water. To bring about such changes requires heat known as *latent heat,* from the Latin 'latens', hidden. This is more readily understood if it is realized that the molecules in a liquid are moving about to a much greater degree than they are in a solid, and in a gas they have an even greater freedom of movement than they have in a liquid. As heat is associated with movement of the molecules, a change from one state to another is accompanied by the storing or releasing of heat. A particular amount of ice always requires the same amount of heat for melting and this same quantity of heat, 144 Btu, is released when 1 lb of water at 32°F freezes into ice at 32°F. This figure of 144 Btu is known as the *latent heat of melting of ice* or the *latent heat of fusion of ice.* Similarly for every 1 lb of water at boiling point, 212°F, that is turned into steam at the same temperature, 970 Btu of heat have to be supplied. This quantity of heat is called the *latent heat of evaporation* and is given up when the steam condenses again to form water. The large amount of heat needed to melt ice, that is its high latent heat, is why ice is so widely used in the fish industry. The latent heat of evaporation is important in every operation in which drying occurs.

Every pound of ice that melts therefore absorbs 144 Btu of heat from somewhere. When fish is iced and some of the ice melts, the heat required to do this is taken from the fish and the fish therefore becomes cooler. The important thing to realize is that the ice must melt if it is to cool the fish. The mere presence of ice is in itself no proof that the fish has been effectively cooled. After it has melted, the cold water produced will have some further cooling effect, although small compared with that of the latent heat of the ice. One pound of ice at 32°F will in melting take up 144 Btu. The water produced will, on warming, for example from 32°F to 35°F, account only for

$$\underset{\text{lb}}{1} \times \underset{\text{deg F}}{3} \times \underset{\text{sp. ht.}}{1} = 3 \text{ Btu}$$

Most ice factories produce ice at temperatures between 23°F and 31°F although it cannot melt until it has warmed up to 32°F. The extra cooling effect of such cold ice can be calculated. The specific heat of ice is 0·5.

Suppose the ice is at 24°F.

The heat required to raise 1 lb of ice from 24°F to 32°F

$$= \underset{\text{deg F}}{(32 - 24)} \times \underset{\text{sp. ht.}}{0{\cdot}5} = 4 \text{ Btu}$$

The heat required to melt 1 lb of ice at 32°F = 144 Btu.

Therefore the total heat required to melt the ice, that is the total cooling effect, is 148 Btu. For all practical purposes in the fish trade the additional 4 Btu provided by cooling 1 lb of ice to 24°F is too small to matter. In any case, ice that leaves the factory at 24°F may be very close to 32°F by the time it is put on the fish.

The amount of ice needed to cool a particular weight of fish can be calculated. Consider, for example, a 2-stone box of cod fillets. Since tap water is generally at about 50°F, the majority of freshly cut fillets are at 50°F. The specific heat of fish is 1. Hence to cool 2 stones (28 lb) of fish from 50°F to 32°F:

$$\underset{\text{lb}}{28} \times \underset{\text{deg F}}{(50 - 32)} \times \underset{\text{sp. ht.}}{1} = 504 \text{ Btu must be removed}$$

Ice will absorb 144 Btu/lb on melting and therefore the weight of ice needed is:

$$\frac{504}{144} = 3\tfrac{1}{2} \text{ lb of ice}$$

This is the weight of ice that will be melted in cooling the fish. In practice when a box of fish is iced for despatch from a port to an inland market some ice will be used in cooling the box itself and more will be melted on the journey. More than the $3\frac{1}{2}$ lb calculated above must therefore be used. How much more will be considered in the next section. The present simple calculation shows that unless at least $3\frac{1}{2}$ lb of ice is melted it is impossible to cool 2 stones of fillets from 50°F to 32°F.

HOW HEAT IS TRANSMITTED

It is now necessary to consider the means by which heat is transferred from one thing to another and also how quickly.

Just as water flows downhill, so heat flows from something at a high temperature to something at a lower temperature. To take water uphill, it must be pumped or carried. Similarly, to make heat go from something cold to something warmer, work has to be done. A refrigerator is a machine which does this, that is it moves heat up a *temperature gradient.* Just as water flows more quickly down a steep slope than a gentle one so heat flows more quickly where there is a big difference of temperature.

It is temperature difference which makes heat go from one place to another, but it may travel by one or all of three entirely different routes, by conduction, convection and radiation.

When a poker is put into the fire, heat travels through the metal and after a time the handle begins to feel warm. This is a common example of the *conduction* of heat. In the fish trade unwanted conduction of heat takes place through the walls of cold stores and ice boxes; conduction is essential on the other hand in freezing and thawing fish.

With *convection,* heat is carried by the movement of a substance. For example, in a traditional smoking kiln, air, which is heated by the fire at the bottom, expands and so becomes lighter and rises, warming the fish in its path. This type of convection is known as *natural convection*, for the movement of the air depends upon change in density when the air is heated. Another example of natural convection is the ordinary domestic hot water system where hot water rises to its tank.

If natural convection is too slow, the movement can be speeded up by a fan or pump. This is known as *forced convection* and occurs in a Torry kiln or in an air blast freezer. In a Torry kiln, heat is transferred from heating coils to the air blowing over them and then from the hot air to the fish. In an air blast freezer the reverse is happening; the fish heats the air which then gives up this heat to the cooling coils.

Heat may travel from one thing to another by *radiation* without warming the intervening space, for example the heat from the sun that reaches the earth. Kippers on a slab in the sun will get hotter than the surrounding air. Electric lamps radiate heat as well as light. Radiation is not as important in the fish trade as conduction or convection, but where it does occur it is usually undesirable. Conduction and convection have both good and bad aspects, and sometimes the object is to stop them and sometimes to encourage them.

It is often useful to know the quantity of heat that moves from one point to another in a given time.

In the ordinary 2-stone wooden fish box the fish is put in first, then a layer of ice and finally the lid nailed on. Some of the ice will be melted in cooling the fish but some of it will also be melted by heat passing by conduction through the lid. It would be useful to know how much heat is actually conducted through the lid in a certain time under any particular conditions.

Figure 14·3 illustrates the problem. A small portion of the lid of the box shown shaded is oblong in section and the area of the top surface is the same

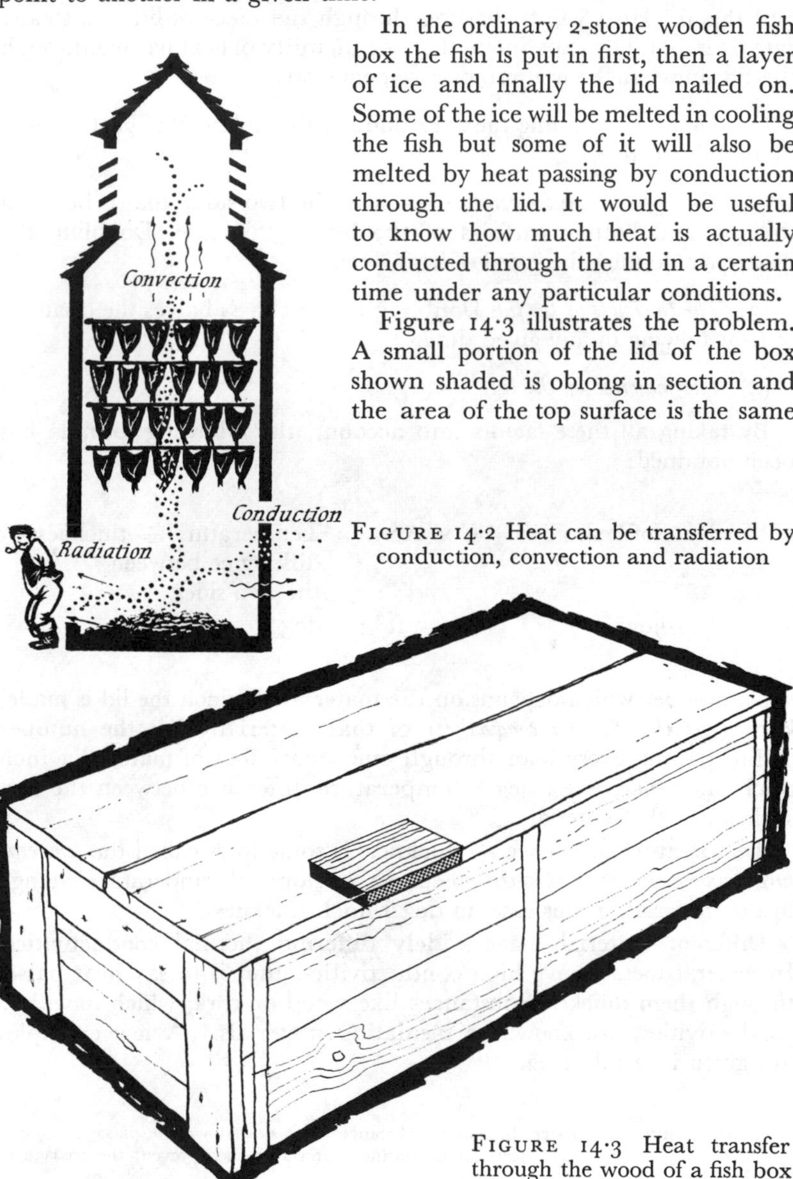

FIGURE 14·2 Heat can be transferred by conduction, convection and radiation

FIGURE 14·3 Heat transfer through the wood of a fish box

as that of the bottom surface. When the box has been standing for some time, and the temperatures have become steady, the under surface of the lid will have cooled almost to ice temperature while the top surface will be at a temperature somewhere between that of the ice and the air. Heat will be passing through the piece of lid at a steady rate. It is found by experiment that the quantity of heat passing through the lid under such circumstances depends on:

(1) *The area.* Doubling the area doubles the quantity of heat flowing in any given time.

(2) *The difference in temperature* between the two sides, that is between the top and bottom surfaces of the lid in this case. Doubling the difference doubles the speed of heat flow.

(3) *The thickness of the lid.* Doubling the thickness halves the quantity of heat flowing in any given time.

(4) *The material of the lid.*

By taking all these factors into account, the following formula has been obtained:

Rate of heat flow	= k ×	Area ×	Temperature difference between the two sides	÷ thickness
Btu/hour		ft^2	deg F	in

k is a number which depends on the material of which the lid is made, known as the *thermal conductivity* of that material. It is the number of Btu passing every hour through one square foot of material 1 inch thick when there is a 1 deg F temperature difference between the two surfaces of the material.

Temperature difference ÷ thickness is sometimes called the *temperature gradient*, because it is the steepness, or slope, of temperature change in the material, in this case, in deg F/inch thickness.

Different materials have widely differing thermal conductivities. In general metals have high conductivities, that is to say heat passes through them quickly. Substances like wood or cork, which have low conductivities, are known as insulating materials.* A few examples are given in Table 14·2.

* Thermal insulating materials, that is substances that slow down the passage of heat, should not be confused with electrical insulating materials which prevent the passage of electricity.

TABLE 14·2

Thermal conductivity (k) *of some common materials*

Material	$k \left(\frac{Btu\ in}{ft^2\ h\ deg\ F} \right)$
Copper	2664
Aluminium	1392
Steel	312
Brick	8
Wood	0·4 to 1·6
Cork	0·25 to 0·3
Glass fibre	0·23 to 0·29
Mineral fibre	0·18 to 0·29
Polyurethane foam	0·18 to 0·29
Expanded ebonite	0·18 to 0·23
Ice	15 to 18
Water	4
Frozen fish	8 to 12
Wet fish	4
Air	0·17

Some reference books give the thermal conductivity through a thickness of 1 *foot* giving values exactly 1/12th of those in Table 14·1, and this point should always be remembered when consulting figures for insulating materials.

Thermal conductivities change a little with temperature but this is unimportant for most purposes.

It is worth noting that the conductivity of frozen fish is greater than that of wet fish; this is important in the operations of freezing and thawing. The very small conductivity of air should also be remembered and the significance of this will be seen in the next section.

Conduction as just considered is what occurs at steady temperatures, that is when the driving force, the temperature gradient, does not alter. In many practical instances, however, things have not had time to settle down to a steady state, and the mere fact that heat is flowing between two points decreases the temperature difference between them. The flow of heat is therefore rapid at first and then gradually gets slower and slower.

Fillets that are all at 50°F when they are packed in a box as shown in Figure 14·4, with a layer of ice both above and below the fish, in theory need 3½ lb of ice to cool them to 32°F; this was shown on page 325. The example did not show, however, how quickly they would cool. The fillet right in the middle will obviously cool most slowly, since it is furthest away from either layer of ice.

If the distance from the middle fillet to an ice layer is, for example, 2 inches and the temperature difference between the fish there and the ice is 18 deg F, there is a temperature gradient of 18 deg F over 2 inches or 9 deg F/inch. When the fish in the middle has cooled to 40°F the temperature difference will be only 8 deg F and the temperature gradient 4 deg F/inch. Hence the rate of cooling will be a lot slower. When this fish is cooled to 35°F the temperature gradient will be only 1·5 deg F/inch and the rate of cooling will be slower still. Thus the rate of cooling gets slower and slower as the temperature gets nearer to 32°F; after six hours the fish in the middle of the box has cooled to 33°F and it will continue to cool more and more slowly. The actual results in this instance are shown in Figure 14·5.

A point of interest at the other end of the graph in Figure 14·5 is that the temperature seems to hesitate before beginning to fall. This can be understood if it is remembered that heat will not flow unless there is a temperature gradient. Right at the beginning all the fillets are at 50°F and heat will not pass from one to the other; only the fillets that are right against the ice will lose heat. Once these cool below 50°F, heat will begin to flow into them from the next fillet and so on. Thus when ice is put on the top of the box there is a definite pause before the fillet in the middle begins to cool. The situation is like that of a long queue of people; when the head of the queue starts to move there is a noticeable pause before the tail starts moving. Similarly the tail continues moving for some time after the head stops.

It is interesting to consider what happens when the box is, for example, 8 inches deep instead of 4 inches so that the distance from the middle of the box to the ice at top or bottom is 4 inches instead of 2 inches. The temperature gradient at the start is then 18 deg F in 4 inches or 4·5 deg F/inch, that is, only half the value for the box 4 inches thick. Heat flow is therefore slower and cooling takes longer. Conversely if the box is shallower than 4 inches, cooling will be quicker. The effect of depth of box on cooling time is very great as will be seen from Table 14·3 which gives the time taken for layers of fish of various thicknesses to cool from 50°F to 35°F.

A single fillet can be cooled very quickly, a large thickness of fillets or fish can be cooled only slowly. Thermal conductivity is a property of the substance itself, in this case, of the fish. Consequently, to cool fish quicker in ice the correct way is to reduce the distance between

TABLE 14·3

Time taken to cool fish in the middle of boxes of various depths iced at top and bottom

Thickness of layer of fillets	*Time to cool from 50°F to 35°F at the middle*
3 inches	2 hours
4	4 hours
5	6½ hours
6	9 hours
10	24 hours
24	5 days

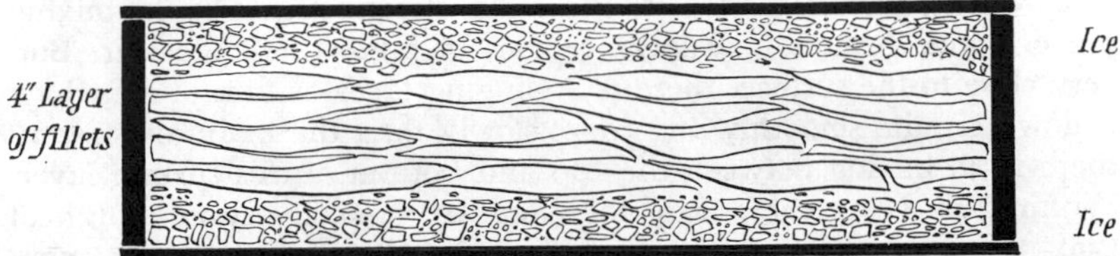

FIGURE 14·4 Cross section of 2-stone box: fillets correctly packed in ice

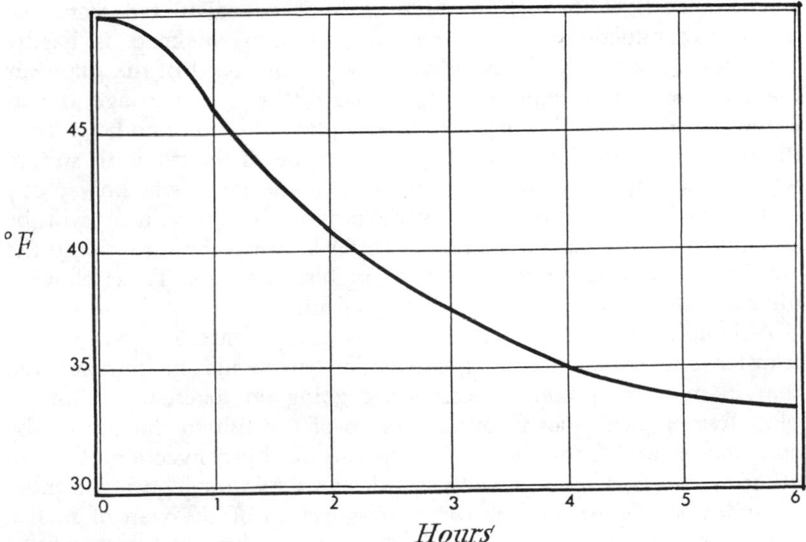

FIGURE 14·5 Cooling curve in centre of 2-stone box of iced fillets

the fish and the nearest piece of ice; that is, to make sure that ice is well distributed amongst fish.

When a packet of fish has just been put into an air blast freezer, a current of cold air passes over the packet and the fish loses heat to the air. The air thus becomes warmer, although not very much warmer, perhaps only a tiny fraction of a degree F, for there is a lot of air. The air then passes on to the cooling unit and loses heat to the refrigerant in the cooling coils. It then passes on to extract more heat from the fish. Thus heat is carried from the fish to the refrigerant by the movement of air, a case of convection and, since there is a fan in the blast freezer, forced convection.

The packet of fish when just put in would be about 50°F and the main stream of air at perhaps minus 40°F. If measurements were made in the air stream very close indeed to the surface of the fish, they would show that there is a very steep temperature gradient there. The reason for this is that in the main stream of air rapid mixing prevents the formation of local regions of hotter or colder air. But, very close to the surface, there is a clinging layer or film of air which is flowing quite smoothly and more slowly than the main air stream; there is no mixing between the top and bottom of this surface layer. Within this layer there can therefore be no convection and heat that passes through it must do so by conduction. The thermal conductivity of air is very low ($k = 0{\cdot}17$) and consequently the very thin surface film is a barrier to heat flow.

The heat flow through the film cannot be calculated from the ordinary conduction formula because the film thickness is hardly ever known, and in any case it varies with the speed of the main air flow. In one set of experiments carried out some years ago it was found that if air was flowing over a glass plate at 9 miles an hour, then the thickness of the layer was 0·034 inches, but if the main air stream was speeded up to 53 miles an hour then the layer was only 0·007 inches thick. The thinner the surface film the more heat will be conducted and the better the heat transfer between the surface and the air. Hence, high air speeds are used in blast freezers. These clinging films are also important in the drying of fish.

Although for purposes of study it is convenient to separate heat transfer into conduction and convection, it should be remembered that in most cases both processes are going on together. In an air blast freezer, heat goes from the centre of the fish to the surface by conduction, and from the surface to the air by convection. Certain other examples, a modern railway fish van for instance, are still more complicated. Figure 14·6 shows a cross-section of the wall of such a van in which there is a sandwich of insulation between two walls of resin-bonded plywood, and the inner wall is lined with aluminium

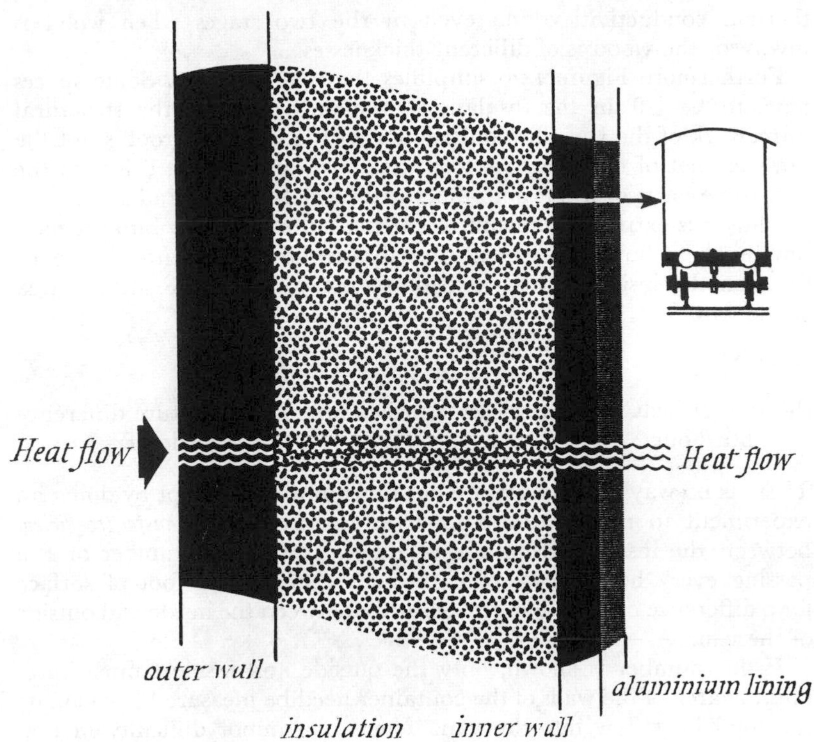

FIGURE 14·6 Transfer of heat through the wall of an insulated rail van

sheet. Heat transfer between the warm air outside the van and cooler air inside it will go on as follows:

From the outside air to the outer wall of the van by convection.
Through the outer wooden wall by conduction.
Through the insulation by conduction.
Through the inner wooden wall by conduction.
Through the aluminium lining by conduction.
From the aluminium lining to the air inside by convection.

There are thus two stages of convection and four of conduction all of which will proceed at different rates. The two convection stages will be different because the air inside the van is almost still and the air outside may be moving quite fast. The conduction stages will all be different because wood, insulation and aluminium all have different thermal conductivities, and even in the two stages when wood is involved, the wood is of different thicknesses.

Furthermore Figure 14·6 simplifies the real situation. Some spaces have to be left in the insulation to make room for the structural framework of the van, the construction of the floor and roof is not the same as that of the sides, the van has doors, and finally, if it is in the sun, some parts of it but not others will receive heat by radiation.

Thus it is extremely difficult to calculate the heat flow into the van, but this heat flow will depend directly on the temperature difference between the inside and the outside of the van and the surface area of the van.

That is:

the rate of heat flow = *a number* × surface area × temperature difference
Btu/hour ft^2 deg F

There is no way of finding out what this number is except by doing an experiment to measure it. Known as the *total heat transfer coefficient* between the inside and outside of the van, it is the number of Btu passing every hour on average through each square foot of surface for a difference of 1 deg F in temperature between the inside and outside of the van.

If this number is known, only the outside and inside temperatures and the area of the walls of the container need be measured to calculate the total heat flow into the van. There is a minor difficulty in that the area of the outside of the container is greater than that of the inside, and it is therefore necessary to know whether the value arrived at for the total heat transfer coefficient of a container refers to the inside or outside area. Values for total coefficients are necessarily approximate, however, and in large containers with relatively thin walls, any error due to taking the wrong area is not very important.

Some values for total heat transfer coefficients are given in Table 14·4.

Because total coefficients are expressed for a square foot it must not be thought that the heat flow is the same for every square foot of surface. The heat flow through the floor of some rail vans is, for example, considerably greater than that through the walls. The values given in the Table are averages for the whole surface. Some people prefer to express heat flow in terms of the whole container instead of for a square foot. They would give the value of a modern insulated rail van as 150 Btu/h deg F instead of giving the area and the total heat transfer coefficient. This value is more useful when considering a particular container, for example when deciding how much ice to use on a particular journey, whereas a figure for the average heat flow through one square foot is useful when comparing containers that differ in size as well as type of insulation, for example old and new types of fish van.

It has been shown that to cool 2 stones of fillets from 50°F to 32°F requires $3\frac{1}{2}$ lb of ice, but how much more ice is needed to keep it cool during the journey? This of course depends very much on the length of the journey, the average outside air temperature and on how the box is stowed and transported.

The example given on page 325 may be continued. Assume that the journey takes 18 hours, that the average air temperature is 50°F, and that the box lies by itself in the guard's van of a passenger train.

The surface area of a 2-stone box is 5·2 ft^2.
The total heat transfer coefficient is 0·5 Btu/ft^{2}h deg F.
The temperature outside the box is 50°F.
The temperature inside the box is 32°F.

Using the formula given on page 328, the heat flow into the box is

$$\underset{\text{coeff.}}{0{\cdot}5} \times \underset{\text{ft}^2}{5{\cdot}2} \times \underset{\text{deg F}}{(50 - 32)} \text{ Btu/hour, that is, } 46{\cdot}8 \text{ Btu/hour.}$$

Hence in the 18-hour journey the total heat entering the box will be

$$\underset{\text{h}}{18} \times \underset{\text{Btu/h}}{46{\cdot}8} \quad \text{or } 844 \text{ Btu}$$

If the fish temperature is to stay at 32°F all this heat must be absorbed by melting ice.

TABLE 14·4

Heat flow into typical fish boxes and rail vans

Container	*Inside surface* (*ft*^{2})	*Total coefficient of inner surface* (*Btu/ft*^{2}*h deg F*)
Six-stone wooden fish box	10	0·5*
Two-stone wooden fish box	5·2	0·5
Six-stone light alloy fish box	10·5	1·2
Old type uninsulated rail van	530–600†	0·55
Modern insulated rail van‡	630–680	0·23

Notes:

* A number of boxes typical of Hull, Grimsby and Aberdeen have been examined and the values are all between 0·4 and 0·6.

† There are a number of slightly different sizes of both old and new type vans.

‡ Not the 'Blue Spot' type.

Since 1 lb of ice requires 144 Btu to melt it, 844 Btu will melt

$$\frac{844}{144} = 5{\cdot}86 \text{ lb of ice or about 6 lb.}$$

The total amount of ice required in theory to keep the fish in this particular box at 32°F is therefore $6 + 3\frac{1}{2} = 9\frac{1}{2}$ lb.

In actual practice, a little more even than this would probably be necessary. In typical commercial procedure a box of fish starts with anything from 4 to 8 lb of ice. This is one of the reasons why so much fish gets to its destination at temperatures well above 32°F.

The amount of ice needed for the journey, as just calculated, is a much more variable and less certain quantity than the amount needed to cool the fish. If the 2-stone box used in the above example had been part of a large consignment stacked together it would have been protected by other boxes and the area available for heat transfer would have been much less than 5·2 square feet. If it had travelled in an insulated container the temperature of the air in the container would have been less than the external temperature of 50°F. In this way it may happen that a light alloy fish box in an insulated vehicle gives better results than a wooden one carried on an open lorry.

INSULATING MATERIALS

Air that is being used to transfer heat, as in an air blast freezer, should have a rapid and turbulent flow, so as to encourage convection. If, on the other hand, there were some way of holding the air perfectly still so that it could only transfer heat by conduction, then it would be a very effective barrier to heat transfer since the conductivity of air is so very low. This in fact is the way in which insulating materials work.

In a double-walled container like the old type of railway fish van, the air in the space between the inner and outer walls is free to move and so to transfer heat by convection. If the space is packed with, for example, cotton wool, the convection is prevented and heat transfer will be by conduction only, through a mixture of air and cotton wool fibres. Any fibrous material such as cotton wool, glass wool or asbestos fibre can be used as a thermal insulator. There is another important class of insulating materials in which air or some other gas is imprisoned as a mass of small bubbles. Expanded ebonite (Onazote) and the many types of foamed plastic insulation are in this class. It does not matter if some gas other than air, for example carbon dioxide or nitrogen, is used for the bubbles, for all gases have low thermal conductivities.

A good insulating material will contain as much air and as little solid as possible, subject always to the consideration that the air must not be able to move about. The better the insulating material the more closely will its thermal conductivity approach that of air, although it will never be quite as small as this because of the solid present. Hence the conductivities of a number of the best insulators are nearly the same (see Table 14·2). In selecting an insulating material the choice is usually made on such grounds as cost, durability and convenience rather than thermal conductivity.

Because of this dependence on the enclosed air, an insulating material will be much less effective if it gets wet and the air is replaced by water, since the thermal conductivity of water is much greater than that of air. The same thing applies even more if the water is liable to get frozen to ice, as in the wall of a cold store. Not only has ice an even higher conductivity than water, but the expansion that occurs on freezing may break up the insulation. Hence, whenever an insulated structure is built, care should be taken to keep the insulation dry.

Finally, it should be noted that the heat flow into any insulated container, particularly a large one, will always be greater than a simple calculation based on thermal conductivity would suggest. This is because various parts of the supporting structure have to go through the insulation, and these may account for much of the heat flow.

WATER ICE VERSUS 'DRY ICE'

In the light of the principles for cooling fish outlined above, the properties of solid carbon dioxide and in particular the way in which its refrigerant properties compare with those of ordinary ice, may be assessed.

Solid carbon dioxide, sold under the names of Drikold and Cardice and often referred to as 'dry ice', is familiar in the fish trade. Unlike ordinary ice, it does not normally melt to a liquid but passes straight from a solid to a gas, just as does snow in frosty weather. It will melt to a liquid if the pressure is above a certain critical value, which is higher than atmospheric pressure. This change occurs at the very low temperature of minus 109°F. The latent heat involved in this process is 241 Btu/lb, that is 1 lb of solid carbon dioxide at minus 109°F turning into 1 lb of carbon dioxide gas also at minus 109°F will take up 241 Btu. In comparing the cooling effect of solid carbon dioxide with ice, there must be added to the latent heat the extra heat that this 1 lb of gas at minus 109°F will take up in warming to 32°F. This requires a further 34 Btu/lb so that the total refrigerating effect of solid carbon dioxide is 275 Btu/lb compared with the refrigerating effect of ice at 32°F which is 144 Btu/lb.

This shows that the total cooling effect of solid carbon dioxide is nearly twice as great as that of ice. Solid carbon dioxide is, however, very much more expensive than ice. Its cost is in the region of 6s. for a 25 lb block, or approximately 27s. a cwt. At the main fish ports, ice costs about 1s. a cwt and at the larger inland towns roughly 4s. to 5s. a cwt. For the same cooling effect, the cost of solid carbon dioxide is therefore anything from three to thirteen times that of ice.

For keeping fish at 32°F, for example if there are a number of boxes of iced fillets in a railway van going from Hull to London, it will be cheaper to use more ice instead of using some solid carbon dioxide 'to preserve the ice' as is sometimes done. If wet fish is being handled, the very low temperature of solid carbon dioxide is a further disadvantage since some of the fish may become frozen.

For the handling of frozen fish the situation is entirely different. Ice is no use at all; putting ice at 32°F on to frozen fish at minus 20°F would warm frozen fish, not cool it. Solid carbon dioxide, if used in sufficient quantity, could cool this fish still further and this would be all to the good. Usually, however, the object of using solid carbon dioxide is to deal with the heat that enters through the walls of an insulated container and would otherwise warm up the frozen fish. Just as the distribution of ice in a box of wet fish has a great influence on how cool the fish is kept, so the distribution of solid carbon dioxide and not merely the quantity used has a great effect on its value in an insulated container. In this type of application, solid carbon dioxide is an alternative to mechanical refrigeration and not an alternative to ice.

WEIGHTS AND MEASURES

In British weights and measures, distance is measured in feet, weight in pounds and time in seconds. This is the so-called *foot, pound, second system* or f.p.s. for short. The f.p.s. unit of heat is the Btu and is based on 1 lb of water. On the Continent, all weights, measures and scientific units as well are expressed in the *centimetre, gramme, second system* or c.g.s. for short. This system is much easier to use because all units are based on multiples and sub-multiples of ten. For example, 1 metre = 10 decimetres = 100 centimetres = 1000 millimetres. To convert from one system to the other involves awkward arithmetic; there are, for instance, 453·59 grammes in a pound and 2·53998 centimetres in an inch. The c.g.s. unit of heat is the *calorie*; one thousand of these make up a *kilocalorie* or large Calorie, spelt with a capital C, which will be familiar from its use in diet sheets.

The relationship of area (length × length) and volume (length × length × length) to the basic units is obvious. In the f.p.s. system these are expressed in square feet (ft^2) and cubic feet (ft^3) respectively. *Density* is merely the mass of unit volume and is expressed as pounds to a cubic foot; that is a weight $\div$ length $\times$ length $\times$ length. Thus, the density of water is 62·4 lb/ft^3 and that of air, 0·08 lb/ft^3. *Specific gravity* is the density of a substance compared to that of water, which is taken as 1·0. Saturated brine, for instance, has a specific gravity at 60°F of 1·204.

PRESSURE OF LIQUIDS AND GASES

A force can be exerted by a liquid or gas, for example by the wind on the sails of a yacht, or a stream on a water wheel. While a billiard cue applies its force to the ball in one place, and the force depends only on the player and not the size of the ball, the wind acts over the whole area of the sail. The total force depends on the area of the sail. When thinking about forces exerted by liquids or gases it is usual to consider the *pressure*, that is the *force on a unit area*, for example lb/in^2. This is the only meaning attached to the word pressure by scientists.

Although pressure means force on a unit area it is not always expressed in this way. Another method is sometimes more convenient. Imagine a vertical iron pipe closed by a cork at the bottom. If water is poured down this pipe a pressure is put on the cork, and the more water that is poured in, the greater is the pressure. For example, a depth of 1 foot of water above the cork gives a pressure of 0·433 lb/in^2. A depth of 3 feet gives a pressure three times as great, 1·299 lb/in^2 and so on, the greater the depth the greater the pressure. The diameter of the pipe makes no difference to the force on a unit area of the cork, although the total force would be greater with a wider pipe. A much heavier, that is denser, liquid like mercury gives a greater pressure; for example a depth of 1 foot of mercury gives a pressure of 5·9 lb/in^2 on the cork. It is sometimes useful to speak of a pressure as '10 inches of mercury' for example; this means that the pressure is the same as that of a column of mercury 10 inches high. The pressure may be due to air or water or anything else, and there need not be any mercury involved. This method is useful with small pressures. It is more convenient to express the air pressure exerted by a fan as 2 inches of water rather than as 0·072 lb/in^2, which is the same thing.

A column of gas can produce a pressure in the same way as a column of liquid and everyone on the Earth is subjected to the pressure of air above them. This pressure of the atmosphere is nearly 15 lb/in^2. It varies slightly with the weather; the use in this connection of the barometer, which registers the air pressure in the atmosphere, is familiar to everybody. Air pressure also decreases with height above sea level; indeed the altimeter of an aircraft is worked by this change of pressure. The standard value of atmospheric pressure, *one atmosphere*, is taken as 14·7 lb/in^2, *or* 29·92 inches of mercury (76 cm of mercury), *or* 33·93 ft of water. All these are the same thing in different units.

Pressure gauges usually read zero at atmospheric pressure, and hence a tyre pressure of 24 lb/in^2 means 24 lb/in^2 *above* atmospheric pressure. Such a pressure is known as a *gauge pressure* and is written 24 lb/in^2 gauge. Sometimes it is necessary to measure pressure starting from zero, that is, a complete vacuum, and these are known as *absolute pressures* which can be abbreviated lb/in^2 abs. To convert from gauge to absolute pressures simply add 14·7, that is, 24 lb/in^2 gauge is 38·7 lb/in^2 abs.

Clearly any pressure below 14·7 lb/in^2 abs means not a pressure in the ordinary sense but a partial vacuum. A perfect vacuum would be 0 lb/in^2 abs. A very low pressure is usually referred to as a high vacuum and it is usual to measure it in millimetres of mercury rather than pounds on a square inch. A millimetre is nearly 1/25th of an inch.

Any substance that can vapourize, that is turn from a liquid or solid into a gas, for example water and petrol, also exercises a pressure, which may be registered on a pressure gauge. Some gases, such as oxygen and hydrogen, are supplied in a highly compressed form in cylinders, and are forced out as a result of the release of pressure when the valve is opened. Other gases, such as ammonia and carbon dioxide, become liquids when compressed into cylinders at ordinary temperatures. Liquid carbon dioxide has to be stored in strong, pressure-resisting tanks at about 280 lb/in^2, that is, about 19 times the ordinary atmospheric pressure of 14·7 lb/in^2, or simply 19 atmospheres.

WATER VAPOUR PRESSURE AND RELATIVE HUMIDITY

The commonest vapour is water vapour. Water vapour is important in fish technology, particularly in freezing, smoking and drying. As the water vapour in the atmosphere is invisible, its existence and effects are not always immediately obvious; they may indeed sometimes only become apparent when it is too late to make the necessary adjustments. It is therefore useful to know something about the way water vapour behaves. When something is dried the molecules of liquid water are transformed into invisible vapour and become mixed with the molecules of oxygen, nitrogen and other water vapour molecules already in the air. Latent heat has to be supplied for evaporation to take place and as a result there is usually a noticeable cooling effect. The amount of water that can be taken up by air depends on its temperature and on how much moisture it already contains, that is its degree of humidity. When the atmosphere is humid, drying is slow. If the surrounding air already contains so much moisture that it can hold no more, further evaporation cannot occur and drying becomes impossible. The air is then said to be *saturated* with water vapour. The warmer the air, the more water vapour it takes to saturate it. If this saturated air is warmed it can take up still more moisture. On the other hand if unsaturated air containing some water vapour is cooled it eventually becomes saturated and if it is then cooled still further, visible water is formed, either liquid as in cloud droplets or dew, or solid as frost.

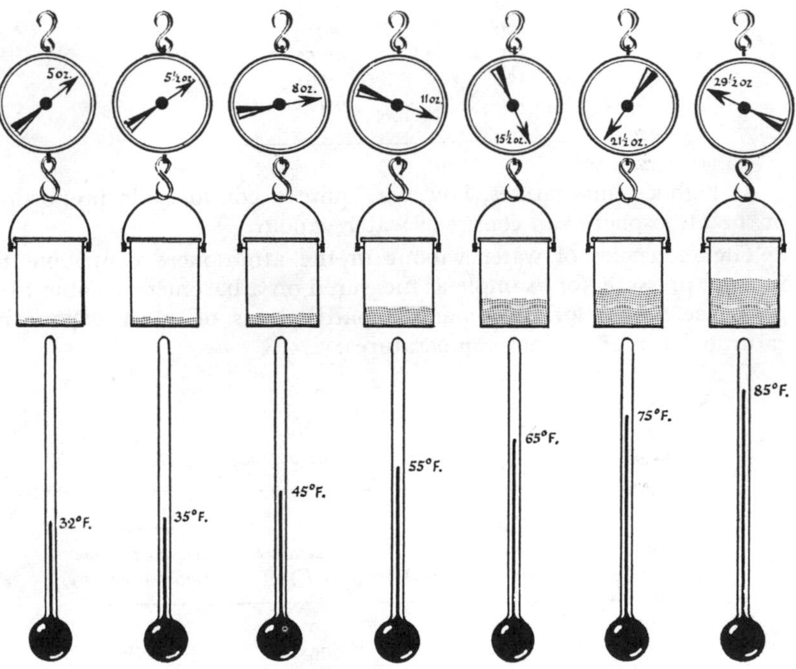

FIGURE 14·7 The weight of water vapour that 1000 cubic feet of air will hold at different temperatures

The water content of air is usually calculated either as grains of moisture in a cubic foot of air (7000 grains = 1 lb), or more usefully, as pounds of water in a pound of dry air (1 cubic foot of air weighs about 1¼ oz). The quantity of water held under conditions of saturation is only a small proportion of the weight of the air itself at ordinary temperatures (see Table 14·5):

TABLE 14·5

Water held by saturated air at various temperatures

Temperature °*F*	*lb water/lb dry air*
40	0·005
50	0·008
60	0·012
70	0·016
80	0·022
90	0·031
100	0·043
110	0·059
120	0·081
130	0·111

At higher temperatures, however, quite a considerable proportion of the atmosphere can consist of water vapour.

The molecules of water vapour in the atmosphere contribute to its total pressure, for example as measured on a barometer. Table 14·6 gives the values for the separate contributions of water vapour in saturated air at various temperatures:

TABLE 14·6

Saturated vapour pressure of water at various temperatures

Temperature (°*F*)	*Saturation vapour pressure* (*inches of mercury*)	*Temperature* (°*F*)	*Saturation vapour pressure* (*inches of mercury*)
35	0·20	70	0·7
40	0·25	75	0·88
45	0·30	80	1·03
50	0·36	85	1·21
55	0·43	90	1·42
60	0·52	95	1·66
65	0·62	100	1·93

It should be realized that the vapour pressure of a pure liquid depends *only* on its temperature. In fact, if drops of water were introduced into a vacuum they would exert the same vapour pressure as they would in air.

The *boiling point* of water, or any other liquid, is the temperature at which its saturation vapour pressure reaches atmospheric pressure, that is normally about 30 inches of mercury. The boiling point therefore varies with atmospheric pressure.

30 000 feet above sea level, on top of Mount Everest, or in an aeroplane, atmospheric pressure is only 8·85 inches of mercury, that is to say only about 30% of normal, because some 70% of the atmosphere is below this elevated level. Water has a vapour pressure of 8·85 inches of mercury at 157°F. At 30 000 ft therefore, water boils at 157°F instead of 212°F.

At pressures above normal atmospheric pressure the boiling point of water is above 212°F (see Table 14·7). In a pressure cooker, autoclave or steam retort as used in canning, water is heated above its usual boiling point in a strong container. Cooking and sterilizing are faster under these conditions.

Refrigerants behave in an equally regular manner.

TABLE 14·7

Boiling points of water at various pressures

Pressure (*lb/in² gauge*)	*Temperature* (*°F*)
0	212·0
1	215·4
2	218·5
3	221·5
4	224·4
5	227·1
10	239·4
15	249·8
20	258·8
30	274·1
40	286·7
50	297·7
75	320·0
100	337·8

Before any drying, of fish for instance, can occur, the water vapour pressure in the space surrounding it must be below the saturation value given in Table 14·6. The *relative humidity* (r.h.) is the actual water vapour pressure expressed as a percentage of the saturation value at that temperature. The r.h. can also be obtained from the weight of water in a definite volume as a percentage of the saturation value in the same volume. Humidity can conveniently be measured by means of a *wet and dry bulb hygrometer*. This consists merely of two thermometers, one of which has its bulb covered by a wet muslin bag. When air is blown over the wet bulb, water evaporates and cools the bulb to an extent that depends on the amount of vapour water already in the air. Table 14·9 gives the relative humidities for the range of air conditions likely to be encountered in the fish trade. Two examples may be given of how this table is used in practice. If the dry bulb reads 85°F and the wet bulb 80°F, that is the difference is 5 deg F, the r.h. is 80%. Relative humidities at dry bulb temperatures intermediate between those listed in Table 14·9 can be estimated from the values on either side.

If the dry bulb is 98°F, and wet bulb is 87·5°F, then the difference is 10·5 deg F. It can be seen from the block of figures marked in Table 14·9 that the r.h. is about 65 to 66%.

When salt or some similar substance is dissolved in water, the vapour pressure of the water at any particular temperature is reduced below the saturation value. Consequently the relative humidity in a confined space above the solution is less than the saturation value for pure water. The more of the substance that is dissolved, that is the stronger the solution, the lower the r.h. that will be maintained. Table 14·8 shows the reduction in r.h. produced by solutions of common salt.

TABLE 14·8

Relative humidity above solutions of common salt

r.h.	*Common salt* (*lb salt/10 gallons water*)
100	0
95	8
90	16
85	24
80	32
75	36 (saturated)*

*Actually 76% r.h.

The lower vapour pressure of solutions of salts and similar substances causes them to absorb water from an atmosphere approaching 100% r.h. This is why it becomes impossible to dry salt fish when the relative humidity of the air rises above about 75%.

TABLE 14·9

Per cent relative humidities at various wet and dry bulb temperatures

Dry bulb temperature (°F)	*Difference between dry and wet bulb temperatures (deg F)* 1	2	3	4	5	6	7	8	9	10	11	12	13	14	15	16	17	18	19	20
20	85	70	55	40	26	12	0													
25	87	74	62	49	37	25	13	11	0											
30	89	78	67	56	46	36	26	16	6	0										
35	91	81	72	63	54	45	36	27	19	10	2	0								
40	92	83	75	68	60	52	45	37	29	22	15	7	0							
45	93	86	78	71	64	57	51	44	38	31	25	18	12	6	0					
50	93	87	80	74	67	61	55	49	43	38	32	27	21	16	10	5	0			
55	94	88	82	76	70	65	59	54	49	43	38	33	28	23	19	14	9	5	0	
60	94	89	83	78	72	68	63	58	53	48	43	39	34	30	26	21	17	13	9	5
65	95	90	85	80	75	70	66	61	56	52	48	44	39	35	33	27	24	20	16	12
70	95	90	86	81	77	72	68	64	59	55	51	48	44	40	36	33	29	25	22	19
75	96	91	86	82	78	74	70	66	62	58	54	51	47	44	40	37	34	30	27	24
80	96	91	87	83	79	75	72	68	64	61	57	54	50	47	44	41	38	35	32	29
85	96	92	88	84	80	77	73	70	66	63	60	57	53	50	47	44	41	39	36	33
90	96	92	89	85	81	78	74	71	68	65	61	58	55	52	49	47	44	41	39	36
95	96	93	89	86	82	79	76	73	69	66	63	61	58	55	52	49	47	44	42	39
100	96	93	89	86	83	80	77	73	70	68	65	62	59	56	54	51	49	46	44	41
105	97	93	90	87	84	81	78	74	71	69	66	63	60	58	55	53	51	48	46	43
110	97	93	90	87	84	81	78	75	73	70	67	65	62	60	57	55	52	50	48	46
115	97	94	91	88	85	82	79	76	74	71	68	66	63	61	58	56	54	52	49	47
120	97	94	91	88	85	82	80	77	74	72	69	67	65	62	60	58	55	53	51	49

15

Fish spoilage

Daughters and dead fish are no keeping wares.
OLD PROVERB

As soon as a fish dies, spoilage begins. Spoilage is the result of a whole series of complicated changes brought about in the dead fish tissue by its own enzymes, by bacteria and by chemical action. The well-known characteristics of spoiling fish are therefore the resultant effects of a host of different changes, few of which are well understood. It is, then, all the more remarkable that the changes in the flesh of any particular species of fresh fish caught on a certain ground and stored in ice, generally follow a definite pattern. It is necessary to know what this spoilage pattern is, in order to be able to assess the freshness of any particular sample, an exercise that is becoming of greater importance in those firms that are establishing quality control on their processing lines.

Millions of bacteria and other micro-organisms, many of them potential spoilers, are present in the surface slime, on the gills and in the intestines of living fish. They do no harm, because the natural resistance of a healthy fish keeps them at bay. Soon after the fish dies, however, bacteria begin to invade the tissues. It is believed that they enter through the gills and kidney, along veins and arteries, and directly through the skin and peritoneum, that is the lining of the belly cavity. It is not known with certainty how long it takes for bacteria to penetrate the skin itself; some evidence indicates, however, that in iced cod penetration begins in the lateral line region about the fourth or fifth day after catching.

Another important series of changes is brought about by the enzymes of the living fish which remain active after its death. Some of these enzyme reactions are almost certainly involved in the flavour changes that occur during the first few days of iced storage, before bacterial spoilage has supervened. In addition to enzymic and bacterial changes, chemical changes, involving oxygen from the air and the fat in the flesh of species such as herring, mackerel and pilchards, may produce rancid odours and flavours. These oxidative changes are of particular importance in frozen fish.

Although the spoilage pattern in any particular species of fish generally follows much the same pattern regardless of the origin of the fish, there are wide differences in the patterns of different groups. The cartilaginous fishes, such as dogfish and skate, are well known to produce large quantities of ammonia even within eight days in ice. But there are wide differences even within the bony fishes. It is, for example, well known that mackerel spoil very rapidly and that redfish (*Sebastes*) will keep better than cod but not so well as halibut. There are, furthermore, wide differences between freshwater and marine bony fishes. In the present state of knowledge, these differences can only be attributed in a general way to differences in some components of the tissues, in enzyme systems and bacterial flora; a great deal more research will be necessary before the underlying reasons for these differences can be explained. Most of the work hitherto carried out has been on cod and closely allied species, and it may therefore be useful to consider the chain of events that occurs, in a well-iced cod, from the moment of death until putrefaction sets in.

CHANGES IN APPEARANCE, ODOUR AND TEXTURE

Freshness is usually judged in the trade entirely by appearance, odour and texture of the raw fish. Since assessment depends upon the senses, these factors are known as *sensory* or *organoleptic*. Few people in the fish industry, however, have had the opportunity of closely observing the day-to-day changes that occur in iced cod, or of trying to describe, in precise terms, how a stale but edible fish differs from a stale and inedible one. Anyone who attempts to write down a specification for a product will know how difficult precise definition can be.

The necessity for accurate definition of the spoilage stages in fish was, however, early realized by Port Health Inspectors. One of the pioneers in this field was Dr. A. G. Anderson, Assistant Medical Officer of Health for Aberdeen. He stated in 1907 that the most important things to look for when assessing the freshness of fish were:

(1) the general appearance of the fish including that of the eyes, gills, surface slime and scales, and the firmness or softness of the flesh

(2) the odour of the gills and belly cavity

(3) the appearance, and particularly the presence or absence of discoloration along the underside, of the backbone

(4) the presence or absence of *rigor mortis* or death stiffening

(5) the manner in which the flesh strips away from the backbone

(6) the appearance of the belly walls

As a result of his observations, Anderson was able to define, in sensory terms, when a fish was stale and unfit for human food and when fresh and wholesome.

Subsequent investigations have developed Anderson's ideas. The changes in odour, texture and appearance in iced cod, and a number of other species as well, from absolute freshness to putridity, have been accurately described.

Five clearly discernible stages in appearance, and ten in the odour of the gills, have been described for cod and similar species stored without crushing in plenty of ice, as shown in Table 15·1. Odour, flavour and texture of the cooked fish have been similarly described, as shown in Table 15·2. Descriptions are also available for a few other species, such as herring, flat fish and *Sebastes* (redfish).

TABLE 15·1

Cod (raw fish)

GENERAL APPEARANCE (5 MARKS)	*Score marks*
Eyes perfectly fresh, convex black pupil, translucent cornea; bright red gills, no bacterial slime, outer slime water white or transparent; bright opalescent sheen, no bleaching.	5
Eyes slightly sunken, grey pupil, slight opalescence of cornea; some discoloration of gills and some mucus; outer slime opaque and somewhat milky; loss of bright opalescence and some bleaching.	3
Eyes sunken; milky white pupil, opaque cornea; thick knotted outer slime with some bacterial discoloration.	2
Eyes: completely sunken pupil; shrunken head covered with thick yellow bacterial slime; gills showing bleaching or dark brown discoloration and covered with thick bacterial mucus; outer slime thick yellow-brown; bloom completely gone; marked bleaching and shrinkage.	0
FLESH INCLUDING BELLY FLAPS (5 MARKS)	
Bluish translucent flesh, no reddening along the backbone and no discoloration of the belly flaps; kidney bright red.	5
Waxy appearance, no reddening along backbone, loss in original brilliance of kidney blood, some discoloration of belly flaps.	3
Some opacity, some reddening along backbone, brownish kidney blood and some discoloration of the flaps.	2
Opaque flesh, marked red or brown discoloration along the backbone, very brown to earthy brown kidney blood, and marked discoloration of the flaps.	0
ODOURS (10 MARKS)	
Fresh seaweedy odours.	10
Loss of fresh seaweediness, shellfish odours.	9
No odours, neutral odours.	8
Slight musty, mousy, garlic peppery, milky or caprylic and like odours.	7
Bready, malty, beery, yeasty odours.	6
Lactic acid, sour milk, or oily odours.	5
Some lower fatty acid odours (for example acetic or butyric acids), grassy, 'old boots', slightly sweet, fruity or chloroform-like odours.	4
Stale cabbage water, turnipy, 'sour sink', wet matches, phosphene-like odours.	3
Ammoniacal (trimethylamine and other lower amines) with strong 'byre-like' (o-toluidine) odours.	2
Hydrogen sulphide and other sulphide odours, strong ammoniacal odours.	1
Indole, ammonia, faecal, nauseating, putrid odours.	0
TEXTURE (5 MARKS)	
Firm, elastic to the finger touch.	5
Softening of the flesh, some grittiness.	3
Softer flesh, definite grittiness and scales easily rubbed off the skin.	2
Very soft and flabby, retains the finger indentations, grittiness quite marked and flesh easily torn from the backbone.	1

Table 15·2

Cod (cooked fish)

approximately 6–8 oz middle cut of fish steamed in glass casserole dishes of 7 inch diameter over boiling water for 35 minutes

ODOUR (10 MARKS)	*Score marks*
Strong seaweedy odours.	10
Some loss of seaweediness.	9
Lack of odour or neutral odours.	8
Slight strengthening of the odour but no sour or stale odour; wood shavings, woodsap, vanillin or terpene-like odours.	7
Condensed milk, caramel or toffee-like odours.	6
Milk jug odours, or boiled potato or boiled clothes-like odours.	5
Lactic acid and sour milk, or 'byre-like' odours.	4
Lower fatty acids (for example, acetic or butyric acids) some grassiness or soapiness, turnipy or tallowy odours.	3
Ammoniacal (trimethylamine and lower amines) odours.	2
Strong ammoniacal (trimethylamine) and some sulphide odours.	1
Strong ammonia and faecal, indole and putrid odours.	0
TEXTURE (5 MARKS)	
Firm thick-white curd, bluish white in appearance, no discoloration.	5
Firm but woolly, lost its bluish whiteness, some yellowing.	3
Softer, cheesy-like, marked discoloration.	2
Sloppy, soapy, very marked browning along the backbone.	1
FLAVOUR (10 MARKS)	
Fresh, sweet flavours characteristic of the species.	10
Some loss of sweetness.	9
Slight sweetness and loss of the flavour characteristic of the species.	8
Neutral flavour, definite loss of flavour but no off flavours.	7
Absolutely no flavour, as if chewing cotton wool.	6
Trace of off flavours, some sourness but no bitterness.	5
Some off flavours, and some bitterness.	4
Strong bitter flavours, rubber-like flavour, slight sulphide-like flavours.	3
Strong bitterness but not nauseating.	1
Strong off flavours of sulphides, putrid, tasted with difficulty.	0

For practical purposes it is not, of course, necessary to identify the stages in spoilage with the accuracy indicated in Tables 15·1 and 15·2. In cod and haddock, properly stowed in ice, four well-defined phases of spoilage can be discerned by most people. These phases roughly correspond to periods of 0 to 6 days, 6 to 10 days, 10 to 14 days, and 14 days and more, in ice. They also correspond approximately to quality gradings used by some of the larger producers.

Phase I first quality (for freshing and freezing).
Phase II second quality (for freshing and smoking).
Phase III for smoking and salting.
Phase IV condemned.

Little serious deterioration, apart from some loss of natural flavour and odour, occurs during Phase I but in Phase II the fish is obviously lacking in odour and flavour. In Phase III not only does the fish begin to taste stale, but its appearance and texture begin to show obvious signs of spoilage; the gills and belly cavity also smell distinctly 'off'. All these changes, which in the later stages are almost entirely due to bacteria, occur at an ever increasing rate until by the fifteenth or sixteenth day of storage, when Phase IV begins, the fish is putrid and generally regarded as inedible.

BACTERIOLOGICAL AND CHEMICAL CHANGES DURING SPOILAGE

Although the underlying changes during spoilage are largely not understood, it is nevertheless possible in a general way to relate the bacteriological and chemical changes to the four phases of spoilage as assessed by sensory means. It is important to know how these changes are interrelated in making any attempt to devise tests for freshness not based on sensory judgements.

It has already been mentioned that bacterial spoilage probably plays little part in the changes that occur in the flesh up to about the sixth day of storage in ice. But the enzymes meanwhile have brought about considerable changes in some of the muscle constituents. In the first instance these changes are associated with *rigor mortis* or the stiffening of the muscle soon after death, which is a noticeable feature of recently-caught fish. As a result of these various changes, some constituents disappear while others are very much altered chemically, so that the characteristic odour and flavour of the fresh fish is lost. Some of these chemical constituents, that are water-soluble and sometimes called extractives, are the first to be acted on by the bacteria which, by the end of Phase II, are present in the muscle in fairly large numbers. The fats and proteins appear at this stage to be resistant to bacterial attack and, indeed, it is only when the so-called extractives have mostly been utilized by bacteria in Phase III that the proteins are attacked to any marked degree.

Extractives are important for a number of other reasons. The actual chemicals present vary in amounts from species to species. Thus, herring and mackerel contain relatively large amounts of certain basic amino acids, notably one called histidine, which incidentally are said to be responsible for the meaty flavour of this group of fishes. Cod and haddock, on the other hand, contain only traces. Skate and dogfish contain large quantities of urea which, again, is virtually absent from cod.

The important substance trimethylamine oxide, which occurs widely in all marine fish, is practically absent from freshwater species. Even in marine fishes, however, quantities vary enormously from one species to another.

Even within a single species there may be considerable variations in the quantities of certain extractives with season, fishing ground and other factors. There is, for example, three times as much of the amino acid taurine in North Sea cod caught in June as there is in January. There is generally about three times as much trimethylamine oxide in Arctic cod as there is in that caught in the North Sea.

These differences in the nature and amount of the extractives in newly-caught fish affect not only initial flavour and odour, but also the subsequent pattern of spoilage. The urea in sharks and rays, for example, is rapidly converted into ammonia by certain marine bacteria with the result that one of the most obvious features of a spoiling fish of this class is its strong smell of ammonia. Bony fish stored under the same conditions do not develop ammonia at such an early stage in spoilage, because very little urea is present.

Although it is known that during Phase II many of the extractives are attacked, unfortunately little is known as yet about the types of compounds manufactured from them by the bacteria. An important exception to this is, however, the substance trimethylamine which results from the breakdown of trimethylamine oxide. Chemical estimation of the amount of trimethylamine in a sample of fish is one of the few promising methods for objectively assessing its degree of freshness.

It has been observed that some amino acids, such as those called leucine, serine and alanine, although present in minute amounts, can be attacked by bacteria to produce strong-smelling substances, as yet unidentified chemically. Some of these odours have been described as being like rotten vegetables, stale cabbage water, onions, and so on. These are all descriptions given to the odour of the gills at certain stages of spoilage of cod (see Table 15·1). It should be remembered that these odoriferous compounds may be present in minute traces of perhaps one part in a million or even less, but may nevertheless be smelt. The chemical separation and identification of such small quantities of substances is a difficult job which scientists have only recently found ways of tackling.

Cod stored in ice seems to alter significantly in appearance, taste and odour between the tenth and twelfth day, Phase III. It is also about this stage that the extractives begin to disappear rapidly and significant changes occur in the bacterial flora. The types of bacteria, at this stage present in enormous numbers, are those most active in spoilage. Over the following few days, with the virtual disappearance of the extractives, the proteins are attacked. Various putrid-smelling compounds, usually associated with rotten fish, occur during this final Phase IV. These include such compounds as hydrogen sulphide, which smells like rotten eggs, mercaptans which smell like stale cabbage-water and rotting vegetation, indole and skatole which smell like excrement and volatile fatty acids that smell like stale sweat.

It should be stressed that this picture relates only to the spoilage of cod and haddock. The spoilage pattern of other species is not the same, but little is known as yet about it in detail.

CHEMICAL TESTS FOR FRESHNESS

It has been mentioned that certain chemical changes in spoiling fish appear to run parallel with changes in odour, texture, appearance and flavour. It is perhaps not surprising, therefore, that various attempts have been made to measure freshness by estimating the quantities of some of these chemicals produced as a result of bacterial and enzymic activity. Since spoilage is a result of many complex changes, and most chemical methods involve the measurement of only one, or at most a few, substances, no chemical method gives the same degree of accuracy in measuring freshness as does a properly trained taste panel. Chemical methods are, on the other hand, usually more rapid and certainly require fewer staff to operate them. Constant vigilance and comparison with samples of known storage history are necessary if a taste panel is to continue to give accurate results. A chemical method, on the other hand, once the technique is mastered, will continue to give accurate results. Unfortunately, chemical tests can generally be carried out only by trained staff in a well-equipped laboratory; only the larger processors could consider setting up such a unit. Nevertheless, many public analysts will undertake certain of these tests. Since buyers of fish are insisting more and more on certain quality standards, it is as well to know what these chemical tests mean and how much reliance can be placed upon the results. These tests have been applied mainly to cod and haddock and what is said below, therefore, refers only to these species.

TRIMETHYLAMINE (TMA)

The TMA test is probably at present by far the most important of all the chemical tests for freshness of cod and haddock. It is used as a routine measure of freshness in some countries, notably Canada, and is beginning to be used by a few firms in Britain.

Trimethylamine oxide, one of the components of marine fish, is broken down by bacteria to give the odoriferous substance trimethylamine, related chemically to ammonia. Consequently really fresh fish has very low TMA values, which rise progressively during spoilage. In about 75% of cases, the TMA test and the taste panel tests agree; for many purposes, therefore, the TMA test gives a good enough indication of the freshness of a particular sample.

The results of a TMA estimation are expressed in one of two ways, either as milligrams of TMA in 100 grams of fish, or milligrams of TMA in 100 millilitres of an extract of the muscle prepared by a standard routine. The approximate equivalents of TMA estimations with the four phases of spoilage and taste panel results are given in Table 15·3.

TABLE 15·3

Correlation of phases of spoilage of cod and haddock with various tests for freshness

Spoilage phase				
Period in ice (days)	0–6	6–10	10–14	14 and over
Taste panel score	10–7½	7½–6	6–4½	4½ and below
TMA mg/100 ml extract	less than 3·0	3·5–7·5	7·5–14·5	14·5 and above
TMA mg/100 g flesh	less than 1·5	1·5–5·0	5·0–14·0	14·0 and above
TVB mg/100 ml extract	up to 16·0	up to 16·0	15·5–29·5	29·5–32·5

These figures give only an approximate indication. Precise conditions of stowage and handling may considerably influence the results.

TOTAL VOLATILE BASES (TVB)

Trimethylamine, as already mentioned, is closely related chemically to ammonia. There are several other similar substances produced during spoilage. It is possible, by a fairly simple analytical procedure, to measure the total quantity of all these substances present in a sample of flesh. Although it might be thought that this test would give a more accurate indication of the freshness of a sample than the TMA test, since a group of substances rather than one is estimated here, this is true only for the later stages of spoilage; it is sometimes used by technologists and the approximate TVB equivalents of the four phases of spoilage in cod and haddock are given in Table 15·3.

AMMONIA

Ammonia, produced in cod, haddock and allied species by the bacterial breakdown of the muscle proteins, appears in quantity at a relatively late stage of spoilage, and this test is therefore not of much application for these species. The test will give an indication only of how poor a sample is that is already obviously stale. It would probably be of use, however, for the estimation of the freshness of cartilaginous fishes such as dogfish and skate.

VOLATILE REDUCING SUBSTANCES (VRS)

This test depends on the fact that most of the substances responsible for off-odours are volatile and most of them can reduce a chemical oxidizing agent such as potassium permanganate. However, the chemical reactions involved with such a complex mixture as the VRS of spoiling fish cannot be interpreted in terms of precise amounts of particular spoilage products. Nevertheless a greater degree of agreement with taste panel results is claimed for this test than is usually obtained with the TMA test. This method is rather more difficult, and the apparatus more complicated and expensive, than that required for TMA estimations. Although experience of the method in this country has been limited, it has been shown to have no advantage over the TMA test for iced fish.

It probably has some use in measuring the freshness of preserved products such as smoked fish. The measurement of VRS of canned fish gives an indication of the freshness of the commodity before canning.

TETRAZOLIUM

This is the so-called 'litmus paper test' for freshness. The basis of the test is that a certain group of colourless chemicals, known as tetrazolium compounds, may be altered by the action of certain bacterial products to give coloured substances. In practice the test has been found to be upset by too many factors to be of any wide application in the fish industry. This is unfortunate since it is a very easy test to undertake. Strips of paper impregnated with certain tetrazolium compounds are laid across the skin of fish. The development of colour in the paper indicates that the fish is beyond a certain stage of spoilage. A refinement is to impregnate the paper with bands of tetrazolium salt of different strengths, so that one paper can be used to indicate more accurately the stage of spoilage of a sample.

In some countries, a form of tetrazolium test is used that involves extracting the mashed flesh with a solution of a tetrazolium salt and then measuring the colour developed in the extract. Although this test possibly gives reasonably reproducible results on the species for which it is used, it does not appear to offer any advantage over the TMA test.

HYPOXANTHINE

The bitter-tasting substance hypoxanthine is a product of some of the enzymic changes occurring in fish soon after death. Since its concentration increases fairly regularly with spoilage, estimation of the amount present in a given weight of flesh gives an objective measure of the degree of spoilage. The estimation can probably be carried out more rapidly than the TMA estimation currently used.

PHYSICAL TESTS FOR FRESHNESS

HENNINGS' METER

An electrical method has recently been developed by Dr. Hennings, formerly employed at the German Fish Technology Research Institute in Hamburg. The equipment employed is compact and a reading can be taken in a few seconds; this makes it suitable for market as well as laboratory use. Although the equipment gives some measure of the expected storage life remaining to a fish, damage to the skin, washing, or even gripping firmly, can lower the reading. It is too early as yet to give a considered opinion on the usefulness of this instrument.

16

Instruments

Measure thrice before you cut once.
Italian proverb

THIS chapter is written for the man who will buy and use instruments and not for the man who will design or repair them. It is concerned with the sort of instrument that can be used for a given job, and with the basic principles on which it works, rather than its details of construction.

Instruments of one kind or another are used in almost every industry. By using the correct instruments in the right places a process may be more closely controlled, so avoiding unintentional variations in the product, for example in the weight of fish cakes. Instrumental control can also reduce costs and supervision of labour. The proper use of instruments is, however, not only an essential for the adequate control of any process; it may in some circumstances be a legal obligation. A boiler used for generating steam, for example, must have gauges to show the pressure of the steam and level of water in the boiler.

Industrial instruments fall into one of two general classes; those that measure and those that control. A measuring instrument may either indicate what the situation is at the moment, as when a thermometer is used to read a temperature, or it may meter what has happened over a period of time, as in the case of a gas or electricity meter. A controlling instrument is in essence a measuring instrument with some sort of switch or regulator built into it to keep conditions within certain limits. The thermostats on a gas oven or electric immersion heater are controllers that keep temperature steady without actually indicating it. All controlling instruments operate within limits, for example a thermostat controlling a heater at 80°F might switch on at 78°F and off at 82°F. A measuring instrument can be used as a guide to hand control; more elaborate and expensive instruments measure, record and control.

In the fish industry, there are many places where instruments could with advantage be used. Instrumentation can in some industries be an expensive business, perhaps 10% of the total cost of the equipment, but many of the instruments that could usefully be employed in the fish industry are relatively inexpensive. Most of the processes carried on in fish factories are of a rather simple nature and involve the control of a few variables, such as temperature, pressure or rate of flow. Well-tried pieces of equipment are available to do these jobs.

BALANCES AND SCALES

Weighing is a frequent operation throughout the fish industry. Even when fish is initially purchased by volume, it is later sold by weight. Retail packages may also be required by the Weights and Measures Act to be of a certain minimum weight.

In addition to this, however, systematic weighing at successive stages of handling and processing can help to check losses, possibly amounting to a few per cent, due to such causes as evaporation or pressure.

One particular balance or weighing machine may be better suited to some types of operation than others and designs are constantly being improved to provide greater speed and accuracy. Whether choosing a balance for a particular job, or for general purposes, it is worth while first to seek the advice of manufacturers. Some firms specialize in particular types of equipment or will design a machine for a particular operation.

Many processes involve a loss in weight. To avoid unnecessary or wasteful loss, it is desirable to keep records of every batch, or of representative samples, taken at regular intervals throughout production. In fish factories, provision should be made for check weighing on a proper statistical basis as an integral feature of the production process. In mechanical fish smoking, for example, there is much to be said for installing a weigh bridge on to which loaded trucks of fish can be wheeled. If the original weight of truck and tenters is known, the weight of fish before and after the smoking operation may be readily determined and the percentage weight loss calculated to see whether the specification has been met.

There are many types of balance in general use in the fish trade, but they mainly fall into one of two groups:

SPRING BALANCES

The spring balance is one of the simplest types of weighing machine. For many purposes it is ideal; it can be light and easily transported, can be used under difficult conditions and will give rapid readings. It has the disadvantages that it is initially not very accurate and that it may in time become progressively even less accurate. The spring balance consists essentially of a coiled spring. The object whose weight is to be ascertained is hung on the bottom of the spring whose top is firmly fixed. The amount by which the spring is extended depends upon the weight applied.

In time the spring may become fatigued, especially if it is frequently overloaded, and slightly high readings may then be obtained. In other words, a tradesman using a spring balance that had been persistently misused might unwittingly give short weight to his customer.

BEAM BALANCES

The simplest type of beam balance is essentially a see-saw. The object to be weighed is placed at one end of the see-saw, and weights are put on to the other end until the beam swings without a tendency to dip on one side more than on the other. Provided that the suspension points are in good condition and well made, these balances are very accurate. Indeed, some of the most accurate balances used in laboratories operate on this principle.

A variant of the simple beam balance is the steelyard. In this instrument, a weight is slid along the beam of the balance until the position is found at which the object to be weighed is first counterbalanced and the whole beam swings freely and evenly. The beam is usually calibrated, so that the weight may be read directly.

Balances are nowadays available that, although they operate on the same principle as the beam balance, nevertheless are so highly refined that their relationship to the beam balance is obscured. These are direct-reading pan balances, which can be very accurate, rapid, robust and simple to use.

TIMERS

Surprisingly few operations in fish processing factories are timed with any accuracy. Brining, for example, is often carried out in a haphazard manner, sometimes with undesirable results. The time for which packets of fish are held in a freezer sometimes seems to depend more upon the total quantity of fish to be frozen than the capacity of the machine to do the job. And yet simple timers, robust enough for a fish factory, but accurate, are available. They may be worked by clockwork or an electric motor, and can be set so that they sound a bell or switch on a warning lamp after any desired period. They can, indeed, be made to switch off equipment at the completion of a process.

THERMOMETERS

Temperature is one of the most important factors in the whole of fish technology. Almost every process is in some way dependent upon temperature. For example, the speed at which wet fish spoils depends on temperature; if the temperature in a kiln is too high, the fish will cook and drop off the tenters; the speed with which a package of fish in a freezer will pass through the critical zone depends upon how long it takes for the temperature at the centre of the package to pass through this critical period.

There are many operations in the fish industry that could be better controlled if temperatures were taken at various times throughout the process. Although some pieces of equipment, such as cold stores and Torry kilns, generally have thermometers fitted to them, it must be admitted that these are not always read and care is not always taken to see that they are in working order.

There are various ways in which temperature may be measured and it is necessary to know what is the best method for any particular set of circumstances. A brief description is given of the various types of equipment that may be employed.

LIQUID EXPANSION THERMOMETERS

A familiar instrument of this kind is the clinical thermometer used in medicine. When the temperature of the bulb, which is filled with mercury, is raised the liquid expands along the fine capillary tube. The amount of this expansion depends upon the extent of the rise in temperature. The actual increase in volume of the mercury is in fact very small so that the capillary tube has to be very fine in order to make it apparent at all. The outside of the thermometer is frequently curved so that, acting as a lens, it makes the fine thread of mercury look bigger.

One of the difficulties in using mercury-in-glass thermometers, and other types of glass thermometers such as alcohol-in-glass, is that they are fragile and very easily broken. But they are simple and generally cheap unless made to be *very* accurate. A glance will usually tell whether or not the thermometer is working; it is quite obvious when the glass is broken. Sometimes, however, the liquid thread only may become broken, and this may not always be so obvious. When reading a liquid-in-glass thermometer, the eye should always be on a level with the surface of the liquid. Another disadvantage of liquid-in-glass thermometers is that it is not always easy to read them at or near the point where the temperature is to be taken; for example, in a cold store. In such a case the mercury-in-steel thermometer is more convenient to use. In this type, the glass is replaced by a steel capillary tube with a steel bulb on the end and the capillary can be as long as required and can be bent round corners. Since it is impossible to see the mercury directly as in a glass thermometer, it is necessary to measure the height of the mercury indirectly and this is usually done by means of a Bourdon gauge similar to that used in a pressure gauge. The Bourdon tube is frequently linked mechanically to a pointer moving on a dial which can be made so large that it can be seen by a short-sighted operator on the other side of the room. Sometimes it is fixed to a pen writing on a clockwork-operated drum and so traces a record of the temperature throughout an operation; for example, temperature in a canning retort.

Thermometers containing mercury are only of any use over the range of temperatures at which mercury is a liquid. Mercury freezes at about minus 38°F and boils at 675°F, but by taking special precautions mercury thermometers may be used up to 950°F. If the temperature is below minus 38°F other liquids must be employed, and alcohol, usually with a pink dye added to it, is most generally employed. It has a useful range from minus 110°F to plus 160°F and other liquids are available for even lower ranges.

The thermometers so far described all share one important disadvantage; they have a bulb that must be warmed or cooled. This takes an appreciable time so that, although they are suitable for measuring fairly steady temperatures such as those in a cold store or smoking kiln, they are too sluggish to measure temperatures that are changing rapidly, such as those of fish in a freezer. The bulb is also frequently large and a mercury-in-steel thermometer with a bulb ½ inch in diameter could not be used to measure, for instance, the temperature of a consumer pack of frozen fish. For temperatures that are changing rapidly and where the sensitive part of the thermometer must be small, other types of measuring instrument are necessary.

SOLID EXPANSION THERMOMETERS

Most materials expand when their temperature is raised but the extent to which they will expand varies from one material to another. Some steels containing nickel and chromium hardly expand at all when their temperature is raised over a fairly wide range. Other metals, such as copper, expand very considerably. This phenomenon is used in bimetallic strip thermometers and in certain types of equipment such as thermostats whose operation is dependent upon temperature. The bimetallic strip consists of two strips of dissimilar metal fastened together. As the temperature rises so one side of the strip expands more than the other and the bimetallic strip begins to curve. The degree of curvature depends upon the amount that the temperature is raised.

ELECTRICAL RESISTANCE THERMOMETERS

This type of thermometer depends upon change in electrical resistance in a wire with change in temperature. The way in which it works can be explained by a simple analogy. The amount of water delivered by a pipe depends upon two things; the diameter of the pipe and the pressure of the water. Suppose that it is desired to deliver a certain quantity of water a minute; then if this is to be delivered through a narrow pipe the pressure must be higher than if it is a wide pipe. If the water pressure necessary for the particular rate of delivery were measured then this could be used as a measure of the resistance to flow of the water. The resistance thermometer works in an analogous manner and in fact measures the resistance to flow of electricity passing along a very fine coil of wire placed in the sensitive tip of the instrument. As the temperature of the tip is raised so the resistance rises.

The advantage of the resistance thermometer is that the sensitive tip can be made very small and, since very little heat is required to warm it up, it will respond rapidly. Robust instruments working on this principle are being made for industrial uses and are finding ready application in the fish industry. Resistance thermometers with a sensitive element fitted into a hypodermic needle are useful for measuring the temperature of wet fish within, for instance, a fish factory. The needle can be rapidly stabbed into the fish or fillet and a reading obtained in a matter of seconds. It is not quite so simple to measure the temperature of frozen fish, since the hypodermic needle may be bent or broken, but if a small hole is first drilled into the frozen fish the sensitive element can then be inserted and the temperature taken without the pack suffering any obvious damage. The method of measuring the temperature of frozen fish is described in Chapter 7.

The electric circuit, meter and battery necessary for the measurement of the change in resistance can be fitted into a small, easily portable box.

Another type of resistance thermometer depends upon the change in resistance with change in temperature of a mixture of metallic oxides; it is finding increasing application in certain industrial instruments. Such a mixture can be sealed into a piece of glass, making what is called a thermistor element which, when connected to a suitable electrical circuit, can give reliable and sensitive readings.

THERMOCOUPLES

If an iron wire and a copper wire are joined together in a closed loop and one of the junctions is heated while the other remains at room temperature, a tiny electric current flows round the loop. Small opposite voltages are produced at each joint and it is the difference between them which causes the current to flow.

Many pairs of metals will produce this effect and the voltage produced at the junctions depends both upon the metals used and the temperature difference between the hot and cold junctions. The phenomenon, known as thermo-electricity, can be used for measuring temperatures. It is desirable that the voltage produced should correspond regularly to the temperature difference; that is to say, twice the voltage should mean twice the temperature difference and so on, for then the measuring scale is evenly spaced. Fortunately it is easy to find pairs of metals that for a limited range come close to this requirement, although different metals may have to be used for different ranges. A thermojunction of iron and copper wire would, for example, be suitable for measuring the temperature of a cold store at minus 20°F but not for the sawdust fires of a smoking kiln which might be at 1000°F.

The voltage produced is very small indeed and sensitive instruments are consequently required for measuring temperatures by this means. Nevertheless the thermocouple has certain considerable advantages. The sensitive part, which is the point where the two wires are soldered

FIGURE 16·1 Using a thermocouple to check freezer performance

or welded together, is extremely small and a thermocouple is consequently even better than a resistance thermometer for measuring temperatures that are changing rapidly. Also, although the apparatus for measuring electrical output may be complicated and expensive, the thermocouple itself is simply two wires which are usually contained within the same outer cover, and is therefore generally cheap and expendable. Hence if the thermocouple becomes frozen into a pack of fish so that it cannot be removed without thawing, it may be simply cut off with pliers. In some circumstances it is an advantage to leave the thermocouple in a pack so that its temperature can subsequently be quickly measured.

There are two basic ways of measuring the voltage produced by a thermocouple. Either some kind of voltmeter, which measures the voltage directly, may be put into the circuit, or another voltage from a battery may be fed into the circuit, thus opposing the thermocouple voltage. The voltage that must be applied to stop any current flowing in the circuit, in fact, gives a more accurate measure of the voltage produced by the thermocouple but the more direct method of using the voltmeter is simpler and sometimes more convenient. In some cases one junction may be at a constant temperature, frequently that of melting ice; or one junction may be at room temperature, and the instrument fitted with some device that gives the same reading as if the junction were at constant temperature. This is known as junction compensation.

Various refinements in the measuring equipment are available; some equipment may measure the voltages of many thermocouples and print the readings on a chart; it may even be adapted to switch heaters or other items of plant on or off at certain temperatures. It may also be specially designed to work in ships or on moving vehicles. In any event, it must use one of the two basic methods of voltage measurement.

PRESSURE GAUGES

The measurement of pressure is not only important in its own right, for example the steam pressure in a fish meal factory, but also because many methods of measuring the flow of both liquids and gases involve measurement of pressure. Most pressure gauges do not measure total pressure but only the difference in pressure between, for example, steam in a boiler and the air outside. This is known as *gauge pressure* and may be measured in pounds a square inch, inches of water or inches of mercury.

An inch of water is that pressure that will force water up an open-ended tube to a height of one inch and keep it there. Similarly with an inch of mercury, but since mercury is about 13½ times as dense as water, one inch of mercury represents a pressure 13½ times as great as one inch of water.

The simplest kinds of pressure gauge consist merely of a tube, frequently bent into the shape of a letter U and filled with a suitable liquid. For prolonged use it is as well to use a liquid that does not evaporate rapidly, for example a light oil or mercury. The liquid chosen depends upon the pressures to be measured. If pressures only slightly above atmospheric are to be measured, then a very light liquid should be employed, but for higher pressures a denser liquid, such as mercury, should be used.

Although pressure-measuring instruments of the U-tube type are simple, cheap, and have no moving parts to go wrong, they are less convenient than instruments having a pointer moving on a dial. Pressure gauges of this latter type are commonly used and frequently depend upon a device known as a Bourdon tube. In principle this is similar to the toy, usually with a whistle built in, that consists of a coiled paper tube. When it is blown, the paper tube unrolls; a light spring incorporated in the structure rolls it up again as soon as blowing ceases. The Bourdon tube consists of a curved tube that is either oval or may be greatly flattened. As the pressure increases inside it, then the tube tends to open out and become less flattened. This in turn tends to straighten out the tube and the extent to which it straightens out depends directly upon the pressure. If one end of the tube is firmly fixed the other can be attached, by means of suitable gears and levers, to the end of a pointer working on a dial. The Bourdon tube is generally C-shaped but it may be, for instance, a flat spiral like the toy in a Christmas cracker. It is normally made to bend only a short distance, otherwise the metal of the Bourdon tube becomes fatigued and it no longer registers pressure accurately. Since the walls of the tube must inevitably be fairly thin, they should never be exposed to pressures higher than the maximum on the scale. Apart from the possible danger of bursting the tube if exposed to too high a pressure there is a great danger of permanently damaging gears and levers. Indeed, even if used for long periods at pressures near the maximum on the scale, the tube may become so distorted that the gauge is no longer reliable.

Bourdon tubes have a wide range of uses. They are used in certain types of distant-reading thermometer, they can be made to register pressures either above or below atmospheric, and it is possible to obtain gauges that will register both on the same dial. A gauge of this type may be used to test whether a can of food has been adequately exhausted during processing. A specially designed instrument with a spike on the end is stuck into the top of the can; the spike makes an air-tight seal and is attached to a gauge which then indicates the pressure in the can.

HYDROMETERS

The density of any substance is the weight of a known volume of it. A gallon of water weighs 10 pounds and the density of water is therefore 10 lb/gallon. It is also 62½ lb/cubic foot. In the metric system a cubic centimetre of water weighs one gram. It is frequently convenient to give the densities of liquids in relation to the density of water, for example one cubic centimetre of benzene weighs 0·8724 grams and one gallon of benzene weighs 8·724 lbs. That is, its density relative to water is 0·8724, and, provided that the densities of water and benzene are measured in the same system of units it does not matter what that system is. This relative density is called the *specific gravity* of the substance.

When specific gravities or densities are to be measured very accurately it is essential to control the temperature within very fine limits since the density of substances varies with temperature. For most purposes in the fish industry, however, very accurate measurement of specific gravity is not necessary.

The densities of liquids, such as beer, battery acid or brine, are generally measured with an instrument called a *hydrometer*. The principle of this instrument depends upon the fact that the denser a liquid, the higher objects will float in it. Sea water, for example, is denser, due to the dissolved salt, than fresh water. A ship sailing down river floats progressively higher in the water as it meets increasingly brackish water and finally the undiluted sea. The traditional method of testing the strength of brine for smoking by means of a potato, a beer bottle or sometimes an egg, depends on this same phenomenon, but unfortunately the method is not sufficiently accurate and too prone to variation; for example, if the potato has been allowed to dry out it will be less dense than if it has just been dug.

The measurement of density and specific gravity is not usually of interest for itself but rather for the information it gives about what is in the liquid. The density of brine is measured with a brineometer not because the weight of a gallon of brine is of much interest to the curer but because it tells him how much salt is in the brine, since increasing the concentration of salt increases the density of the brine.

A *brineometer* or salinometer is simply a hydrometer made specially for use with salt solutions. In strong brine it floats with almost the entire stem above the surface, but in weak brine most of the stem is submerged. Brineometers are marked in degrees; a saturated solution of salt is 100° brine and 0° brine is pure water.

A saturated brine contains about 36 lb of salt in every 100 lb, or 10 gallons, of water. An 80° brine commonly used in fish smoking therefore contains $36 \times \frac{80}{100} = 28{\cdot}8$ or roughly 29 lb of salt in 10 gallons of *water*. This is not the same thing as 29 lb of salt in 10 gallons of *brine*.

Since the density of a liquid changes with temperature, the brineometer scales are only correct at one particular temperature, usually 60°F. For use in the fish industry the errors due to using brineometers at *slightly* different temperatures are too small to matter. This is because the density of water and solutions of substances in water change only slightly with the change of temperature and also, in the case of brine, the amount of salt that will dissolve in water does not alter very much with changes in temperature. For example, 100 lb of water will dissolve about 34 lb of salt at 32°F and 35 lb at 80°F.

FIGURE 16·2 Checking air flow in a Torry kiln with a wool streamer

Salt is, however, unusual in this respect since most substances are much more soluble in hot water than cold.

There are other ways in which the salt content of a brine can be measured. One method which is used, although no suitable instrument is available commercially, is to measure the electrical resistance of a solution of the brine. The resistance of the brine depends on its strength. Standard equipment could be used for carrying out this measurement. One firm has designed a special meter for rapid and direct measurement of the percentage of salt in smoked fish.

FLOWMETERS

Although the fish industry is not very often concerned with measuring the flow of liquids it is quite frequently interested in the flow of gases, especially air. Air flow is particularly important in blast freezers, blast thawers and smoking kilns.

No method of measuring air speed is accurate in the region of a fan where the airstream is spinning round and round as well as going forward. Air is invisible and the path is sometimes unexpected. It is sometimes useful to be able to follow its movement. This can be done quite simply by using streamers of light thread with a piece of paper tied at the free end, held at arm's length in the air current. Care must be taken to see that the streamers are not shielded by the body. An elaborate method uses a chemical smoke produced by a fuming liquid called titanium tetrachloride.

ANEMOMETERS

One of the most direct methods of finding the speed of a current of air is to measure the speed with which it rotates a fan. Instruments working on this principle are known as anemometers. The fan must be very light and operating on bearings with very low frictional resistance. As the fan rotates it drives a delicate counting device which must also operate with very little friction. The speed of the airstream is found by knowing the number of revolutions of the fan in a certain time, the instrument having been previously graduated in airstreams with known flow rates. But however carefully the machine is made and however small the frictional resistance of the fan and its associated counting mechanism, some energy is inevitably lost in working it, so that the anemometer cannot be used for measuring very low flow rates of about 50 feet a minute or less.

Anemometers also have certain other disadvantages. For example, they must be placed in the airstream but they cannot be used at high temperatures or in smoke, because these damage the mechanism. Also the usual types of instrument do not give a direct reading; it is necessary to take readings at the beginning and end of a certain interval of time which must be measured accurately with a stop watch. The air speed can then be calculated. Direct-reading instruments of this type have however come on to the market fairly recently but they are expensive. The delicate mechanism is also easily damaged.

VELOMETERS

Fortunately there are other methods for measuring air speeds and one very useful technique is to place a single pivoted vane in the airstream and measure the force that is exerted upon it. A velometer is an instrument of this type in which the vane moves in a specially shaped channel, and the force upon it is opposed by a small spiral spring. Fixed to the vane is a pointer which moves across the scale. A velometer can be put into the airstream or connected to it by inlet and outlet tubes which must be specially designed. This latter method is often more convenient but although it gives some protection to the instrument from hot gas or smoke, it should be remembered that the hot gas still passes through it. One advantage of the velometer over the anemometer is that it is a direct-reading instrument, but its delicate mechanism is still easily damaged.

PITOT TUBES

An instrument that measures gas flow by the determination of pressure difference is the *Pitot tube*. With such an arrangement the air speed affects the height of liquid according to a known mathematical relationship. That is to say, a flow rate may be determined by measuring a pressure difference. A convenient type of pitot tube for use in kilns or blast tunnels has a detecting end, known as a head, and measures the difference between the *impact pressure* and the *static pressure* by an inclined-tube manometer (Figure 16·3). This type of manometer is very useful for measuring small pressures in the region of one inch of water, such as are often encountered in air flow measurement in the fish industry. The pitot head must, of course, face into the airstream but the accuracy with which it is pointed is not critical. A properly designed pitot head may be inclined at an angle of up to 20° to the direction of flow before the error in measurement exceeds 2%. It is important however that the pitot tube itself should not be in any way dented, distorted or dirty.

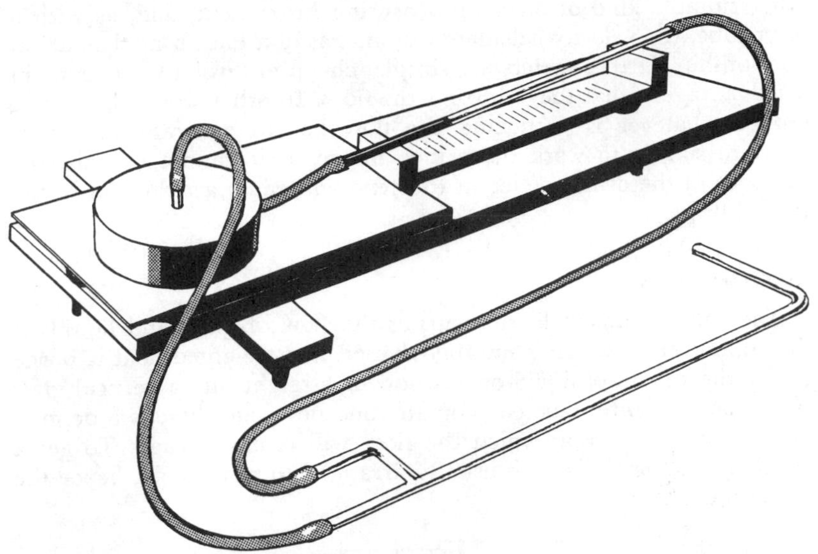

FIGURE 16·3 Inclined tube manometer with pitot static tube

ORIFICE PLATES

All the flowmeters so far described measure the flow at one particular place. If it is desired to know the air flow in a blast freezer, for example, it is necessary to make a number of measurements at different points. The air velocity may be greater in the centre than it is near the walls, and any type of obstruction may produce changes of velocity near to it. Thus a single measurement may be misleading and it is necessary to take the average of a number of readings.

Anemometers, velometers, and pitot tubes are all useful in making occasional tests but are not very suitable for permanent installation. If a flow-measuring device installed as part of a plant is required, so that flow rates may be permanently indicated on an instrument panel, there are a number of devices available, most of which depend on measuring the difference in pressure in front of and behind an obstruction fixed in the gas or liquid stream. Only one such device will be described here, the oldest and most common, the *orifice meter.* After the stream of gas or liquid in the pipe has been forced through the hole in the orifice plate it does not immediately spread out again to fill the pipe but forms a neck. By the side of this neck the pressure is lower than in the stream before the orifice. The pressure difference can be measured with a U-tube, an inclined tube manometer or any other suitable kind of pressure-measuring instrument, and, as with a pitot tube, the velocity calculated by means of a known mathematical relationship. Orifice meters are simple, cheap and robust. Their main disadvantage is that they obstruct the flow. In other words the pump, fan or whatever is producing the flow has to generate additional pressure simply to work the orifice meter. If it cannot do this, the presence of the orifice meter in the pipe will cause a reduction in the rate of flow.

ROTAMETERS

Another instrument for measuring the flow of gases and liquids is the rotameter in which a suitably shaped and weighted float is borne up by the stream of liquid or gas flowing straight into a vertical glass tube that is slightly tapered. For any one flow rate there is a definite position in the tube at which the float will remain steady. To get a steadier reading the float is usually arranged to spin round, hence the name rotameter.

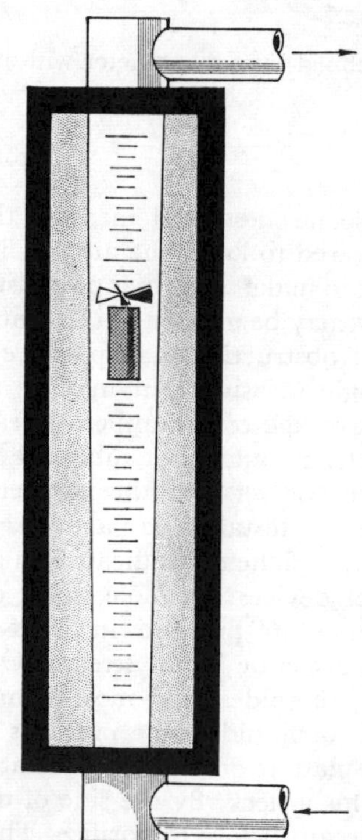

FIGURE 16·4 Rotameter variable area flowmeter

HYGROMETERS AND MOISTURE METERS

It is often desirable to measure the moisture content or humidity of the air in smoking or drying operations; it may also be necessary to know the moisture content of the commodity at the end of the process.

HYGROMETERS

The moisture content of air can be measured by various forms of hygrometer. The crudest type is the *horsehair hygrometer*. Many substances, such as horsehair, catgut, and paper, absorb water vapour from the atmosphere to an extent that depends upon its relative humidity. As these substances pick up water they expand and may be made to operate a needle on a scale or even make an old woman with an umbrella come out of her toy house. Instruments based on this principle are almost impossible to calibrate and are totally unreliable. Also the sensitive material may be readily damaged by exposure to, for example, wood smoke in a kiln. In any event the materials suffer from ageing effects to an unknown extent.

The most reliable method of measuring relative humidity employs two thermometers. The bulb of one is enclosed in a wet muslin sleeve and the other is uncovered. When air is blown over the two thermometers water evaporates from the wet bulb and consequently cools it to a temperature below that of the dry bulb. The difference in temperature between the two thermometers is known as the wet bulb depression, and depends on the humidity of the surrounding air. This instrument is known as a *wet and dry bulb hygrometer* or sometimes a *psychrometer*.

Relative humidities for ranges of temperature can be obtained from Table 14·9 on page 345, and tables and charts for readings outside this range are available. Various precautions must be taken to ensure reliable results. The muslin sleeve round the wet bulb must be kept moist with clean water and, after remoistening, a reading should not be taken for half a minute or so in order to allow the temperature to settle down again. The temperatures of the two thermometers should also be read within a few seconds of each other. The wet bulb should be placed downwind of the dry bulb and not the other way round, otherwise evaporation from the wet bulb may cool the air and the dry bulb reading will then be too low. Air speed over the wet bulb must be greater than about 3 feet a second otherwise a true reading will not be obtained. For practical purposes there is no upper limit to this air speed. The draught over the thermometer may be provided by a small fan, as in the Assmann psychrometer. It is possible to arrange a continuous direct feed to keep the wet bulb moist all the time, but the muslin sleeve must then be replaced by a clean one from time to time if accurate readings are to be obtained. It is also possible to record the wet and dry bulb temperatures separately on a chart

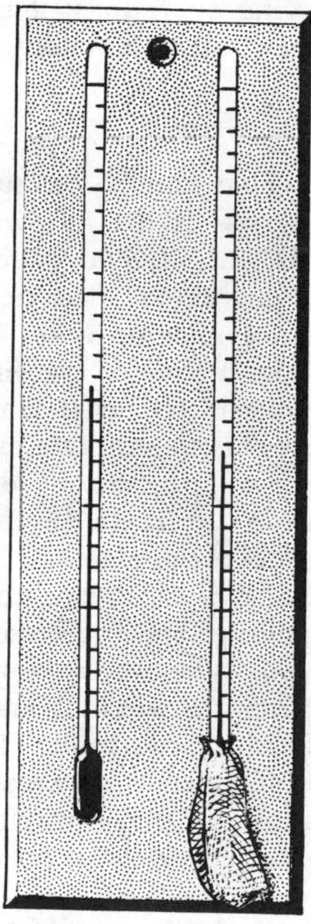

FIGURE 16·5 Wet and dry bulb hygrometer

using, for example, mercury-in-steel thermometers. Although wet and dry thermocouples can be used similarly, there is no industrial instrument of this type on the market.

Instruments for measuring humidity based on a variety of other principles, chiefly on changes of electrical conductivity with moisture content, are on the market, but none is sufficiently robust or free from fault to be recommended for the industrial purposes described in this book.

MOISTURE METERS

A standard method for measuring the water content of a product is to take a known weight of it, dry it in a dish overnight in a laboratory oven and weigh again next day when it has cooled down in a dry atmosphere. This, although reliable, is a slow process. A commercial instrument employing this principle, with heat obtained from an infra-red lamp, can be used to determine the moisture content of a sample of fish meal in about 20 minutes. A quick method for a substance such as fish flesh containing much more water than fish meal is to heat a known weight of it with a liquid such as toluene, which does not mix with water, and collect the water driven off. This method, however, still takes about an hour and the drying process cannot usually be held up as long as this to see if the water content is correct. Moreover, toluene is very flammable and for this reason should be used only by properly trained technicians in a properly equipped laboratory.

Various types of moisture meter have, therefore, been devised for arriving at a quick estimate of the water content of a product. Which particular method is used depends very much on the type of product. A powder, such as fish meal, can obviously be tested in ways that whole fish cannot. One common type of meter depends on the variations in electrical capacity or electrical conductivity in substances resulting from variations in moisture content. Water has an electrical capacity 20 times as great as that of most other substances. Unfortunately, although suitable for measuring the water content of seeds and grains in the range 5%–20% of water, such instruments are upset by the presence of salt. The more water there is in a product the better it conducts electricity and this fact is applied in the Marconi moisture meter, but this does not seem to give reliable results on fish meal.

Another method of measuring moisture content is to mix a known amount of the material with ordinary calcium carbide such as used in acetylene lamps. Carbide reacts with the water present producing acetylene gas and a certain amount of heat, so that if the reaction is carried out in a sealed vessel a pressure is generated which can be read on a Bourdon gauge, suitably graduated to read percentage moisture content. This method seems to be useful in obtaining rough estimates of the moisture content of certain crops in the field, but does not seem to have been used with success in the fish meal industry.

SMOKE METERS

The optical density of smoke can be measured by determining how much light from a lamp of known brightness will penetrate a certain fixed distance. Although this method measures only the droplets and solid particles in the smoke, which are known to be considerably less important in smoking than the invisible vapours, it is nevertheless found in practice that it gives a useful measure of the thickness of the smoke. The intensity of light can be measured by a device known as a photo-electric cell and all smoke meters incorporate one of these somewhere in their design. The simplest type merely gives the density of the smoke as a reading on a dial, but other types have been devised that can measure smoke on a meter or can draw a graph of the smoke density in a kiln during a smoking operation. In the simplest type, in which the smoke density is recorded on a dial, light is focused by a lens system on to a photo-electric cell at a fixed distance from the lens. Both lens and photo-electric cell are sealed into smoke-tight boxes. In order to prevent tar from the smoke being deposited on the surface of the photo-electric cell and the lens system, electrically heated glass windows are fixed immediately in front of each. Smoke in the kiln passes freely between the lens system and the photo-electric cell and the electric current produced by the photo-electric cell is measured on a suitable meter which reads directly the optical density per foot. A number of refinements of design are necessary to get an accurate answer, but the instrument can nevertheless be made both robust and reliable. It has the advantage that the sensitive part of the instrument can be placed anywhere in a kiln and joined to the indicating part of the instrument by a long length of cable. The lamp and photo-cell must be in the kiln but the meter could, if required, be in the manager's office.

Another type of smoke meter, which is now available commercially, will not only indicate the smoke density on a dial, but also meters the smoke in the same way as an electricity meter shows how much electricity has been used.

A third type graphs the smoke density and has certain advantages over the other types mentioned although it is much more expensive. It also has the disadvantage that the detecting and measuring parts are fixed together and the whole instrument must therefore be mounted on the kiln.

17

How to find out more

One gives nothing so freely as advice.

DUC DE LA ROCHEFOUCAULD

THIS book has been written by scientists working at the Torry Research Station and the Humber Laboratory for those employed in the various branches of the British fish industry. Inevitably, much has been omitted and much else abbreviated. The reader may not therefore always be able to find from this book alone all that he needs to know about a specific point.

The Torry Research Station and the Humber Laboratory have been set up to help to bring the fruits of science to the British fish industry. The results of research frequently appear in scientific and trade journals and also in special publications of Her Majesty's Stationery Office, and these are mostly written for some particular section of the industry. Enquiries about the problems of fish handling and processing are welcomed at both establishments. These should be addressed to the Director, The Torry Research Station, PO Box 31, 135 Abbey Road, Aberdeen, Scotland (Telephone: Aberdeen 24258), or to the Officer-in-Charge, Humber Laboratory, Wassand Street, Hull, East Yorkshire, England (Telephone: Hull 27879).

Index

Index